Green Solutions for Degradation of Pollutants

Edited by

Neha Agarwal
Department of Chemistry
Navyug Kanya Mahavidyalaya
University of Lucknow
Lucknow, India

Vijendra Singh Solanki
Department of Chemistry
Institute of Science and Research
IPS Academy
Indore, India

&

Sreekantha B. Jonnalagadda
School of Chemistry and Physics
University of KwaZulu-Natal, Westville Campus
Durban, South Africa

Green Solutions for Degradation of Pollutants

Editors: Neha Agarwal, Vijendra Singh Solanki, Sreekantha B. Jonnalagadda

ISBN (Online): 978-981-5238-96-9

ISBN (Print): 978-981-5238-97-6

ISBN (Paperback): 978-981-5238-98-3

First published in 2024.

need for a court order if at any point you breach any terms of this License Agreement. In no event will any delay or failure by Bentham Science Publishers in enforcing your compliance with this License Agreement constitute a waiver of any of its rights.

3. You acknowledge that you have read this License Agreement, and agree to be bound by its terms and conditions. To the extent that any other terms and conditions presented on any website of Bentham Science Publishers conflict with, or are inconsistent with, the terms and conditions set out in this License Agreement, you acknowledge that the terms and conditions set out in this License Agreement shall prevail.

Bentham Science Publishers Pte. Ltd.
80 Robinson Road #02-00
Singapore 068898
Singapore
Email: subscriptions@benthamscience.net

CONTENTS

PREFACE

The rapid industrialization, urbanization and technological advancement have generated various contaminants on a global level. The presence of these contaminants in different environmental matrices, either from natural or anthropogenic activities, represents a threat to the natural environment and living entities. Therefore, much control and research are required to eradicate and minimize the negative impacts of these pollutants from the contaminated environment. Though conventional treatment approaches and advanced techniques are effective in the removal of pollutants from the environment, these techniques are highly expensive, energy consuming and non-environment friendly in nature. In this view, there is a need for eco-friendly and sustainable solutions with minimal negative post-environmental impact.

The book "Green Solutions for Degradation of Pollutants" is a compilation of chapters on environment friendly techniques of remediation of pollutants. Green solutions are basically a collection of techniques and practices that are based on the generation of non-toxic end products, renewable energy sources and other factors that mitigate the negative impacts caused by human activities. The book will be highly useful for students, researchers, environmentalists, academicians, environmental microbiologists, life sciences and nanosciences experts, waste treatment industries, and for a well-read audience. It will also serve as a learning resource for researchers and students in environmental science, microbiology, nanotechnology, freshwater ecology, and microbial biotechnology.

Agarwal *et al.*, in chapter one, have discussed the transport fate and accumulation of emerging environmental pollutants and critically assessed their toxic impacts on the environment and living beings. They have also highlighted the possible solutions that could be used to remove these contaminants in a sustainable manner.

Amrit Mitra, in chapter two, has given a comprehensive description of microbial potential for biodegradation of organic pollutants, their removal mechanisms, and distribution of pollutants in environmental matrices, biodegradation pathways and the efficacy of biodegradation for complete mineralization.

Shankar *et al.*, in chapter three, have given an overview of the green synthesis of metal nanoparticles using plant extracts, the pollutants degradation mechanism, and their environmental and biological applications in detail.

Amrit Mitra, in chapter four, has given a detailed account of the current advancements in green bioremediation methods, how various contaminants are broken down by microorganisms and what the future holds for bioremediation in terms of lowering global pollution levels.

Mishra *et al.*, in chapter five, have thrown light on carbon dots as a new group of zero-dimensional luminescent nanomaterials, their synthesis, classification, properties and applications in environmental pollution control and environmental protection measures.

Kumar *et al.*, in chapter six, have given a detailed account of green synthesis of nanoparticles using plant extracts and multiple applications of these nanoparticles in environmental remediation along with their biological applications. They have also discussed why green synthesis is more advantageous than classical chemical synthesis.

Saivenkatesh *et al.*, in chapter seven, have discussed the nanoparticles synthesized by microalgae, their characterization methods, and their multiple applications with a special focus on environmental remediation. They have also highlighted the challenges involved in using microalgae-derived NPs along with their future perspectives.

Shahi *et al.*, in chapter eight, have summarized the green synthesis methods of various nanomaterials, their remediation methodology, mechanism of action, and prospective applications in environmental remediation. Additionally, they have also highlighted the efficient removal and valorization of waste materials using nanobioremediation.

Kumar *et al.*, in chapter nine, have talked about the remediation of heavy metal-contaminated soil using phytoremediation as a sustainable approach. According to them, modern phytoremediation methods may be used for large-scale decontamination of contaminated soil in a sustainable manner.

Bais *et al.*, in chapter ten, have given a detailed discussion on the immense potential of nanomaterials in the bioremediation of polluted water. They have given a comprehensive comparison of nanobioremediation with other conventional bioremediation methods to make water environmentally non-hazardous.

Srivastava *et al.*, in chapter eleven, have addressed the biogenic and green synthesis of palladium nanoparticles to remove different types of pollutants from wastewater.

Jangid *et al.*, in chapter twelve, have given a comprehensive review of different domains of nanotechnology in the treatment of wastewater and also thoroughly covered various fundamental aspects of nanotechnology such as types, synthesis, applications and future directions for a green and sustainable environment.

Manoj Kumar *et al.*, in chapter thirteen, have given a comprehensive overview of eco-friendly initiatives, methods and preventive measures to remove microplastic and nanoplastics from the global environment. They have also underlined the potential of green nanomaterials to solve the growing problem of plastic pollution and emphasized the importance of sustainable and environmentally friendly solutions.

Neha Agarwal
Department of Chemistry
Navyug Kanya Mahavidyalaya
University of Lucknow
Lucknow, India

Vijendra Singh Solanki
Department of Chemistry
Institute of Science and Research
IPS Academy
Indore, India

&

Sreekantha B. Jonnalagadda
School of Chemistry and Physics
University of KwaZulu-Natal, Westville Campus
Durban, South Africa

ACKNOWLEDGEMENTS

With a profound sense of reverence and gratitude, we express our hearty indebtedness to Benthem Science Publishers for their specialized support and encouragement given to us for completing this project and publishing it in the form of an edited book.

Our sincere gratitude goes to all the authors who contributed their time and expertise to this book. Secondly, we wish to acknowledge the valuable contributions of the reviewers regarding the improvement of quality, coherence, and content presentation of chapters. Without their support, this book would not have become a reality. Their research and input were essential for the successful completion of this work.

We wish to pay heartfelt gratitude to our mentors for their contribution to this work, and we most gratefully acknowledge the constant encouragement and support given by our friends and family members for their unconditional and whole-hearted support to complete this work successfully.

Finally, we would like to thank The Great Almighty who has always been there to bless us to achieve success in all our endeavors. This achievement would not have been possible without His mercy.

List of Contributors

Amrit K. Mitra	Department of Chemistry, Government General Degree College, Singur, Hooghly, West Bengal, India
Anupma Singh	Department of Chemistry, DDU Govt. P.G. College, Sitapur, Lucknow, India
Anamika Srivastava	Department of Chemistry, Banasthali Vidyapith, Banasthali, Rajasthan, India
Annu Yadav	Department of Chemistry, Banasthali Vidyapith, Banasthali, Rajasthan, India
Azhar Ullah Khan	Department of Chemistry, School of Life and Basic Sciences, Jaipur National University, Jaipur, India
Deepankshi Shah	Department of Environmental Science, Parul Institute of Applied Sciences, Parul University, Vadodara, Gujarat, India
E. Rajalakshmi	Department of Chemistry, Bishop Heber College, Tiruchirappalli, Tamil Nadu, India
Gitanjali Arora	Department of Chemistry, Banasthali Vidyapith, Banasthali, Rajasthan, India
J. Princymerlin	Department of Chemistry, Bishop Heber College, Tiruchirappalli, Tamil Nadu, India
Jaya Dwivedi	Department of Chemistry, Banasthali Vidyapith, Banasthali, Rajasthan, India
Keshav Lalit Ameta	Centre for Applied Chemistry, School of Applied Material Sciences, Central University of Gujrat, Gandhinagar, Gujrat, India
M. Nanda	Bioresource Product Research Laboratory, Department of Botany, School of Life Science, Guru Ghasidas Vishwavidyalaya (A Central University), Bilaspur, Chhattisgarh, India
Manoj Kumar	Department of Hydro and Renewable Energy, Indian Institute of Technology Roorkee, Roorkee, Uttarakhand, India
Manish Srivastava	Department of Chemistry, University of Allahabad, Prayagraj, Uttar Pradesh, India
M.K. Gupta	Department of Chemistry, H. R. College, Jai Prakash University, Amnour, Chapra, India
Mohd. Tariq	Department of Life Sciences, Parul Institute of Applied Sciences, Parul University, Vadodara, Gujarat, India
N. Maurya	Department of Chemistry, Kamla Rai College, Gopalganj, Jai Prakash University, Chapra, India
Nakul Kumar	Gandhinagar Institute of Science, Gandhinagar University, Gandhinagar, Gujarat, India
Neha Agarwal	Department of Chemistry, Navyug Kanya Mahavidyalaya, University of Lucknow, Lucknow, India
Neetu Singh	Department of Physics, Government Degree College, Kuchalai, Sitapur, Lucknow, India
Nirmala Kumari Jangid	Department of Chemistry, Banasthali Vidyapith, Banasthali, Rajasthan, India
Pankaj Kumar	Department of Environmental Science, Parul Institute of Applied Sciences, Parul University, Vadodara, Gujarat, India

Rajendra	Department of Chemistry, Banasthali Vidyapith, Banasthali, Rajasthan, India
Ramesh Kumar	Department of Environmental Science, School of Earth Sciences, Central University of Rajasthan, Ajmer, Rajasthan, India
Rekha Sharma	Department of Chemistry, Banasthali Vidyapith, Banasthali, Rajasthan, India
Ruchi Shrivastava	Department of Chemistry, Institute of Science and Research, IPS Academy, Indore, Madhya Pradesh, India
Sreekantha B. Jonnalagadda	School of Chemistry and Physics, University of KwaZulu-Natal, Westville Campus, Durban, South Africa
Sankara Rao Miditana	Department of Chemistry, Government Degree College, Puttur, Tirupathi, Andhra Pradesh, India
Saivenkatesh Korlam	Department of Botany, SVA Government Degree College, Srikalahasti, Tirupati, Andhra Pradesh, India
S. Mishra	Department of Chemistry, Jai Prakash University, Chapra, India
Snigdha Singh	Department of Environmental Science, Parul Institute of Applied Sciences, Parul University, Vadodara, Gujarat, India
Sunil Soni	School of Environment and Sustainable Development, Central University of Gujarat, Gandhinagar, India
S. Padmavathi	Department of Botany, SVA Government Degree College, Srikalahasti, Tirupati, Andhra Pradesh, India
S. Agrawal	Bioresource Product Research Laboratory, Department of Botany, School of Life Science, Guru Ghasidas Vishwavidyalaya (A Central University), Bilaspur, Chhattisgarh, India
S.K. Shahi	Bioresource Product Research Laboratory, Department of Botany, School of Life Science, Guru Ghasidas Vishwavidyalaya (A Central University), Bilaspur, Chhattisgarh, India
Shivraj Gangadhar Wanale	School of Chemical Sciences, Swami Ramanand Teerth Marathwada University, Nanded, Maharashtra, India
Shipra Choudhary	Department of Microbiology and Biotechnology, Meerut Institute of Engineering & Technology, Meerut, Uttar Pradesh, India
Shraddha Bais	Department of Chemistry, Institute of Science and Research, IPS Academy, Indore, Madhya Pradesh, India
Shruti	Department of Chemistry, Banasthali Vidyapith, Banasthali, Rajasthan, India
S. Ambika	Department of Chemistry, Bishop Heber College, Tiruchirappalli, Tamil Nadu, India
Vijendra Singh Solanki	Department of Chemistry, Institute of Science and Research, IPS Academy, Indore, India
Virendra Kumar Yadav	Department of Life Sciences, Hemchandracharya North Gujarat University, Matarvadi Part, Gujarat, India
Vimala Bind	Department of Zoology, Navyug Kanya Mahavidyalaya, University of Lucknow, Lucknow, India
V. J. Maodiswari	Department of Botany, Bishop Heber College, Tiruchirappalli, Tamil Nadu, India

Y. Manojkumar   Department of Chemistry, Bishop Heber College, Tiruchirappalli, Tamil Nadu, India

CHAPTER 1

Emerging Pollutants in Aquatic Environment: Critical Risk Assessment and Treatment Options

Neha Agarwal[1,*]**, Vijendra Singh Solanki**[2]**, Sreekantha B. Jonnalagadda**[3]**, Keshav Lalit Ameta**[4]**, Neetu Singh**[5]**, Anupma Singh**[6] **and Vimala Bind**[7]

[1] *Department of Chemistry, Navyug Kanya Mahavidyalaya, University of Lucknow, Lucknow, India*

[2] *Department of Chemistry, Institute of Science and Research, IPS Academy, Indore, India*

[3] *School of Chemistry and Physics, University of KwaZulu-Natal, Westville Campus, Durban, South Africa*

[4] *Centre for Applied Chemistry, School of Applied Material Sciences, Central University of Gujrat, Gandhinagar, Gujrat, India*

[5] *Department of Physics, Government Degree College, Kuchalai, Sitapur, Lucknow, India*

[6] *Department of Chemistry, DDU Govt. P.G. College, Sitapur, Lucknow, India*

[7] *Department of Zoology, Navyug Kanya Mahavidyalaya, University of Lucknow, Lucknow, India*

Abstract: The chemical compounds that have been identified as dangerous to the environment, ecosystem and human health are classified as Emerging Pollutants (EPs). EPs include a variety of compounds such as dyes, pesticides, antibiotics, drugs, endocrine disruptors, hormones, industrial wastes and chemicals, and microplastics. These pollutants are malignant and non-biodegradable in nature, so they are responsible for the unhealthy and unsustainable environment. The occurrence of these pollutants has raised global concerns not only in various environmental matrices (air, water, and soil) but also in biological systems due to their toxic nature. These pollutants get accumulated in the environment and ecosystem and cause intensified environmental problems, global warming, deterioration of soil quality, the greenhouse effect, and ecological imbalance. Consequently, they affect the quality of life and the maintenance of the environment on a global level. Recent research indicates that if this trend is continued, situations will worsen in the near future. Sustainable solutions, such as bioremediation, nano-bioremediation, microbial degradation *etc.*, are becoming increasingly important for the removal of these EPs as an efficient tool for sustainable development and pollution control. Therefore, the main aim of this chapter is to assess the current threats and future challenges associated with emerging pollutants so that focus can be drawn on sustainable green solutions for a greener and healthier environment.

* **Corresponding author Neha Agarwal**: Department of Chemistry, Navyug Kanya Mahavidyalaya, University of Lucknow, Lucknow, India; Tel: +91-9454784074; E-mail: nehaagarwal4074@gmail.com

Keywords: Bioremediation, Emerging contaminants, Ecosystem, Environment, Green solutions, Nano-bioremediation, Non-biodegradable, Pollutants, Pollution control, Sustainable.

INTRODUCTION

With the rapid technological advancements and industrialization, the environmental quality has deteriorated, which is an alarming sign for sustainability. Different categories of emerging contaminants (ECs), like heavy metals, pesticides, pharmaceuticals, endocrine disrupting agents, personal care products, dyes, detergents, plastics, *etc.*, are causing menace at a global level as they adversely affect the environment, ecosystem and living beings [1 - 3]. Among different types of pollution, water pollution is an important subclass that severely affects global life. Water is a vital component of life; it has been contaminated due to high industrialization in recent decades and severely affects the quality of life [4]. For the last few decades, EPs have attracted worldwide attention, and many attempts have been made to mitigate the release and accumulation of EPs into the environment to prevent the dangerous impact on the environment. Many studies have been conducted to monitor progress in this field. For instance, in a study performed by Barbosa *et al.*, various treatment techniques were reviewed with their removal efficacy of EPs that concluded the future research perspective for a risk-free environment. They also reviewed the interaction of microplastics with pollutants and concluded that marine microplastic debris may dangerously affect human health [5]. Another review done by Taheran *et al.* emphasized that if EPs are present in scarce concentrations, conventional sewage treatment processes are not capable of treating them efficiently [6]. In fact, chemical and physical methods that are used to treat effluents do not degrade these pollutants completely, but rather change their forms, which are more toxic to the environment and human health, even in low concentrations [7]. Literature studies also suggest that current information on mechanisms available for water remediation needs to be updated to avoid future risks to the ecosystem and environment [8, 9].

Due to high costs, difficult techniques and improper efficiency involved, the issue of emerging pollutants has become a challenge. Therefore, there is an urgent need to protect the environment and living beings by developing sustainable methods for the removal of these pollutants [10]. Bioremediation is the most promising technology over conventional methods of wastewater treatment because it is an eco-friendly and cost-effective technique with the possible recovery of elements and for solving environmental problems [11, 12]. Nanotechnology has also emerged as a promising technology, which has shown great potential in various fields along with the treatment of pollutants [13]. Currently, bionanotechnology is attracting great attention in the remediation of pollutants as green solutions, which are eco-friendly, cost-effective and easy-to-handle tools for the bioremediation of wastewater and other categories of environmental pollutants. This chapter presents a concept to assess the occurrence, fate and risk assessment of emerging pollutants and also provides an overview of sustainable solutions for water resource management.

EMERGING POLLUTANTS IN AQUATIC ENVIRONMENT: TRANSPORT, FATE AND BIOACCUMULATION

As a result of uncontrolled urbanization, industrial development, healthcare activities and other anthropogenic activities, there is a rapid increase in EPs on a global level [14]. The synthetic persistent organic chemicals that adversely affect the ecosystem and human health but are not monitored in the environment are known as EPs. Different routes and fate of EPs are shown in Fig. (**1**).

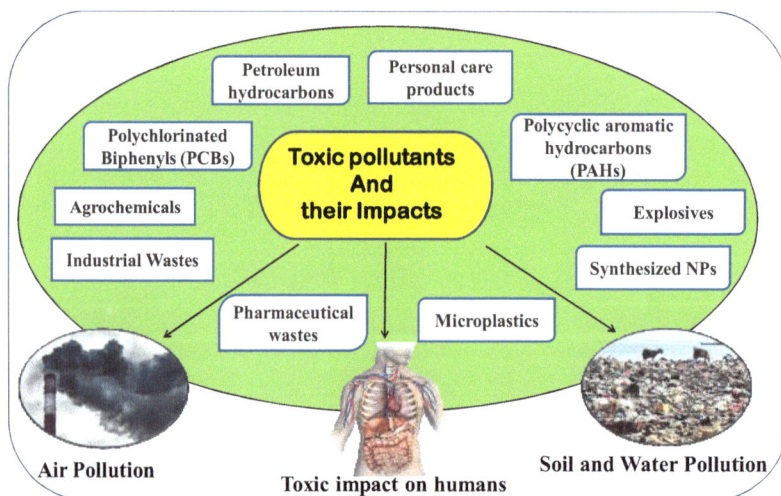

Fig. (1). Different types of toxic EPs and their impacts.

Many studies have been conducted on the route and fate of EPs in aquatic environments [15]. However, EPs can enter into an aqueous environment through various direct and indirect routes and can get bioaccumulated through food chains and food webs, causing serious health hazards to living beings. Therefore, many studies have focused on their fates and bioaccumulation [16 - 18]. In aquatic environments, the concentrations of EPs can vary over a wide range from ng/L to g/L. Their toxicological effects on living organisms may result in acute and chronic toxicity, endocrine disruption, resistance to antibiotics and human health hazards [19]. According to a recent study, EPs, such as pharmaceuticals, pesticides, and phosphorus-based flame retardants, have been reported in marine bivalves in municipal wastewater and landfill leachate effluent discharges in Hong Kong [20]. For the first time, the presence of ninety-nine EPs was reported in the gonads of sea urchin *Paracentroyus lividus* by Rocha *et al.* (2018) [21]. Another study gave the first evidence of the presence of benzotriazoles (BTs) degradation products (BTTPs) in urban aquifers that may severely deteriorate the groundwater quality [22]. Another group of pollutants, known as "contaminants of emerging concern", are also released into the environment, surface and groundwater resources [23, 24]. Although some EPs have existed in the environment for many years, which might be very harmful to our ecosystems, their occurrence has been analyzed only recently [25].

The most prominent classes of EPs are dyes, pesticides, pharmaceuticals, disinfection by-products, industrial chemicals, and plastics. For example, pharmaceuticals represent a subclass of emerging pollutants due to their uncontrolled use to treat a wide variety of diseases and their diverse physico-chemical and biological characteristics [26]. Pharmaceutical compounds, after excretion in the original form or as metabolites, can be found in different varieties such as urban wastewater, hospital sewers, and surface waters [27, 28].

The World Health Organization has also declared that resistance to antibiotics is the biggest threat to global health and the environment. Heavy metals also have the property of bioaccumulation and environmental persistence as EPs, which enter the aquatic systems through various routes and affect the ecosystem and human health negatively [29]. Plastic waste, after accumulation in the environment, is broken down into micro and nano plastics, gradually forming nano-plastics of less than 5 mm in size [30]. Once accumulated, microplastics migrate and diffuse into the environment and carry other environmental pollutants like antibiotics and heavy metals [31, 32]. Since these substances have a potential impact on aquatic life and human health, and there is a lack of knowledge regarding their environmental implications and analytical and sampling techniques, urgent action is required to tackle this problem at multiple levels.

CHALLENGES AND RISKS ASSOCIATED WITH EMERGING POLLUTANTS

Although EPs frequently occur in various environmental matrices on a global level, the knowledge of their hazards and ecological risks is not sufficient. EPs, even in low concentrations of ng/L, can have adverse effects on living beings, such as genotoxicity, carcinogenicity, hormonal interference in fishes, endocrine disruption, and immune toxicity [33]. Endocrine disruptors are highly toxic to wildlife, altering the reproductive behavior, and sexually dimorphic neuroendocrine system, and also to human beings by creating problems in the cardiovascular system and causing abnormal neural behaviors. They are also linked to diabetes and obesity. Similarly, perfluorinated compounds get bioaccumulated in fish and fishery products and have adverse effects on developmental and reproductive systems [34]. Engineered nanoparticles also have negative toxicological impacts and are very harmful to human health, resulting in cytotoxicity, oxidative stress, carcinogenic effects, inflammatory effects in the lungs, genotoxicity, and augmented intestinal collagen staining [35]. It is essentially important to understand that EPs are not isolated in the environment but in complex mixtures of contaminants [36, 37]. The mixture of ECs can have additive or multiplicative ecotoxicological effects [38]. The joint toxic effects may result in antagonistic interactions, which can lead to a cock tail effect [39].

Many studies had confirmed that many EPs are not dangerous for the environment if their concentrations detected in soil and water are very low [40]. However, relevant concentrations (ng/L or μg/L) can alter and have a negative impact on ecological interactions [41, 42]. According to several studies, it has been found that after entering the environment, these pollutants are transformed into metabolic by-products under different environmental conditions, such as degradation in the presence of light, oxidation and reduction, and microbial decomposition, but the risk analysis of these pollutants remains insufficient [43]. Many studies have demonstrated the effects of EPs on animal behavior and altered microbial communities and their function even in trace concentrations [44, 45]. EPs may also create resistance to antibiotics [46]. The effects of pesticides and pharmaceuticals on fluvial biofilms in a Mediterranean river were also studied, and it was observed that autotrophic biomass increased peptidase and decreased the photosynthetic efficiency when biofilms were shifted to highly polluted areas of EPs. In low concentrations also, heavy metals can affect and damage multiple organs such as kidneys, lungs, liver, esophagus, skin, and stomach, and can also cause neurodegenerative diseases and disorders [47, 48]. Heavy metals can also cause oxidative damage and endocrine disruption by accumulating in several organs in aquatic organisms, which can also affect their survival and growth [49]. Consequently, the potential ecological impacts of EPs require the development of efficient technologies that can easily remove them from water and other environmental matrices.

POSSIBLE SOLUTIONS FOR DEGRADATION AND REMOVAL OF EPS

As a result of the increasing risks due to the continuous occurrence and accumulation of EPs in the environment, their treatment and eradication have become necessary but cumbersome [50]. EPs that commonly occur in an aqueous environment are difficult to remove by applying conventional treatment technologies, such as physical and chemical methods, but these are not degraded completely and change their forms [51]. These modified forms are highly toxic and can cause damage even in trace concentration [52]. Over the conventional methods, bioremediation is considered the most promising technology for cleaning up the environment due to its eco-friendly and cost-effective nature. This technique can recover useful elements and can solve environmental problems [53]. Some of the commonly used strategies to mitigate the emission of EPs in different environmental matrices are given below.

Conventional Treatment Techniques and Advanced Oxidation Processes

Water pollution by EPs is a serious problem due to their continuous discharge and accumulation through various routes into the environment. Conventional treatment techniques such as membrane bioreactor and activated sludge have been used to remove biodegradable contaminants but failed to completely remove these EPs from wastewater [54]. Therefore, advanced oxidation processes (different photochemical and chemical processes, as mentioned in Fig. (**2**)) were used to treat wastewater. Both the traditional treatment methods to treat wastewater are effective to some extent and are still used today. However, the rising water scarcity requires the reuse of water after absolute filtration. The primary and secondary treatments are not very effective in meeting the standard of reusable water that can be used for domestic and industrial purposes [55]. Hence advanced treatment methods are required after the secondary treatment that helps in further removing the toxic materials [56].

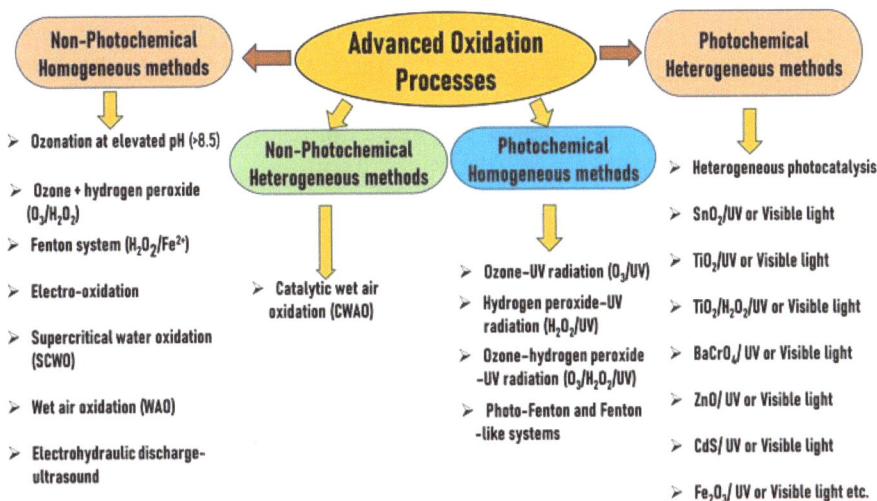

Fig. (2). Various advanced oxidation processes for the treatment of pollutants.

To solve this problem, activated sludge and conventional wastewater treatment processes can be used in combination with advanced oxidation processes such as ozonation, photodegradation, and biodegradation, which increase the efficiency of the treatment to a greater extent. However, the major drawback of this combination of processes is the high energy consumption and high costs involved. For the degradation of EPs, it is very important to know the oxidation potential of conventional and advanced processes in wastewater treatment plants [57, 58]. For instance, some EPs could be degraded by chlorine, such as methyl indole, chlorophene and nortriptyline, while benzotriazole and N, N-diethyl-m-toluamide were found to be recalcitrant and were not altered by chlorine [58, 59]. Another study reported that chlorine and ozone could degrade part of the EPs present in water and confirmed that EPs, which are easily oxidized by chlorine, are also oxidized by ozone with the same efficiency. Conventional and advanced oxidation processes such as chemical precipitation, ion exchange, and electrochemical removal, as discussed above, may remove some EPs from wastewater and can reduce their concentration in potable water but have many drawbacks, including high-energy consumption, incomplete removal, production of toxic sludge, and high operational and maintenance cost, which can result in improper and inadequate application of these technologies. Therefore, there is a need to develop effective and environmentally friendly solutions that include biological and nanotechnology approaches for the effective removal of these contaminants from the global environment [60, 61].

Advanced oxidation processes (AOPs) have shown a promising effect for treating contaminated water and also for the removal of naturally occurring toxins, impurities of emerging concern, pesticides, and other harmful contaminants, *etc.* AOPs include several methods for creating hydroxyl radicals and some other reactive oxygen species like superoxide anion-radical and hydrogen peroxide. However, hydroxyl radicals are still the most common species that enhance the effectivity of AOPs. Most of the organic compounds react with the hydroxyl radicals to form a carbon-centered radical. Further, this carbon-centered radical reacts with the oxygen molecule to generate a peroxyl radical, which undergoes further reactions and ultimately produces oxidation products such as ketones, aldehydes, and alcohols [62]. Hydroxyl radical is also able to detach an electron from the electron-rich substrates to create a radical cation, which is quickly hydrolyzed in an aqueous environment that leads to the formation of an oxidized product. It has been observed that the oxidation products are often less toxic and more receptive to bioremediation. Advanced oxidation processes involve UV/H_2O_2, UV/O_3, Fenton, sonolysis, nonthermal plasmas, radiolysis, photocatalysis, and supercritical water oxidation processes. Sonolysis and radiolysis of aqueous media can form hydroxyl radicals when the chemical oxidants are not present in the water. On the other hand, photochemical methods like photo-Fenton-type processes require the presence of a catalyst or precursor to generate the hydroxyl radical [63]. Sonolysis produces the hydroxyl radicals at or near a gas−liquid interface, while the radiolysis process of aqueous media generates those hydroxyl radicals that are considered to be homogeneous for the timescales [64, 65].

The sonolysis method is not cost-effective as the operating cost is very high for large-scale water treatment, while the radiolysis treatments are low cost as the operation cost is quite low in comparison to sonolysis methods. Fenton and photo-Fenton-type processes have also grabbed significant attention for the treatment of water [66]. However, the consumption of Fe(II) and the requirement for the removal of generated iron sludge during Fenton-type advanced oxidation processes have restricted its application for the treatment of water. These restrictions can be controlled by photo-Fenton processes that effectively utilize solar irradiation to recreate the Fe (II) species that leads to hydroxyl radical production. The formation of the hydroxyl radical using various homogeneous and heterogeneous AOPs involves distinct reaction dynamics that consequently lead to different reaction pathways. A more comprehensive understanding of the structure-reactivity relationships for the groups of compounds for individual treatment processes is based on kinetic data for the identification of an effective AOP.

Bioremediation for Achieving Environmental Sustainability

It is essential to incorporate the ecological and biological components to attain the aims of environmental sustainability that have been lacking in conventional and advanced oxidation techniques. Environmentally friendly solutions are sometimes neglected in favor of technical solutions. As a result, for a sustainable ecosystem, biological treatment methods must be implemented. As one of the most favorable biotechnological applications, bioremediation uses microbial enzymes to break down harmful organic pollutants into less toxic compounds. The widespread use of genetically engineered microorganisms (GEMs) can also help to eliminate toxic organic pollutants such as naphthalene, benzene, petroleum, and other organic compounds [67]. Waste management can be done efficiently through bioremediation because persistent organic pollutants that are hard to break down can be successfully remediated through bioremediation. Bioremediation is the process of removing contaminated materials from the environment using bacteria, algae, fungi, plants and yeast [68]. Different enzymes produced by these microorganisms speed up biochemical reactions that break down pollutants through metabolic pathways [69, 70]. Enzymes play a very crucial role in the process of metabolism at every stage [71]. These enzymes must act on the pollutants for their bioremediation, and optimum environmental parameters are required for speedy microbial growth and degradation during biodegradation [72]. Several factors, such as soil type, physical, chemical, and biological factors, source of carbon and nitrogen, and type of microorganisms, affect the process of bioremediation [12, 72].

Different microorganisms can degrade EPs under aerobic and aerobic conditions. Different aerobic species of bacteria such as *Rhodococcus, Mycobacterium, Bacillus, Pseudomonas, and Sphingomonas* can degrade a variety of complex organic compounds such as pesticides, hydrocarbons, and polyaromatic compounds [73]. These microorganisms mineralize these contaminants and are used as a source of carbon and energy [74]. Bacterial species such as *Pseudomonas*, *Aeromonas,* and sulfate-reducing bacteria can be bioremediate EPs under anaerobic conditions. Microbial degradation of azo dyes occurs under anaerobic environmental conditions by the oxidation of organic substrates [75]. Bioremediation can be used in multiple ways; some of the commonly used methods are mentioned in Fig. (**3**).

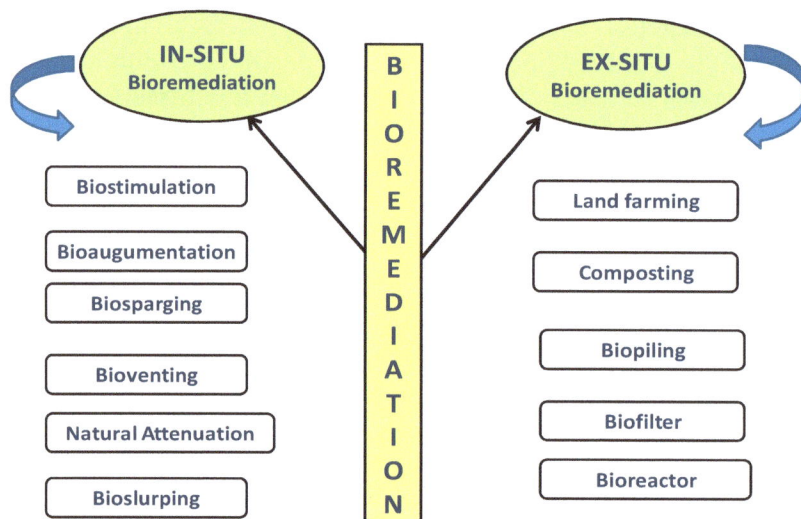

Fig. (3). Different bioremediation techniques.

Bioremediation can combat serious environmental issues in an environment-friendly and economical manner, and it has many advantages over traditional and physicochemical methods, such as cost and energy efficiency, specificity, selectivity, minimal requirement, *etc.* However, certain limitations are associated with bioremediation, such as the degradation of a toxic compound is time consuming. Moreover, its applications are restricted to severely contaminated sites with hazardous and toxic pollutants [76 - 79]. Therefore, given the benefits and drawbacks of every method and to tackle remediation problems, remediation methods can be integrated for better results. Nanobioremediation is one of the latest methods which have drawn a lot of attention from researchers in the past few decades. The benefits of nanotechnology and the advantages of bioremediation are successfully integrated into nanobioremediation.

Phytoremediation for Achieving Environmental Sustainability

Bioremediation with plants is known as phytoremediation. It is a natural biological process that degrades harmful EPs and recalcitrant xenobiotics that cause pollution in the environment. It is an eco-friendly method that can be successful. The plants absorb heavy metals and remove toxins from the water and soil in a cost-effective way. Contaminated soils can be cleaned up using plant extracts to remove pollutants and lower their bioavailability in soil [80]. Varieties of processes are employed in phytoremediation depending on the quantity and form of the pollutant [81]. Common methods of phytoremediation are extraction, sequestration, rhizofilteration, phytostabilization and transformation for removing heavy metals. Pollutants from the roots and shoots are removed as an important part of phytoremediation. Plants are better candidates for phytoremediation because they ideally absorb Cu, Zn, Cr and Ni. According to a study, microorganisms in the rhizosphere can increase the availability of heavy metals and their uptake by plants [82]. The accumulation of heavy metal in plants depends upon the metal, its solubility, translocation, and plant species [83]. Metals, pesticides, crude oils, explosives and other toxic pollutants have been lowered through phytoremediation processes around the globe.

Nanotechnology as a Sustainable Solution

Nanotechnology has numerous applications as it has high removal efficiency, less time period, and is economical in comparison to other technologies of environmental remediation. Nanotechnology has given a new perspective to wastewater treatment [84]. Depending on their shape, size, structure, and composition, different varieties of nanomaterials such as nanofibres, nanodots, nanotubes, nanoshells, nanocomposites and nanoclusters are used for eliminating contaminants from different environmental matrices [85]. Green synthesis of nanoparticles has multiple sources of synthesis, such as bacteria, fungi, algae and plant extract. The use of green synthesized nanoparticles for the treatment of EPs and other pollutants makes nanotechnology a promising alternative to the current forms of treatment [13].

In the green synthesis of nanoparticles, several biological factors, such as pH, reaction medium, and temperature, influence the properties of the formed nanoparticles. Different organisms can generate different compositions of metallic nanoparticles with different sizes, distributions and morphologies, such as spherical, triangular, cubic, and rod shape [86]. Large-scale production of nanoparticles is mainly governed by the choice of bacteria and methods of synthesis [87]. Biodegradable waste can also be incorporated into the process of synthesizing nanoparticles, which not only saves our environment but also prevents the exhaustion of any natural resource [88]. When bioremediation is combined with nanotechnology to achieve remediation, it is known as nano-bioremediation. Nano-bioremediation is more efficient, less time-consuming and environment friendly. Through an integrated approach, the disadvantages of individual technologies can be removed and provide better results. Recently, a group of workers combined both technologies for the removal of chlorinated aliphatic hydrocarbons and confirmed the integrated potential of nano-bioremediation by efficiently removing a wide range of chlorinated aliphatic hydrocarbons [89]. Polychlorinated biphenyls (PCBs) have been investigated by Le *et al.* (2015) by the integrated approach, and they found that the treatment of PCBs with Pd/Fe NPs followed by bioremediation with B. xenovorans could effectively transform PCBs into less toxic compounds [90]. Carbon nanotubes (CNTs) and carbon dots, along with bioremediation, are also successfully used for contaminant removal. Though highly efficient and frequently used, the toxicity of NPs for microorganisms is well seen in the literature. Therefore, efforts should be made to develop integrated approaches that are non-toxic and sustainable.

The sustainability of the environment is necessary for the survival of living beings. To sustain life, we must conserve our environment, ecosystem and habitats. The presence of toxic pollutants in the global environment in different forms cannot be denied. The long-term impacts of these pollutants seriously impact our environment, ecosystems, humans, and biota. Because of the limitations of available traditional and advanced remediation techniques, we should switch to green substitutes for the rehabilitation of the environment. Biological solutions, including plants and microorganisms, can facilitate the conservation and restoration of ecosystems in a sustainable manner. To relieve our planet from the undesirable anthropogenic concerns that generate huge amounts of pollutants, an integrated approach is the need of the hour. Though physiochemical and traditional treatments are effective, they have certain limitations and are unsustainable. Biological treatments are economical but not so effective and are non-consistent. However, to accomplish environmental sustainability goals, biological techniques, in combination with the latest techniques, must be used to their full potential to degrade pollutants in a green manner. Moreover, future technologies must be both effective and eco-friendly as well. Also, these technologies should be capable of removing the numerous types of emerging pollutants with low cost and energy consumption, as the existing system for testing emerging pollutants is a highly energy-consuming process. So, it is necessary to enhance energy potency to reduce the usage of energy.

Furthermore, the efficacy of the treatment needs to be adjusted for the concentrations of emerging pollutants in the aquatic environment.

REFERENCES

[1]   Bunke D, Moritz S, Brack W, Lopex Herraez D, Posthuma L, Nuss M. Developments in society and implications for emerging pollutants in the aquatic environment. Environ Sci Eur 2019; 31(1): 32.

[2]   Caliman FA, Gavrilescu M. Pharmaceuticals, personal care products and endocrine disrupting agents in the environment - A review. Clean 2009; 37(4-5): 277-303.
[http://dx.doi.org/10.1002/clen.200900038]

[3]   Gavrilescu M, Demnerová K, Aamand J, Agathos S, Fava F. Emerging pollutants in the environment: present and future challenges in biomonitoring, ecological risks and bioremediation. N Biotechnol 2015; 32(1): 147-56.
[http://dx.doi.org/10.1016/j.nbt.2014.01.001] [PMID: 24462777]

[4]   Bharti PK, Kumar P, Singh V. Impact of industrial effluents on ground water and soil quality in the vicinity of industrial area of Panipat city, India. J Appl Nat Sci 2013; 5(1): 132-6.
[http://dx.doi.org/10.31018/jans.v5i1.294]

[5]   Barboza LGA, Dick Vethaak A, Lavorante BRBO, Lundebye AK, Guilhermino L. Marine microplastic debris: An emerging issue for food security, food safety and human health. Mar Pollut Bull 2018; 133: 336-48.
[http://dx.doi.org/10.1016/j.marpolbul.2018.05.047] [PMID: 30041323]

[6]   Taheran M, Naghdi M, Brar SK, Verma M, Surampalli RY. Emerging contaminants: Here today, there tomorrow! Environ Nanotechnol Monit Manag 2018; 10: 122-6.
[http://dx.doi.org/10.1016/j.enmm.2018.05.010]

[7]   Rajan M. cyanobacteria as a potential source of phycoremediation from textile industry effluent. J Bioremediat Biodegrad 2014; 2014: 1-4.

[8]   Rasheed T, Bilal M, Nabeel F, Adeel M, Iqbal HMN. Environmentally-related contaminants of high concern: Potential sources and analytical modalities for detection, quantification, and treatment. Environ Int 2019; 122: 52-66.
[http://dx.doi.org/10.1016/j.envint.2018.11.038]

[9]   Geissen V, Mol H, Klumpp E, et al. Emerging pollutants in the environment: A challenge for water resource management. Int Soil Water Conserv Res 2015; 3(1): 57-65.
[http://dx.doi.org/10.1016/j.iswcr.2015.03.002]

[10]   Ferreiro C, Gómez-Motos I, Lombraña JI, et al. Contaminants of emerging concern removal in an effluent of wastewater treatment plant under biological and continuous mode ultrafiltration treatment. Sustainability 2020; 12(2): 725.
[http://dx.doi.org/10.3390/su12020725]

[11]   Sahu O. Reduction of organic and inorganic pollutant from waste water by algae. Int Lett Nat Sci 2014; 13: 1-8.
[http://dx.doi.org/10.56431/p-8aq47u]

[12]   Agarwal N, Solanki VS, Gacem A, et al. Bacterial laccases as biocatalysts for the remediation of environmental toxic pollutants: a green and eco-friendly approach—A Review. Water 2022; 14(24): 4068.
[http://dx.doi.org/10.3390/w14244068]

[13]   Agarwal N, Solanki VS, Pare B, Singh N, Jonnalagadda SB. Current trends in nanocatalysis for green chemistry and its applications- a mini-review. Curr Opin Green Sustain Chem 2023; 41(2): 100788.
[http://dx.doi.org/10.1016/j.cogsc.2023.100788]

[14]   Petrie B, Barden R, Kasprzyk-Hordern B. A review on emerging contaminants in wastewaters and the

environment: Current knowledge, understudied areas and recommendations for future monitoring. Water Res 2015; 72: 3-27.
[http://dx.doi.org/10.1016/j.watres.2014.08.053] [PMID: 25267363]

[15] Gomes I B, Maillard J Y, Simoes L, *et al.* Emerging contaminans affect the microbome of ater systems-strategies for the mitigation. NPJ Clean Water 2020; 3(39).
[http://dx.doi.org/10.1038/s41545-020-00086-y]

[16] Gogoi A, Mazumder P, Tyagi VK, Tushara Chaminda GG, An AK, Kumar M. Occurrence and fate of emerging contaminants in water environment: A review. Groundw Sustain Dev 2018; 6: 169-80.
[http://dx.doi.org/10.1016/j.gsd.2017.12.009]

[17] Salthammer T. Emerging indoor pollutants. Int J Hyg Environ Health 2020; 224: 113423.
[http://dx.doi.org/10.1016/j.ijheh.2019.113423] [PMID: 31978722]

[18] Tang Y, Yin M, Yang W, Li H, Zhong Y, Mo L, *et al.* Emerging pollutants in water environment: Occurrence, monitoring, fate, and risk assessment. Water Environment Research. John Wiley and Sons Inc. 2019; 91: pp. 984-1.

[19] Hlavínek P, Žižlavská A. Occurrence and removal of emerging micropollutants from urban wastewater. Water Management and the Environment: Case Studies. Cham;: Springer International Publishing 2018; pp. 231-54.
[http://dx.doi.org/10.1007/978-3-319-79014-5_11]

[20] Burket SR, Sapozhnikova Y, Zheng JS, Chung SS, Brooks BW. At the intersection of urbanization, water, and food security: determination of select contaminants of emerging concern in mussels and oysters from hong kong. J Agric Food Chem 2018; 66(20): 5009-17.
[http://dx.doi.org/10.1021/acs.jafc.7b05730] [PMID: 29526083]

[21] Rocha AC, Camacho C, Eljarrat E, *et al.* Bioaccumulation of persistent and emerging pollutants in wild sea urchin *Paracentrotus lividus.* Environ Res 2018; 161: 354-63.
[http://dx.doi.org/10.1016/j.envres.2017.11.029] [PMID: 29195184]

[22] Trček B, Žigon D, Zidar VK, Auersperger P. The fate of benzotriazole pollutants in an urban oxic intergranular aquifer. Water Res 2018; 131: 264-73.
[http://dx.doi.org/10.1016/j.watres.2017.12.036] [PMID: 29304380]

[23] Agarwal N, Solanki VS, Awasthi RR, Bind V, *et al.* Release and accumulation of pharmaceuticals in the environment: critical risk assessment for environment, ecosystem and human health in book pharmaceuticals: Boon or Bane. Newyork: Nova Science Publishers 2022; pp. 1-23.

[24] Khan S, Naushad M, Govarthanan M, Iqbal J, Alfadul SM. Emerging contaminants of high concern for the environment: Current trends and future research. Environ Res 2022; 207: 112609.
[http://dx.doi.org/10.1016/j.envres.2021.112609] [PMID: 34968428]

[25] Rivera-Utrilla J, Sánchez-Polo M, Ferro-García MÁ, Prados-Joya G, Ocampo-Pérez R. Pharmaceuticals as emerging contaminants and their removal from water. A review. Chemosphere 2013; 93(7): 1268-87.
[http://dx.doi.org/10.1016/j.chemosphere.2013.07.059] [PMID: 24025536]

[26] Basheer AA. New generation nano-adsorbents for the removal of emerging contaminants in water. J Mol Liq 2018; 261: 583-93.
[http://dx.doi.org/10.1016/j.molliq.2018.04.021]

[27] Yehya T, Favier L, Kadmi Y, Audonnet F, Fayad N, Gavrilescu M, *et al.* Removal of carbamazepine by electrocoagulation: investigation of some key operational parameters. Enviro Eng Management J (EEMJ) 2015; 14(3) 2015; 14: (3).

[28] WW, Nödler K, Farinelli A, Blum J, Licha T. Environmental Engineering and Management. Environmental Engineering 2018; 17.

[29] Meng T, Wang C, Florkowski WJ, Yang Z. Determinants of urban consumer expenditure on aquatic products in Shanghai, China. Aquac Econ Manag 2023; 27(1): 1-24.

[http://dx.doi.org/10.1080/13657305.2021.1996480]

[30]   Thompson RC, Olsen Y, Mitchell RP, Davis A, Rowland SJ, John AWG, *et al.* Lost at sea: where is all the plastic? Science 2004; 304(5672): 838.

[31]   Karbalaei S, Hanachi P, Walker TR, Cole M. Occurrence, sources, human health impacts and mitigation of microplastic pollution Environm Sci Pol Res. Springer Verlag; 2018; 25: pp. 36046-63.
[http://dx.doi.org/10.1007/s11356-018-3508-7]

[32]   Moreno-Jiménez E, Leifheit EF, Plaza C, *et al.* Effects of microplastics on crop nutrition in fertile soils and interaction with arbuscular mycorrhizal fungi. J Sustain Agric Environ 2022; 1(1): 66-72.
[http://dx.doi.org/10.1002/sae2.12006]

[33]   Miraji H, Othman OC, Ngassapa FN, Mureithi EW. Research Trends in Emerging Contaminants on the Aquatic Environments of Tanzania. Scientifica. Hindawi Limited; 2016.

[34]   Naidu R, Arias Espana VA, Liu Y, Jit J. Emerging contaminants in the environment: Risk-based analysis for better management. Chemosphere 2016; 154: 350-7.
[http://dx.doi.org/10.1016/j.chemosphere.2016.03.068] [PMID: 27062002]

[35]   Water. Vasilachi IC, Asiminicesei DM, Fertu DI, Gavrilescu M. Occurrence and fate of emerging pollutants in water environment and options for their removal.Switzerland;: MDPI AG 2021; 13.

[36]   Peng Y, Gautam L, Hall SW. The detection of drugs of abuse and pharmaceuticals in drinking water using solid-phase extraction and liquid chromatography-mass spectrometry. Chemosphere 2019; 223: 438-47.
[http://dx.doi.org/10.1016/j.chemosphere.2019.02.040] [PMID: 30784750]

[37]   Snow DD, Cassada DA, Bartelt-Hunt SL, *et al.* Detection, occurrence and fate of emerging contaminants in agricultural environments. Water Environ Res 2015; 87(10): 868-1937.
[http://dx.doi.org/10.2175/106143015X14338845155101] [PMID: 26420073]

[38]   Brennan G, Collins S. Growth responses of a green alga to multiple environmental drivers. Nat Clim Chang 2015; 5(9): 892-7.
[http://dx.doi.org/10.1038/nclimate2682]

[39]   Di Poi C, Costil K, Bouchart V, Halm-Lemeille MP. Toxicity assessment of five emerging pollutants, alone and in binary or ternary mixtures, towards three aquatic organisms. Environ Sci Pollut Res Int 2018; 25(7): 6122-34.
[http://dx.doi.org/10.1007/s11356-017-9306-9] [PMID: 28620858]

[40]   Richmond EK, Grace MR, Kelly JJ, Reisinger AJ, Rosi EJ, Walters DM. Pharmaceuticals and personal care products (PPCPs) are ecological disrupting compounds (EcoDC). Elementa. University of California Press; 2017; 5.

[41]   Subirats J, Timoner X, Sànchez-Melsió A, *et al.* Emerging contaminants and nutrients synergistically affect the spread of class 1 integron-integrase (intI1) and sul1 genes within stable streambed bacterial communities. Water Res 2018; 138: 77-85.
[http://dx.doi.org/10.1016/j.watres.2018.03.025] [PMID: 29573631]

[42]   Wang J, Wang J, Zhao Z, *et al.* PAHs accelerate the propagation of antibiotic resistance genes in coastal water microbial community. Environ Pollut 2017; 231(Pt 1): 1145-52.
[http://dx.doi.org/10.1016/j.envpol.2017.07.067] [PMID: 28886881]

[43]   Čese M, Heath D, Krivec M, Košmrlj J, Kosjek T, Heath E. Seasonal and spatial variations in the occurrence, mass loadings and removal of compounds of emerging concern in the Slovene aqueous environment and environmental risk assessment. Environ Pollut 2018; 242: 143-54.

[44]   Meador JP, Yeh A, Gallagher EP. Adverse metabolic effects in fish exposed to contaminants of emerging concern in the field and laboratory. Environ Pollut 2018; 236: 850-61.
[http://dx.doi.org/10.1016/j.envpol.2018.02.007] [PMID: 29471284]

[45]   Yeh A, Marcinek DJ, Meador JP, Gallagher EP. Effect of contaminants of emerging concern on liver

mitochondrial function in Chinook salmon. Aquat Toxicol 2017; 190: 21-31.
[http://dx.doi.org/10.1016/j.aquatox.2017.06.011] [PMID: 28668760]

[46]   Wang Y, Lu J, Mao L, *et al.* Antiepileptic drug carbamazepine promotes horizontal transfer of plasmid-borne multi-antibiotic resistance genes within and across bacterial genera. ISME J 2019; 13(2): 509-22.
[http://dx.doi.org/10.1038/s41396-018-0275-x] [PMID: 30291330]

[47]   Zamora-Ledezma C, Negrete-Bolagay D, Figueroa F, Zamora-Ledezma E, Ni M, Alexis F, *et al.* Heavy metal water pollution: A fresh look about hazards, novel and conventional remediation methods. Environ Technol Innov 2021; 22: 101504.
[http://dx.doi.org/10.1016/j.eti.2021.101504]

[48]   Cabral Pinto MMS, Marinho-Reis AP, Almeida A, *et al.* Human predisposition to cognitive impairment and its relation with environmental exposure to potentially toxic elements. Environ Geochem Health 2018; 40(5): 1767-84.
[http://dx.doi.org/10.1007/s10653-017-9928-3] [PMID: 28281140]

[49]   Guerra A, Etienne-Mesmin L, Livrelli V, Denis S, Blanquet-Diot S, Alric M. Relevance and challenges in modeling human gastric and small intestinal digestion. Trends Biotechnol 2012; 30(11): 591-600.
[http://dx.doi.org/10.1016/j.tibtech.2012.08.001] [PMID: 22974839]

[50]   Sanchez W, Egea E. Health and environmental risks associated with emerging pollutants and novel green processes. Environ Sci Pollut Res Int 2018; 25(7): 6085-6.
[http://dx.doi.org/10.1007/s11356-018-1372-0] [PMID: 29417484]

[51]   Sophia AC, Lima EC. Removal of emerging contaminants from the environment by adsorption. Ecotoxicol Environ Saf 2018; 150: 1-17.
[http://dx.doi.org/10.1016/j.ecoenv.2017.12.026] [PMID: 29253687]

[52]   Thanigaivel S, Vinayagam S, Gnanasekaran L, Suresh R. Matias Soto-Moscoso, Chen Wei-Hsin, Environmental fate of aquatic pollutants and their mitigation by phytoremediation for the clean and sustainable environment: A review. Enviro Res, 240(1), 2024, 117460.
[http://dx.doi.org/10.1016/j.envres.2023.117460]

[53]   Hlihor RM, Gavrilescu M, Tavares T, Favier L, Olivieri G. Bioremediation: An overview on current practices, advances, and new perspectives in environmental pollution treatment. Biomed Res Int. 2017; 2017:6327610.
[http://dx.doi.org/10.1155/2017/6327610]

[54]   Babuponnusami A, Muthukumar K. A review on Fenton and improvements to the Fenton process for wastewater treatment. J Environ Chem Eng 2014; 2(1): 557-72.
[http://dx.doi.org/10.1016/j.jece.2013.10.011]

[55]   Sharma NK, Philip L. Combined biological and photocatalytic treatment of real coke oven wastewater. Chem Eng J 2016; 295: 20-8.
[http://dx.doi.org/10.1016/j.cej.2016.03.031]

[56]   Saravanan A, Deivayanai VC, Kumar PS, *et al.* A detailed review on advanced oxidation process in treatment of wastewater: Mechanism, challenges and future outlook. Chemosphere 2022; 308(Pt 3): 136524.
[http://dx.doi.org/10.1016/j.chemosphere.2022.136524] [PMID: 36165838]

[57]   Gomes J, Costa R, Quinta-Ferreira RM, Martins RC. Application of ozonation for pharmaceuticals and personal care products removal from water. Sci Total Environ 2017; 586: 265-83.
[http://dx.doi.org/10.1016/j.scitotenv.2017.01.216] [PMID: 28185729]

[58]   Sichel C, Garcia C, Andre K. Feasibility studies: UV/chlorine advanced oxidation treatment for the removal of emerging contaminants. Water Res 2011; 45(19): 6371-80.
[http://dx.doi.org/10.1016/j.watres.2011.09.025] [PMID: 22000058]

[59]   Acero JL, Benitez FJ, Real FJ, Roldan G, Rodriguez E. Chlorination and bromination kinetics of emerging contaminants in aqueous systems. Chem Eng J 2013; 219: 43-50.
       [http://dx.doi.org/10.1016/j.cej.2012.12.067]

[60]   Bethke K, Palantöken S, Andrei V, *et al.* Functionalized cellulose for water purification, antimicrobial applications, and sensors. Adv Funct Mater 2018; 28(23): 1800409.
       [http://dx.doi.org/10.1002/adfm.201800409]

[61]   Kumar R, Sharma P, Manna C, Jain M. Abundance, interaction, ingestion, ecological concerns, and mitigation policies of microplastic pollution in riverine ecosystem: A review. Sci Total Environ 2021; 782(782): 146695.
       [http://dx.doi.org/10.1016/j.scitotenv.2021.146695]

[62]   Cooper WJ, Cramer CJ, Martin NH, Mezyk SP, O'Shea KE, Sonntag C. Free radical mechanisms for the treatment of methyl tert-butyl ether (MTBE) *via* advanced oxidation/reductive processes in aqueous solutions. Chem Rev 2009; 109(3): 1302-45.
       [http://dx.doi.org/10.1021/cr078024c] [PMID: 19166337]

[63]   Wang Z, Chen C, Ma W, Zhao J. Photochemical Coupling ofIron Redox Reactions and Transformation of Dissolved OrganicMatter (DOM). J Phys Chem Lett 2012; 3: 2044-51.
       [http://dx.doi.org/10.1021/jz3005333]

[64]   Mason TJ. Sonochemistry—Beyond Synthesis. Educ Chem 2009; 46: 140-4.

[65]   Mozumder A. Radiation chemistry: background, current status and outlook. J Phys Chem Lett 2011; 2(23): 2994-5.
       [http://dx.doi.org/10.1021/jz2012758]

[66]   Sun C, Chen C, Ma W, Zhao J. Photodegradation of organic pollutants catalyzed by iron species under visible light irradiation. Phys Chem Chem Phys 2011; 13(6): 1957-69.
       [http://dx.doi.org/10.1039/C0CP01203C] [PMID: 21082142]

[67]   Singh S, Singh S, Kushwaha R. Bioremediation of hydrocarbons and xenobiotic compounds.Bioremediation: Challenges and Advancements. Sharjah, United Arab Emirates;: Bentham Science Publishers 2022; pp. 1-48.

[68]   Enerijiofi KE. Bioremediation for Environmental Sustainability. 2021.

[69]   Nannipieri P, Kandeler E, Ruggiero P. Enzyme activities and microbiological and biochemical processes in soil. Enzy Environ. CRC Press;: Boca Raton, FL, USA: 2002; pp. 1-33.

[70]   Khalid F, Hashmi MZ, Jamil N, Qadir A, Ali MI. Microbial and enzymatic degradation of PCBs from e-waste-contaminated sites: a review. Environ Sci Pollut Res Int 2021; 28(9): 10474-87.
       [http://dx.doi.org/10.1007/s11356-020-11996-2] [PMID: 33411303]

[71]   Malik S, Dhasmana A, Kishore S, Kumari M. Bioremediation and Phytoremediation Technologies in Sustainable Soil Management Apple Academic Press. New York, NY, USA;: Microbes and Microbial Enzymes for Degradation of Pesticides 2022; pp. 95-127.
       [http://dx.doi.org/10.1201/9781003281207-5]

[72]   Ali U, Mudasir S, Farooq S, Nazir R. Factors affecting bioremediation. J Res Dev 2015; 15: 102-9.

[73]   Bala S, Garg D, Thirumalesh BV, Sharma M, Sridhar K, Inbaraj BS, Tripathi M. Recent strategies for bioremediation of emerging pollutants: a review for a green and sustainable environment. Toxics 2022; 10(8):484.
       [http://dx.doi.org/10.3390/toxics10080484]

[74]   Medfu Tarekegn M, Zewdu Salilih F, Ishetu AI. Microbes used as a tool for bioremediation of heavy metal from the environment. Cogent Food Agric 2020; 6(1): 1783174.
       [http://dx.doi.org/10.1080/23311932.2020.1783174]

[75]   Garg SK, Tripathi M. Microbial strategies for discoloration and detoxification of azo dyes from textile effluents. Res J Microbiol 2017; 12(1): 1-19.

[http://dx.doi.org/10.3923/jm.2017.1.19]

[76]   Shekher Giri B, Geed S, Vikrant K, *et al.* Progress in bioremediation of pesticide residues in the environment. Environ Eng Res 2021; 26(6): 200446.
[http://dx.doi.org/10.4491/eer.2020.446]

[77]   Azubuike CC, Chikere CB, Okpokwasili GC. Bioremediation techniques–classification based on site of application: principles, advantages, limitations and prospects. World J Microbiol Biotechnol 2016; 32(11): 180.
[http://dx.doi.org/10.1007/s11274-016-2137-x] [PMID: 27638318]

[78]   Agarwal N, Yadav RS, Solanki VS. The Latest Trends In Bioremediation of Pharmaceutical Contaminants: Limitations and Future Prospects. In: Agarwal N, Ed. Pharma: Boon or Bane. Newyork: Nova Science publishers, 2023; pp. 221-44.

[79]   Zelmanov G, Semiat R. Iron(3) oxide-based nanoparticles as catalysts in advanced organic aqueous oxidation. Water Res 2008; 42(1-2): 492-8.
[http://dx.doi.org/10.1016/j.watres.2007.07.045] [PMID: 17714754]

[80]   Berti WR, Cunningham SD. "Phytostabilization of metals," in Phytoremediation of Toxic Metals: Using Plants to Clean-up the Environment. New York, NY;: John Wiley & Sons, Inc. 2000; pp. 71-88.

[81]   Wei Z, Van Le Q, Peng W, *et al.* A review on phytoremediation of contaminants in air, water and soil. J Hazard Mater 2021; 403: 123658.
[http://dx.doi.org/10.1016/j.jhazmat.2020.123658] [PMID: 33264867]

[82]   Sheoran V, Sheoran AS, Poonia P. Role of hyperaccumulators in phytoextraction of metals from contaminated mining sites: a review. Crit Rev Environ Sci Technol 2010; 41(2): 168-214.
[http://dx.doi.org/10.1080/10643380902718418]

[83]   Lasat MM. Phytoextraction of toxic metals: a review of biological mechanisms. J Environ Qual 2002; 31(1): 109-20.
[http://dx.doi.org/10.2134/jeq2002.1090] [PMID: 11837415]

[84]   Sanghvi G, Thanki A, Pandey S, Singh NK. Engineered bacteria for bioremediation. Biorem Pollut. The Netherlands: Elsevier Amsterdam; 2020; pp. 359-74.

[85]   Goutam SP, Saxena G, Singh V, Yadav AK, Bharagava RN, Thapa KB. Green synthesis of TiO_2 nanoparticles using leaf extract of *Jatropha curcas* L. for photocatalytic degradation of tannery wastewater. Chem Eng J 2018; 336: 386-96.
[http://dx.doi.org/10.1016/j.cej.2017.12.029]

[86]   Magalhães-Ghiotto GAV, Oliveira AM, Natal JPS, Bergamasco R, Gomes RG. Green nanoparticles in water treatment: A review of research trends, applications, environmental aspects and large-scale production. Environ Nanotechnol Monit Manag 2021; 16: 100526.
[http://dx.doi.org/10.1016/j.enmm.2021.100526]

[87]   Samadi N, Golkaran D, Eslamifar A, Jamalifar H, Fazeli MR, Mohseni FA. Intra/extracellular biosynthesis of silver nanoparticles by an autochthonous strain of *Proteus mirabilis* isolated from photographic waste. J Biomed Nanotechnol 2009; 5(3): 247-53.
[http://dx.doi.org/10.1166/jbn.2009.1029] [PMID: 20055006]

[88]   Koenig JC, Boparai HK, Lee MJ, O'Carroll DM, Barnes RJ, Manefield MJ. Particles and enzymes: Combining nanoscale zero valent iron and organochlorine respiring bacteria for the detoxification of chloroethane mixtures. J Hazard Mater 2016; 308: 106-12.
[http://dx.doi.org/10.1016/j.jhazmat.2015.12.036] [PMID: 26808236]

[89]   Le TT, Nguyen KH, Jeon JR, Francis AJ, Chang YS. Nano/bio treatment of polychlorinated biphenyls with evaluation of comparative toxicity. J Hazard Mater 2015; 287: 335-41.
[http://dx.doi.org/10.1016/j.jhazmat.2015.02.001] [PMID: 25679799]

[90]   Diao M, Yao M. Use of zero-valent iron nanoparticles in inactivating microbes. Water Res 2009; 43(20): 5243-51.
[http://dx.doi.org/10.1016/j.watres.2009.08.051] [PMID: 19783027]

CHAPTER 2

A Critical Review of Microbial Potential for Biodegradation Mechanism of Organic Pollutants

Amrit K. Mitra[1,*]

[1] *Department of Chemistry, Government General Degree College, Singur, Hooghly, West Bengal, India*

Abstract: The rise in environmental pollution is a major issue of concern in the current times. Due to globalization and the Industrial Revolution in the 20th century, pollution has become a problem for the world's population. Numerous factors, including unchecked human activity, careless use of petroleum products, industrial waste emissions, poor waste management, release of toxic organic by-products, and increased use of pesticides, insecticides and fertilizers have contributed to increased pollution and its detrimental effects on the planet Earth. For all forms of life, organic molecules are known to have the potential to be carcinogenic and poisonous. To reduce organic pollutants and dispose of industrial waste properly, several techniques have been put forth and put into action, but some of them are either not relevant or have not produced the expected outcomes. For the past few decades, research has been focused on finding biological methods of degradation of complex organic contaminants. Numerous microbial species obtained from polluted native environments have been shown to digest hazardous, complex organic chemicals and can be used to effectively biodegrade contaminated areas. The development of recombinant DNA technologies has revitalized the area of bioremediation by enabling the emergence of microorganisms and entire microbial communities that contain novel genes and enzymes with improved efficiencies. This chapter discusses the significance of isolating efficient indigenous microbial species, different factors that affect the distribution of pollutants in the soil matrix, biodegradation pathways, and physiological factors that affect the efficiency of biodegradation for complete mineralization. In addition to these, efforts to enhance the biodegradation potential of microbes through multiple pathways have also been highlighted.

Keywords: Anthropogenic pollutants, Bioaugmentation, Biodegradation, Bioremediation, Biostimulation, Mineralization, Organic Pollutants, Pollutant management.

---

* **Corresponding author Amrit K. Mitra:** Department of Chemistry, Government General Degree College, Singur, Hooghly, West Bengal, India; Tel: +91-33-2630-0126; +91-9432164011; Fax: +91-33-2630-0126; E-mail: ambrosia12june@gmail.com

Neha Agarwal, Vijendra Singh Solanki and Sreekantha B. Jonnalagadda (Eds.)

INTRODUCTION

The organic substances of human origin cause severe environmental pollution. Since the beginning of the Industrial Revolution, the biosphere has experience significant increase in pollution [1 - 4]. Petroleum-based hydrocarbons, as well as various pesticides used in agriculture and pest management, are examples of common organic pollutants of public concern [5, 6]. Textiles, hydrocarbon oils, soaps, detergents, and other useful materials were among the chemically generated commodities that expanded substantially in the late 1800s and early 1900s [7]. The impacts of these substances on the environment are determined by several processes that differ in the properties of each component. Halogenated chemical pollutants, including polychlorinated biphenyls (PCBs), dichloro-diphenyl-trichloroethane (DDT), dechlorane plus, and dibromide-phenyl-ethane are of great concern due to their longevity, bioaccumulation and potential exposure to humans and animals [8, 9]. Other industrial by-products, for instance, phthalates (plasticizers found in bottles), toys, and personal care products, also act as organic pollutants. Polybrominated diphenyl esters (PBDEs) that are utilized in a variety of consumer goods are now found in the environment. There are numerous industrial uses for chlorinated ethanes, chlorinated ethenes and chlorinated benzenes as solvents and degreasers, as well as biocides and their precursors [10, 11]. They pose serious threats to both human and environmental health due to their acute and chronic toxicity, persistence and bioaccumulation. The EPA (Environment Protection Agency) has listed several organochlorines as priority contaminants, emphasizing the potential environmental danger they pose [12]. The primary factor in water and soil contamination is the release of hydrocarbons into the environment. It is well recognized that petroleum-based hydrocarbon contaminants negatively impact both terrestrial and aquatic life as well as soil productivity. The extensive use of pesticides, drugs (such as non-steroidal anti-inflammatory medications (NSAIDs), and antibiotics) and other chemicals, their unplanned disposal, and subsequent presence in different ecosystems are matters of keen concern [2]. These compounds are now more prevalent due to their widespread use in soil, water and sediments. Consequently, there is an increased understanding of the risks imposed by these organic pollutants and their removal from the environment.

The microorganism's ability to break down and detoxify these contaminants is known as microbial degradation or biodegradation. Different varieties of microorganisms, such as bacteria, fungi, and protozoa, decompose diverse substrates biologically [13, 14]. After being released into the environment, organic pollutants can undergo degradation, sorption-desorption, volatilization, uptake by plants, runoff into surface waters and transfer into groundwater. Through biotic or abiotic degradation and transformation, these organic pollutants are assimilated into the environment. These mechanisms either mineralize organic contaminants into a carbon field or turn them into degradation products [15]. Microorganisms can adapt themselves to changing environmental conditions. Due to recent developments that have allowed for extensive and high-throughput research of ecologically relevant microorganisms, interest in the microbial breakdown of contaminants has increased. This adds to our understanding of their biodegradation mechanisms and metabolic pathways [4, 16 - 19]. This chapter intends to offer a cutting-edge review of the microbial potential in the degradation of organic pollutants for the green remediation of the environment.

ORIGIN AND OCCURRENCE OF ORGANIC POLLUTANTS

Organic pollutants (OPs) are those chemical substances that have a blatantly adverse effect on the environment. OPs are a class of exceedingly poisonous synthetic organic molecules that can persist in the environment under natural conditions for a long time. To safeguard both human health and the environment, the international community has developed tools to detect the existence and regulation of these pollutants [16]. The Stockholm Convention of 2013 was the most well-intentioned plan, which attempted to eliminate organic pollutants and, if that is impossible, to restrict emissions and discharges [16].

A considerable amount of OPs are released into the soil as a result of the rapid population growth, increased fuel consumption and production of industrial chemicals, fertilizers, pesticides and medications. Based on their environmental half-lives, OPs are divided into two classes *i.e.*, persistent organic pollutants (POPs) and non-POPs. Non-POPs can be broken down into simple, non-polluting components like carbon dioxide and nitrogen by chemical reactions or natural microbes [20 - 22]. One illustration of a non-persistent contaminant is organic waste. Due to their high persistence and toxicity in soil, the POPs pose significant health concerns to humans through food chains [23 - 27]. These OPs become highly toxic, poisonous compounds and pose threats to living beings when they cross the acceptable limits. Detergents, petroleum hydrocarbons, plastics, organic solvents, pesticides, insecticides and dyes are the main sources of release of these organic compounds into the environment [23, 24]. Since the 20th century, the POPs have drawn attention due to their highly hazardous bio-accumulative qualities. Owing to their possible toxicological characteristics, POPs have been deemed to be more harmful and are further listed as follows:

1. They stay intact for many years and are vulnerable to long-distance transportation.
2. Since they are widely dispersed in the environment, including soil, water, and most significantly in air, they build up in the fatty tissue of living things,

including humans, and are more abundant at the top of the food chain.

3. Harmful to both human beings and animals
4. Primarily differ in the number of chlorine substitutions and last for a long time in the environment, especially in soils, sediments, and air
5. POPs undergo a variety of reactions in municipal and industrial waste, landfill effluents and agricultural operations that confirm their presence. POPs are also found in the biota of these remote locations

Organometallic compounds, oxygen, phosphorus, nitrogen-based organic compounds and hydrocarbons are the three main groups of organic pollutants. Hydrocarbons and associated chemicals (dioxins, DDT, and polycyclic aromatic hydrocarbons (PAHs)) are the main group of agricultural pollutants [28]. Anthropogenic sources continuously pollute the biosphere by using a wide variety of harmful substances that they manufacture in various ways. Military waste, farming practices and agricultural practices are the three main sources of organic pollutants that include gasoline, PAHs, dioxins, organophosphates, triazines and carbamates [3]. Owing to their low polarity, hydrocarbons are lipophilic, poorly water-soluble and persistent [29, 30].

Oxygen, nitrogen and phosphorus-containing organic compounds are water-soluble and have a low environmental persistence. This could be a result of the presence of relatively strong polar bonds created by the attachment of nitrogen, oxygen, or phosphorus atoms to carbon and other atoms, which give the linked compounds a high degree of polarity. The least significant group of chemicals in terms of their effects on the environment is the organometallic category, which may have organic combinations, such as lead and tin, with organic elements based on carbohydrates [16]. POPs are a class of compounds that cause environmental difficulties due to their bioaccumulation ability and environmental longevity as well as consumption patterns. They are transmitted across long distances on the Earth and can linger for a long time in the environment. As POPs, two categories of important substances exist: (I) PAHs and (II) halogenated hydrocarbons. The aforementioned group includes organo-chlorines such as DDT [16, 29 - 31].

MICROBIAL DEGRADATION OF ORGANIC POLLUTANTS

Bioremediation is a biological process that transforms wastes into compounds that can be used and reused by other species [32 - 35]. Microorganisms play an important role which is necessary for a precarious alternative approach to problems. The organic substance is degraded or transformed by enzymatic reactions of microorganisms. It works on two levels: co-metabolism and growth. The organic contaminant, which is the source of carbon and energy for growth, is completely mineralized as a result of this process. Co-metabolism is the metabolic process of an organic substance where a growth substrate is present, which is the main source of carbon and energy.

Bacteria, fungi and yeasts are among the microorganisms engaged in the biodegradation process. Reports on the participation of algae and protozoa in biodegradation are scarce [36]. Even though there are numerous biodegradation processes, carbon dioxide is often the end product of the disintegration. The term 'biodegradable material' refers to organic material that can be broken down by microbes. Some microbes have naturally occurring catabolic enzymes that break down, change, or collect a wide range of substances, such as hydrocarbons, PAHs, radionuclides and metals. Microorganisms can degrade organic pollutants in both aerobic and anaerobic environments. The bulk of organic contaminants does, however, biodegrade more quickly and completely in aerobic environments [36 - 38]. The activation and incorporation of oxygen is the primary enzymatic reaction performed by oxygenases and peroxidases. The primary mechanism of aerobic hydrocarbon breakdown is depicted in Fig. (**1**). Organic pollutants are gradually transformed into intermediates of the central intermediary metabolism by peripheral degradation routes. Gluconeogenesis produces the sugars needed for different biosynthesis and development processes [1].

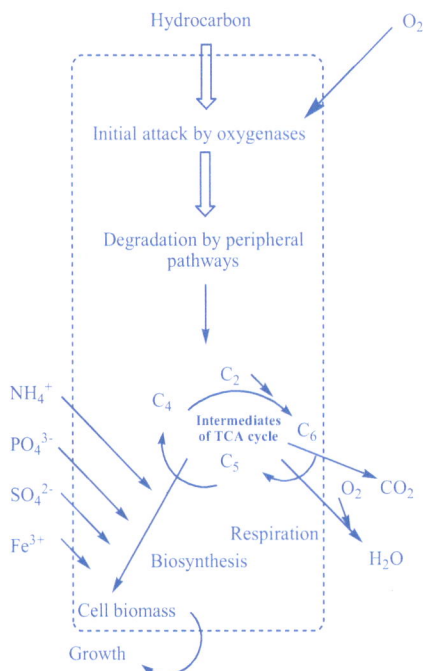

Fig. (1). The basic principle of aerobic hydrocarbon breakdown by microorganisms [4].

This chapter covers different types of microorganisms that help in the breakdown of various contaminants and a deep understanding of the process of biodegradation. It will also focus on 'bioremediation' by explaining the processes like biostimulation, bioaugmentation and natural attenuation that use microorganisms' degradation abilities in the process of bioremediation [39 - 41]. The use of genetically engineered microorganisms (GEMs), which is also an effective method of degradation of contaminants, is one goal of this chapter. The second goal of this chapter is to illustrate the significance of some GEMs in this procedure and to discuss challenges that must be overcome before GEMs can offer an efficient clean-up procedure at a lower cost. Natural attenuation occurs during the bioremediation of pollutants using the biodegradation abilities of microorganisms. GEM is utilized to enhance the capacity of microorganisms for biodegradation. Many factors can affect the effectiveness of this method; however, hazards are also involved in using GEM in the field (Fig. **2**) [17, 42].

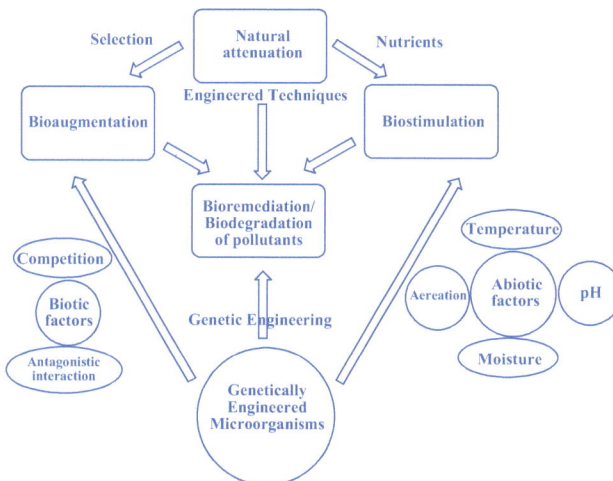

Fig. (2). Representative summary of bioremediation [17, 42].

Biodegradation of Petroleum Hydrocarbons

Biodegradation of petroleum-based hydrocarbon is a complicated process that is affected by the variety and quantity of hydrocarbons present, as well as their chemical composition [1]. The type and size of a hydrocarbon molecule influence its susceptibility to microbial degradation. Short-chain alkanes ($< C_{10}$) typically damage the lipid membrane architecture of microorganisms, making them poisonous to bacteria. Volatilization is the most common method of elimination [43]. While higher alkanes with long chains are highly resistant to microbial biodegradation, intermediate chain length alkanes (C_{10}-C_{24}) often degrade swiftly [1, 4, 43]. According to their varying susceptibilities to microbial attack, hydrocarbons are categorized in decreasing susceptibility order as follows: *n*-alkanes > branched alkanes > low molecular weight aromatics > cyclic alkanes [10].

The first step in the aerobic degradation of hydrocarbons by bacteria is the introduction of molecular oxygen into the hydrocarbon. Prokaryotes use dioxygenase, which converts aromatic hydrocarbons to *trans*-dihydrodiols that are further oxidized to dihydroxy compounds. In the case of benzene, the product is catechol [1]. Mono-oxygenases are used by eukaryotic microorganisms to convert benzene to benzene 1,2-oxide, which is then converted into dihydroxy dihydro benzene due to the addition of water. The subsequent oxidation of this results in catechol, which undergoes *ortho* or *meta* cleavage to produce muconic acid or hydroxymuconic semialdehyde, respectively [8].

Biochemistry of Oxidation of Hydrocarbon

Whether they have short or long chains, alkanes go through beta-oxidation, followed by terminal oxidation, to produce the corresponding aldehydes, alcohols, and fatty acids. However, additional evidence reveals that branched alkanes typically experience di-terminal and sub-terminal *n*-alkane or fatty acid oxidation [44, 45]. Although the enzyme oxidation mechanism of an alkane is not well understood, hydroxylation, dehydrogenation, and hydroperoxidation are the recognized pathways [45]. While the aromatic fraction of petroleum is highly resilient to microbial attack, *n*-alkanes and cycloalkanes parts of petroleum degrade quickly. Aerobic degradation normally begins by oxidizing a terminal methyl group to produce a primary alcohol in the case of n-alkanes with two or more carbon atoms [46]. This is then further oxidized to the equivalent aldehyde before being transformed into a fatty acid (Fig. **3**, Pathway a).

Acetyl-CoA is produced by activating fatty acids and then processing them through a β-oxidation pathway. In some instances, the di-terminal hydroxylation of fatty acids results in the oxidation of both ends of the alkane molecule, producing a di-hydroxy fatty acid, which is then transformed into a dicarboxylic acid, resulting in the formation of acetyl CoA [13, 14]. The tricarboxylic acid (TCA) cycle is where the acetyl CoA enters to be broken down into carbon dioxide and water. It is observed that n-alkanes also undergo sub-terminal oxidation [46, 47]. A Baeyer-Villiger monooxygenase (BVM) then oxidizes the secondary alcohol-derived product to yield the matching ketone, which is then transformed into an ester. An esterase (E) hydrolyzes the ester, producing an alcohol and a fatty acid that go into the process of oxidation (Fig. **3**, Pathway b). In some microorganisms, both terminal and sub-terminal oxidation can occur [46].

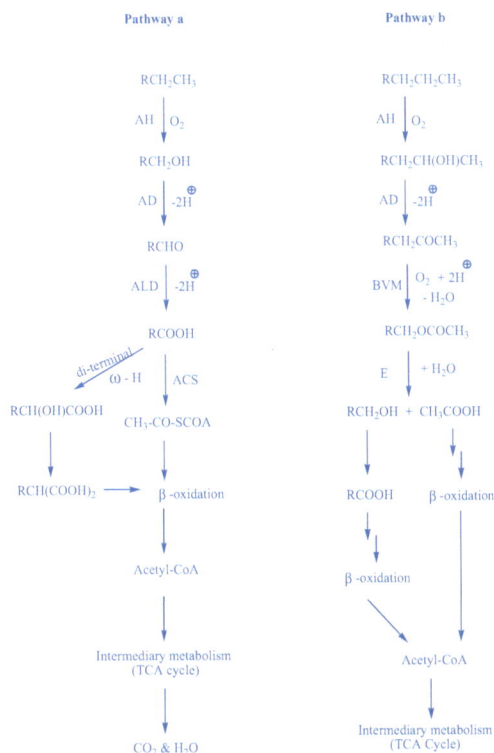

Fig. (3). Terminal oxidation of alkane pathway (Pathway a) and sub-terminal oxidation of *n*- alkane pathway (Pathway b) [4, 46]. (AH: alkane hydroxylase; AD: alcohol dehydrogenase; ALD: aldehyde dehydrogenase; ACS: acyl – CoA synthetase; ω-H: ω-hydroxylase).

Anaerobic Degradation of Alkanes

Quantitatively speaking, saturated hydrocarbons make up the largest portion of all petroleum hydrocarbons. For the last few decades, considerable progress has been made in understanding the role of microorganisms in the anaerobic biodegradation of alkanes. Several physiological groupings of microorganisms can use them as carbon and energy sources when they are cultivated under nitrate-reducing, chlorate-reducing, and methanogenic environments. Petroleum (also known as crude oil) is a complicated mixture made up of hydrocarbons and thousands of other distinctly different chemical elements. Crude oil's hydrocarbon content can be divided into four primary fractions: saturated hydrocarbons, resins, aromatic hydrocarbons and asphaltenes. These substances are less reactive due to the absence of functional groups and have limited solubility in water. Nonetheless, the first hydrocarbon assault continues to be a crucial stage in the later metabolism of these apolar substrates. Also, the initial step's integration of molecular oxygen into hydrocarbon molecules is an oxygen-dependent biochemical process that is advantageous energetically.

Oil biodegradation takes place largely in anoxic conditions in the subsurface by sulfate-reducing bacteria or other anaerobes employing several electron acceptors as an oxidant [43]. Hydrocarbon substrates (such as toluene) are converted to oxidative intermediates in the presence of nitrate before further degradation [8]. It has been demonstrated that anaerobic degradation of petroleum hydrocarbons occurs at extremely low rates in natural environments and its ecological relevance has generally been regarded as being minimal [8]. However, it has been demonstrated that anaerobic conditions are necessary for the microbial degradation of oxidized aromatic compounds like halogenated aromatic compounds and benzoates like halobenzoates and polychlorinated biphenyls [8]. Numerous bacterial strains have been identified that can utilize alkanes as a source of carbon under anoxic conditions [46]. However, the recycling of hydrocarbons in the environment also benefits from the anaerobic breakdown of alkanes. For instance, *Azoarcus* sp. utilizes C_6-C_8 alkanes, whereas *Desulfobacterium* sp. metabolizes C_{12}-C_{20} alkanes [46 - 48]. For some strains, anaerobic breakdown pathways for the alkanes have been studied [46]. There seem to be two main tactics in use (Fig. **4**); one includes adding a fumarate molecule to the alkane, activating it at a sub-terminal position and producing an alkyl-succinate derivative (Pathway a). The production of glycyl radical (an organic radical intermediate) is the possible mechanism by which this reaction takes place [48]. The fumarate molecule is combined with one of the alkane's terminal carbon atoms in the second reaction pathway (Pathway b), which has only been discussed about propane (Fig. **4**) [46, 49, 50].

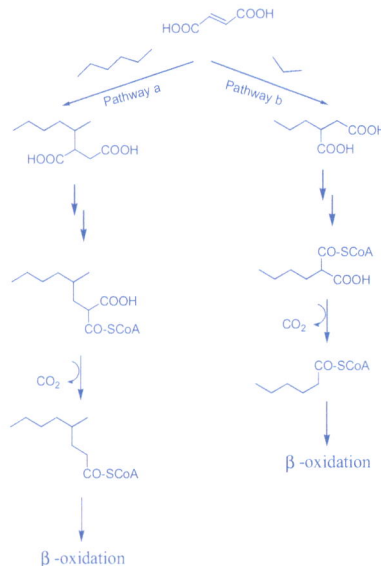

Fig. (4). Anaerobic breakdown of alkane *via* the addition of fumarate molecule [46].

Biodegradation of Organic Chemicals (Pesticides)

Various chemicals have accumulated as a result of rapid industrialization and rising population. An incredible effort has been made to apply new methods to decrease or eradicate these toxins from the environment due to the frequent and broad usage of man-made 'xenobiotic' compounds. Typical pollution treatment techniques, such as landfilling, recycling, pyrolysis and incineration, have had negative effects on the environment and can result in the creation of harmful intermediates [51]. Moreover, some techniques are more expensive and occasionally challenging to use, especially the use of pesticides in large agricultural areas [52]. Exploiting microorganisms' capacity to remove contaminants from contaminated places is one promising treatment strategy. This process is known as bioremediation and is an alternative treatment method that is efficient, minimally harmful, cost-effective, adaptable and environmentally benign.

As known to us, pesticides are natural substances that are used to prevent, eradicate, repel, or mitigate pest populations to boost agricultural productivity. Chemicals that are used to control pests are also referred to as pesticides. Today, it is a routine practice to apply pesticides to agricultural soil, and integrated pest management (IPM) heavily relies on these practices. A few pesticides persist in the soil and produce pollutants, resulting in intermittent pollution of an aquatic environment. Microorganisms (bacteria and fungi) with the potential to degrade pesticides can be found in several pesticide-contaminated locales. When these pesticides are administered to crops, the soil absorbs them. However, the effluent from the pesticide industry, as well as sewage sludge, activated sludge, wastewater, natural waters, sediments and places near the pesticide companies, are also rich sources of microorganisms that can break down pesticides. *Pseudomonas sp.* is the most effective bacterial genus for the breakdown of hazardous substances [53]. Numerous fungal species, including *Aspergillus fumigatus*, *Aspergillus niger*, *Aspergillus terreus*, and *Rhizopus microsporus*, were isolated from contaminated soils in Algeria [54]. *Trichoderma viridae* fungi have also been linked to the breakdown of the insecticides endosulfan and methyl parathion [55]. These microorganism's internal or extracellular enzymes play a significant part in the breakdown of these organic contaminants. These chemicals are persistent in the soil due to the absence of microbial systems that contain the enzymes that break down pesticides. In these circumstances, the external addition of microflora that degrades pesticides is advised [56]. Additionally, the biotransformation of such chemical contaminants depends on a wide range of environmental parameters like temperature, pH, nutrients, redox potentials, contact duration, water content *etc.*, in addition to the existence of microorganisms with the proper degrading enzymes [57]. The microbial metabolism of pollutants is influenced by several factors, including physiological, ecological, biochemical and molecular factors [53, 58, 59].

Pesticide Microbial Degradation Mechanism

Pesticide microbial degradation could take place during three phases. In Phase I, oxidation, reduction, or hydrolysis are used to change a parent compound's initial characteristics to create a water-soluble and typically less toxic derivative. In the second phase, pesticide metabolite is conjugated to an amino acid or sugar, increasing its water solubility and decreasing its toxicity in comparison to the original pesticide. Phase II metabolites are transformed into secondary conjugates, which are non-toxic, in the third phase. The production of intracellular or extracellular enzymes, such as hydrolytic enzymes, peroxidases, oxygenases, *etc.*, by fungi and bacteria takes place in these activities [54].

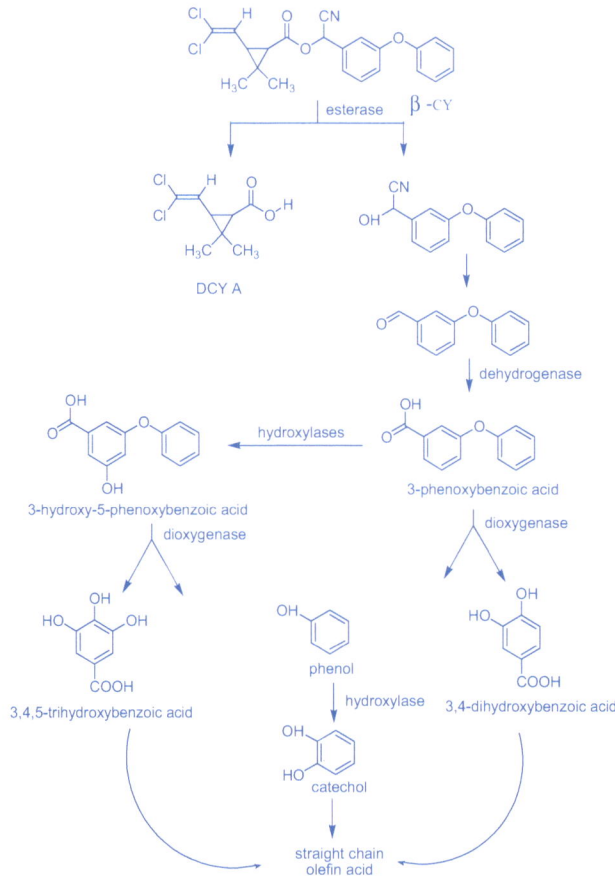

Fig. (5). Degradation pathway of beta-CY by *Aspergillus niger* YAT [61].

The process by which organic molecules are converted into non-toxic inorganic compounds by soil bacteria is known as mineralization [60]. Numerous pesticides are natural chemical analogs, and certain microorganisms have the enzymes needed to break them down. They might serve as a supply of microbial nutrition before being broken down by microorganisms into inorganic substances, carbon dioxide and water. It should be emphasized that, in most instances, the degradation of pesticides in the microbial body requires the synergistic effect of a series of events. Using co-metabolism and mineralization, Deng *et al.* discovered that *Aspergillus niger* YAT could completely break down beta-CY (β-CY) and its intermediates [61]. Fig. (**5**) depicts the process by which Aspergillus niger YAT degrades β-CY.

Factors affecting Organic Pollutant Biodegradation

Due to their ability to adapt to harsh environmental conditions, microorganisms can break down a variety of organic contaminants. As a result, microbes play a significant role in site restoration. However, a variety of parameters, such as the concentration of pollutants and the physicochemical properties of the environment, affect the effectiveness of biodegradation [17, 62]. The key environmental elements that affect the microbial breakdown of organic contaminants are given in Table **1**.

Table 1. Environmental factors responsible for the microbial breakdown of organic contaminants.

Nutrients	For Smooth Microbial Development to take Place, A Good Amount of Oxygen and Nutrients must be Present in a Usable form and in the correct proportions. Low amounts of one or more vitamins and amino acids are necessary for many bacteria and fungi. The metabolism of cells also needs phosphorus and nitrogen [4, 17].
Temperature	For the functioning of biological enzymes, the optimal temperature is required. As the temperature drops by 10°C, the rate of biodegradation reduces to half [17].
pH	A pH range of 6.5 to 8.5 is ideal for biodegradation, though it can occur under a wide variety of pH conditions. Additionally, the majority of heterotrophic bacteria prefer a pH of 7.0, whereas fungi are highly resistant to acidic environments [4, 17].
Moisture	Moisture also affects the pace of metabolism of pollutants by influencing the availability of soluble materials, osmotic pressure and pH of the environment.
Oxygen	Another crucial factor is oxygen because it affects how microbes assimilate nutrients and produce energy. For bacteria and fungi to use organic contaminants like hydrocarbons, they need an electron sink. Oxygen is the most frequent electron sink in the subsequent steps as well. Nitrate or sulfate reduction may enable further biodegradation of partially oxygenated intermediates in the absence of molecular oxygen [8].

(Table 1) cont.....

Nutrients	For Smooth Microbial Development to take Place, A Good Amount of Oxygen and Nutrients must be Present in a Usable form and in the correct proportions. Low amounts of one or more vitamins and amino acids are necessary for many bacteria and fungi. The metabolism of cells also needs phosphorus and nitrogen [4, 17].
Soil Type	The organic compound that can be adsorbed on a solid surface depends on the kind of soil and the amount of organic matter in the soil. In a comparable process called absorption, a pollutant permeates the soil matrix in its entirety. Both adsorption and absorption decrease the contaminant's availability to the majority of bacteria, which consequently decreases the pace at which the chemical is metabolized. The movement of fluids and the migration of contaminants in groundwater may be affected by variations in the porosity of zones of the aquifer matrix. When soils are highly saturated with water, the matrix's capacity to transfer gases like CO_2, O_2, and CH_4 is reduced
Oxidation-Reduction Potential	It is possible to gauge the system's electron density by looking at a soil's oxidation-reduction potential. When compounds are oxidized, electrons are transported to other more oxidized compounds known as electron acceptors. This process is how biological energy is produced. Low electron density (Eh > 50 mV) denotes oxidizing and aerobic circumstances, while high electron density denotes reducing, anaerobic conditions [17].

ENHANCING MICROORGANISMS BIODEGRADATIVE CAPABILITIES

Popular environmental pollutants that are toxic, genotoxic, mutagenic and carcinogenic include aromatic hydrocarbons. The bioremediation of petroleum hydrocarbons using microbes as a source of carbon for microbial growth was very effective. For example, Pseudomonas species isolated from petroleum hydrocarbon-contaminated soil have been demonstrated to break down hydrocarbons. Microorganism's capacity for biodegradation may be increased through biostimulation, bioaugmentation, or using GEMs.

Biostimulation

It is the process of environmental modification to stimulate the existing bacteria that are capable of bioremediation. This can be achieved by providing different types of nutrients, electron donors and acceptors, such as oxygen, nitrogen, phosphorus and carbon (*e.g.*, in the form of molasses), which are ordinarily present in inadequate amounts to control microbial activity. Biostimulation is performed to encourage natural microorganisms to biotransform a variety of soil pollutants [63]. The goal of biostimulation is to increase the activity of native microorganisms that can degrade contaminants from the soil environment. Perfumo *et al.* described that the population of naturally occurring microorganisms that are capable of bioremediation rises when nutrients, oxygen, or other electron donors and acceptors are added to the coordinated site [64]. According to Margesin *et al.*, biostimulation is a sort of natural remediation that can accelerate pollutant breakdown by enhancing factors including aeration, nutrient addition, pH control and temperature regulation [65]. According to their opinion, biostimulation is an effective remediation method for removing petroleum pollutants from soil, but it necessitates an evaluation of both the inherent degradation capabilities of the native microflora and the ambient factors affecting the kinetics of the *in situ* process. Microorganisms can quickly transform trichloroethene and perchloroethene into ethane when lactate is added during biostimulation [66]. The increased rate of pentachlorophenol (PCP) dechlorination by the soil bacteria was due to the biostimulation of native microorganisms by the addition of lactate and anthraquinone-2,6-disulfonate (AQDS) [67]. According to Chen *et al.*, glycerol seems to exhibit the most favorable metabolic properties against phenol toxicity on the native *Rhizobium Ralstonia taiwanensis*, resulting in improved phenol degradation efficiency [68]. In contaminated soils, biostimulation was found to be more effective than natural attenuation of biodiesel [69]. The addition of commercial fertilizers with nitrogen and phosphate to contaminated soil improved the abundance of microbes that break down hydrocarbons as well as the overall degradation of petroleum hydrocarbons [70]. Another study that used poultry manure as an organic fertilizer found that biodegradation was improved when only poultry manure was present, but the efficiency of biodegradation was affected by the addition of alternative carbon substrates for surfactants [71]. However, high concentrations of nutrients can prevent the efficiency of the biodegradation process [72].

Bioaugmentation

The use of bacterial cultures to hasten the breakdown of pollutants is known as bio-augmentation. As an alternative method for the bioremediation of oil-contaminated ecosystems, bioaugmentation (the addition of oil-degrading microorganisms) has been advocated since the 1970s to enhance native populations. The justification for this strategy is that native microbial populations would not be able to degrade the broad range of possible substrates found in complex mixes like petroleum [73]. To aid the native microorganisms in their biodegradative activities, a technique known as bioaugmentation involves introducing microorganisms with the potential for biodegradation into a contaminated environment [74]. The microflora in the contaminated soil sediments is particularly well adapted to the high concentration of organic contaminants. Soils that have recently become contaminated with hydrocarbons can be remediated using microorganisms that have been isolated from contaminated soil sediments. It has been discovered that priming with 2% bio-remediated soil can speed up the biodegradation of fuel oil-treated soil's PAH components. The inclusion of GEMs designed for the degradation of the hydrocarbons in contaminated soil may occasionally be necessary to achieve this. With the use of bioaugmentation, it is possible to supply particular bacteria in large quantities to finish the biodegradation process. Through the years, bioaugmentation has been used to successfully remediate contaminants such as pesticides, petroleum compounds and an increasing number of harmful organic chemicals [75, 76]. When the indigenous hydrocarbon-degrading population is low and when seeding may reduce the lag period to begin the bioremediation process, bioaugmentation may be considered [77]. The seed microorganisms should be able to degrade the majority of petroleum components, maintain genetic stability during storage, survive in foreign and unfavorable environments, compete with native microorganisms and move through the pores of the sediment to the contaminants [78]. The enzymatic preferences and abilities of various microbial species for the breakdown of oil molecules vary. Some bacteria may break down branched, linear, or cyclic alkanes. Some favor mono- or polynuclear aromatics, while others degrade alkanes and aromatics simultaneously.

When biostimulation and bioattenuation fail, the bioaugmentation strategy is thought to be used [17]. Numerous studies have demonstrated that both biotic and abiotic factors have an impact on the effectiveness of bioaugmentation. The outcomes of bioaugmentation are also significantly influenced by biotic variables, such as competition for scarce carbon sources between native and exogenous microorganisms, antagonistic relationships and predation by protozoa and bacteriophages [79]. The utilization of inocula made up of microbial strains or microbial consortia that have been well suited to decontaminate the site is essential for the success of bioaugmentation treatments. The effectiveness of foreign microorganisms (those in inocula) depends on their capacity to outcompete native microbes, predators and other abiotic variables. When selecting the microorganisms, it is important to take into account factors affecting the growth of the microorganisms, such as the chemical makeup and concentration of the pollutants, the accessibility of the contaminant to the microorganisms, the size and makeup of the microbial population and the physical environment. A potential method to hasten bioremediation is the combination of bioaugmentation and biostimulation [17].

Genetically Engineered Microorganisms (GEMs)

It is becoming more common to use some genetically modified microbes to modify how well they can consume harmful pollutants like hydrocarbons and pesticides. In the late 1980s and early 1990s, this method was first mentioned. The potential to examine the genetic diversity and metabolic variability of microbes is the foundation for the ability to 'engineer' microorganisms to increase degradation capabilities [80]. Microorganisms that have been genetically transformed through the use of genetic engineering techniques are referred to as GEMs [81]. By producing GEMs, genetic engineering has enhanced the usage and removal of hazardous wastes in laboratory settings [8]. With improved degradation capabilities covering a variety of chemical contaminants, GEMs have demonstrated biodegradation potential in soil, water and activated sludge environments [81]. Scientists are currently looking for novel genetically modified microbes to improve their capacity to digest specific substances like hydrocarbons and insecticides. Recombinant DNA techniques have reportedly been shown to significantly speed up the breakdown of hazardous waste in lab settings. In some circumstances, the GEMs have a stronger capability to degrade several dangerous chemicals. Genetic engineering technology is currently being widely used as a potential tool in the bioremediation process. Analysis of diverse gene and metabolic diversity in microorganisms occurs during bioremediation [80]. Such microorganisms are advantageous in biodegradation, biotransformation, biosorption and bioaccumulation due to their genetic makeup. In their chromosomal and extrachromosomal DNA, these microorganisms have been found to contain the fundamental blueprint for the genes that code for biodegradative enzymes. By locating these genes and introducing them into a specific host through an appropriate vector under the strict control of relevant promoters, recombinant DNA techniques expedite the development of an organism's ability to metabolize a xenobiotic compound. This is dependent on the ability to modify and exchange genetic data. Recombinant DNA technology-related biotechnology has recently evolved into a fine-tuning of the bioremediation technology by improving pollutant-degrading microbes through strain alterations and genetic modification of specific regulatory and metabolic genes, which are very crucial in developing effective, secure and affordable techniques for bioremediation. Bioremediation can be used to remove undesired compounds from different environmental matrices because it is the most effective method of degrading dangerous pollutants.

The initial stage in creating a Genetically Modified Microorganism (GMM) is choosing the appropriate gene or genes. After that, the DNA fragment that will be used to make the clone is placed into a vector and delivered to host cells. Recombinant cells are the modified bacteria. The creation of several gene copies and the selection of cells harboring recombinant DNA come next. The last phase involves searching for clones with the desired DNA inserts and biological traits. Four kinds of genes have been identified that are responsible for the degradation of various environmental contaminants, including toluene, chlorobenzene acids, other halogenated insecticides and toxic wastes. The first one is the OCT plasmid, which breaks down lower alkanes; the second is the XYL plasmid, which breaks down toluenes and xylenes; the third one is the CAM plasmid, which breaks down camphor; and last is the NAH plasmid, which breaks down naphthalene [30]. Aromatic hydrocarbons such as camphor, salicylate and naphthalene can all be broken down by Pseudomonas putida that carries plasmids as well as hybrid plasmids created by the recombining process [82]. It can also grow quickly on crude oil because it could break down hydrocarbons more quickly than any other single plasmid [83]. This GEM is a superbug (oil-eating bug) [17]. Through genetic engineering or plasmid breeding, molecular approaches can quickly create microbes with enhanced catalytic capabilities that can break down any environmental contaminant [82]. The issues brought on by strain competition in a mixed culture would not exist if GEM strains were used as an inoculum during seeding. Field testing of these organisms could be postponed due to safety and the potential ecological harms that may occur but there is great debate on the release of GEMs into the environment [84]. Table **2** showcases the interaction between microorganisms and hydrocarbons, oils and dyes.

Table 2. Interaction among microorganisms and hydrocarbons, oils and dyes.

Microorganisms	Compound	References
Penicillium chrysogenum	Monocyclic aromatic hydrocarbons such as benzene, toluene, ethyl benzene, xylene, and phenol compounds	[85, 86]
Pseudomonas alcaligenes, Pseudomonas mendocina, Pseudomonas putida, Pseudomonas veronii, Achromobacter, Flavobacterium, Acinetobacter	Polycyclic aromatic hydrocarbons, petrol and diesel	[87, 88]
Pseudomonas putida	Monocyclic aromatic hydrocarbons [*e.g.* benzene and xylene]	[87, 89]
Phanerochaete chrysosporium	Monocyclic aromatic hydrocarbons [*e.g.* benzene and xylene]	[90]
Coprinellus radians	PAHs, methylnaphthalenes and dibenzofurans	[91]
Alcaligenes odorans, Bacillus subtilis, Corynebacterium propinquum, Pseudomonas aeruginosa	phenol	[92]
Tyromyces palustris, Gloeophyllum trabeum, Trametes versicolor	hydrocarbons	[93]
Aspergillus niger, Aspergillus fumigatus, Fusarium solani and Penicillium funiculosum	hydrocarbons	[94]
Candida viswanathii	Phenanthrene, benzopyrene	[95]
cyanobacteria, green algae and diatoms and *Bacillus licheniformis*	naphthalene	[96, 97]

(Table 2) cont.....

Microorganisms	Compound	References
Gleophyllum striatum	striatum Pyrene, anthracene, 9-methylanthracene, Dibenzothiophene Lignin peroxidase	[98]
Acinetobacter sp., Pseudomonas sp., Ralstonia sp. and Microbacterium sp,	aromatic hydrocarbons	[99]
Fusarium sp.	oil	[98]
Alcaligenes odorans, Bacillus subtilis, Corynebacterium propinquum, Pseudomonas aeruginosa	oil	[45]
Bacillus cereus A	diesel oil	[100]
Aspergillus niger, Candida glabrata, Candida krusei and Saccharomyces cerevisiae	crude oil	[101]
B. brevis, P. aeruginosa KH6, B. licheniformis and B. sphaericus	crude oil	[102]
Pycnoporus sanguineous, Phanerochaete chrysosporium and Trametes trogii	industrial dyes	[103]
Penicillium ochrochloron	industrial dyes	[104]
Exiguobacterium indicum, Exiguobacterium aurantiacums, Bacillus cereus and Acinetobacter bauma	azo dyes effluents	[105]
Bacillus fi rmus, Bacillus macerans, Staphylococcus aureus and Klebsiella oxytoca	vat dyes, Textile effluents	[106]

CONCLUSION AND FUTURE PERSPECTIVES

An appealing and commonly accepted method of waste reduction is the use of microorganisms in the biodegradation of organic pollutants. Bioremediation is the most effective and green method for cleaning up contaminated organic pollutant sites since it is environmentally friendly and sustainable. There is no spillover habitat loss quality either. Our environment's microorganisms have a stronger potential to convert hazardous substances into less dangerous products. Carbon dioxide, microbial biomass, and water are harmless byproducts of biodegradation that are environmentally safe. The recovery of polluted places could be accelerated using bioaugmentation, biostimulation, or GEMs.

For biodegradative investigations, suitable environmental conditions should be maintained. Also, the growing body of bacterial genomic data offers new chances to comprehend the genetic and molecular underpinnings of the breakdown of organic contaminants. Microbial enzymes play a significant role in the biodegradation of organically polluted soils like those polluted with diesel, petroleum, PAHs, *etc.* These enzymes will be made available in the future, which is anticipated to usher in a new era of microbiology that will enable various environmental cleanup methods.

ACKNOWLEDGEMENTS

The author would like to acknowledge the Department of Science & Technology and Biotechnology, Government of West Bengal, India, for financial assistance (Memo No. 860(Sanc.)/STBT-11012(25)/5/2019-ST SEC dated 03/11/2023). The author would also like to acknowledge the guidance of Prof. Arup Kumar Mitra of St. Xavier's College, Kolkata, for introducing him to the field of 'Microbiology' during his undergraduate days at St. Xavier's College, Kolkata.

REFERENCES

[1]   Das N, Chandran P. Microbial degradation of petroleum hydrocarbon contaminants: an overview. Biotechnol Res Int 2011; 2011: 1-13.
[http://dx.doi.org/10.4061/2011/941810] [PMID: 21350672]

[2]   Saadoun IM. Impact of oil spills on marine life. Emerging Pollutants in the Environment - Current and Further Implications 2015; 10: 60455.

[3]   Connell DW, Wu RS, Richardson BJ, Lam PK. Chemistry of organic pollutants, including agrochemicals. Environ Ecol Chem 2009; 6: 181.

[4]   Mbachu AE, Chukwura EI, Mbachu NA. Role of microorganisms in the degradation of organic pollutants: a review. Energy Environ Eng 2020; 7(1): 1-1.
[http://dx.doi.org/10.13189/eee.2020.070101]

[5]   Marinescu M, Dumitru M, Lăcătuşu AR. Biodegradation of petroleum hydrocarbons in an artificial polluted soil. Res J Agric Sci 2009; 41(2): 157-62.

[6]   Head IM, Swannell RPJ. Bioremediation of petroleum hydrocarbon contaminants in marine habitats. Curr Opin Biotechnol 1999; 10(3): 234-9.
[http://dx.doi.org/10.1016/S0958-1669(99)80041-X] [PMID: 10361073]

[7]   Gianfreda L, Rao MA. Potential of extra cellular enzymes in remediation of polluted soils: a review. Enzyme Microb Technol 2004; 35(4): 339-54.
[http://dx.doi.org/10.1016/j.enzmictec.2004.05.006]

[8]   Jain PK, Gupta VK, Gaur RK, Lowry M, Jaroli DP, Chauhan UK. Bioremediation of petroleum oil contaminated soil and water. Res J environ toxic 2011; 5(1): 1.

[9]   Miyoshi K, Kamegaya Y, Matsumura M. Electrochemical reduction of organohalogen compound by noble metal sintered electrode. Chemosphere 2004; 56(2): 187-93.
[http://dx.doi.org/10.1016/j.chemosphere.2004.02.013] [PMID: 15120565]

[10]   Dijk JA. Bacterial oxidation of low-chlorinated compounds under anoxic conditions Microbiology. Wageningen University and Research; 2005.

[11]   Mehboob F. Anaerobic microbial degradation of organic pollutants with chlorate as electron acceptor. Microbiology. Wageningen University and Research 2010.

[12]   Van Eekert M. Transformation of chlorinated compounds by methanogenic granular sludge. Environ Tech. Wageningen University and Research 1999.

[13]   Mrozik A, Piotrowska-Seget Z, Labużek S. Bacterial degradation and bioremediation of polycyclic aromatic hydrocarbons. Pol J Environ Stud 2003; 12(1).

[14]   Siddiqa A, Faisal M. Microbial degradation of organic pollutants using indigenous bacterial

strains.Handbook of Bioremediation. Academic Press 2021; pp. 625-37.
[http://dx.doi.org/10.1016/B978-0-12-819382-2.00039-9]

[15]   Kotani T, Yurimoto H, Kato N, Sakai Y. Novel acetone metabolism in a propane-utilizing bacterium, Gordonia sp. strain TY-5. J Bacteriol 2007; 189(3): 886-93.
[http://dx.doi.org/10.1128/JB.01054-06] [PMID: 17071761]

[16]   Mir RA, Gulfishan M. The biodegradation of organic pollutants. Eur J Mol Clin Med 2020; 7(10): 3552-62.

[17]   Joutey NT, Bahafid W, Sayel H, El Ghachtouli N. Biodegradation: involved microorganisms and genetically engineered microorganisms. Biodegr-life Sci 2013; 1: 289-320.

[18]   Yadav D, Singh S, Sinha R. Microbial degradation of organic contaminants in water bodies. Techno Advanc. Resources, Strategies and Scarcity: Pollutants and Water Management 2021; 16: pp. 172-209.

[19]   Huang DY, Zhou SG, Chen Q, Zhao B, Yuan Y, Zhuang L. Enhanced anaerobic degradation of organic pollutants in a soil microbial fuel cell. Chem Eng J 2011; 172(2-3): 647-53.
[http://dx.doi.org/10.1016/j.cej.2011.06.024]

[20]   Xu P, Tao B, Ye Z, *et al.* Simultaneous determination of three alternative flame retardants (dechlorane plus, 1,2-bis(2,4,6-tribromophenoxy) ethane, and decabromodiphenyl ethane) in soils by gas chromatography–high resolution mass spectrometry. Talanta 2015; 144: 1014-20.
[http://dx.doi.org/10.1016/j.talanta.2015.07.031] [PMID: 26452921]

[21]   Sarnaik S, Kanekar P. Bioremediation of colour of methyl violet and phenol from a dye-industry waste effluent using *Pseudomonas* spp. isolated from factory soil. J Appl Bacteriol 1995; 79(4): 459-69.
[http://dx.doi.org/10.1111/j.1365-2672.1995.tb03162.x]

[22]   Devi PI, Thomas J, Raju RK. Pesticide consumption in India: A spatiotemporal analysis. Agric Econ Res Rev 2017; 30(1): 163-72.
[http://dx.doi.org/10.5958/0974-0279.2017.00015.5]

[23]   Meng F, Huang Q, Yuan G, Cai Y, Han FX. The beneficial applications of humic substances in agriculture and soil environments. New trends in remov of heavy metals from Indust Wastewater. Elsevier 2021; pp. 131-60.
[http://dx.doi.org/10.1016/B978-0-12-822965-1.00007-6]

[24]   Alharbi OML, Basheer AA, Khattab RA, Ali I. Health and environmental effects of persistent organic pollutants. J Mol Liq 2018; 263: 442-53.
[http://dx.doi.org/10.1016/j.molliq.2018.05.029]

[25]   Jones KC, de Voogt P. Persistent organic pollutants (POPs): state of the science. Environ Pollut 1999; 100(1-3): 209-21.
[http://dx.doi.org/10.1016/S0269-7491(99)00098-6] [PMID: 15093119]

[26]   Kelly BC, Ikonomou MG, Blair JD, Morin AE, Gobas FA. Food web specific biomagnification of persistent organic pollutants. Science 2007; 317(5835): 236-9.

[27]   Sharma BM, Bharat GK, Tayal S, Nizzetto L, Čupr P, Larssen T. Environment and human exposure to persistent organic pollutants (POPs) in India: A systematic review of recent and historical data. Environ Int 2014; 66: 48-64.
[http://dx.doi.org/10.1016/j.envint.2014.01.022] [PMID: 24525153]

[28]   Goodhead RM, Tyler CR. Endocrine-disrupting chemicals and their environmental impacts. Water 2009; 13(10): 1347.

[29]   El-Shahawi MS, Hamza A, Bashammakh AS, Al-Saggaf WT. An overview on the accumulation, distribution, transformations, toxicity and analytical methods for the monitoring of persistent organic pollutants. Talanta 2010; 80(5): 1587-97.
[http://dx.doi.org/10.1016/j.talanta.2009.09.055] [PMID: 20152382]

[30]   Muir DCG, Howard PH. Are there other persistent organic pollutants? A challenge for environmental

chemists. Environ Sci Technol 2006; 40(23): 7157-66.
[http://dx.doi.org/10.1021/es061677a] [PMID: 17180962]

[31]   Harrad S. Persistent organic pollutants. John Wiley & Sons; 2009; p. 19.

[32]   Vidali M. Bioremediation. An overview. Pure Appl Chem 2001; 73(7): 1163-72.
[http://dx.doi.org/10.1351/pac200173071163]

[33]   Omokhagbor Adams G, Tawari Fufeyin P, Eruke Okoro S, Ehinomen I. Bioremediation,
biostimulation and bioaugmention: a review. Int J Environ Bioremediat Biodegrad 2020; 3(1): 28-39.
[http://dx.doi.org/10.12691/ijebb-3-1-5]

[34]   Iwamoto T, Nasu M. Current bioremediation practice and perspective. J Biosci Bioeng 2001; 92(1): 1-8.
[http://dx.doi.org/10.1016/S1389-1723(01)80190-0] [PMID: 16233049]

[35]   Eweis JB, Ergas SJ, Chang DP, Schroeder ED. Bioremediation principles. McGraw-Hill Book
Company Europe 1998.

[36]   Alexander M. Biodegradation and bioremediation. Gulf Professional Publishing 1999.

[37]   Sivan A. New perspectives in plastic biodegradation. Curr Opin Biotechnol 2011; 22(3): 422-6.
[http://dx.doi.org/10.1016/j.copbio.2011.01.013] [PMID: 21356588]

[38]   Bradley PM. History and ecology of chloroethene biodegradation: a review. Bioremediat J 2003; 7(2): 81-109.
[http://dx.doi.org/10.1080/713607980]

[39]   Bento FM, Camargo FAO, Okeke BC, Frankenberger WT. Comparative bioremediation of soils
contaminated with diesel oil by natural attenuation, biostimulation and bioaugmentation. Bioresour
Technol 2005; 96(9): 1049-55.
[http://dx.doi.org/10.1016/j.biortech.2004.09.008] [PMID: 15668201]

[40]   Roy A, Dutta A, Pal S, *et al.* Biostimulation and bioaugmentation of native microbial community
accelerated bioremediation of oil refinery sludge. Bioresour Technol 2018; 253: 22-32.
[http://dx.doi.org/10.1016/j.biortech.2018.01.004] [PMID: 29328931]

[41]   Xu Y, Lu M. Bioremediation of crude oil-contaminated soil: Comparison of different biostimulation
and bioaugmentation treatments. J Hazard Mater 2010; 183(1-3): 395-401.
[http://dx.doi.org/10.1016/j.jhazmat.2010.07.038] [PMID: 20685037]

[42]   Sayler GS, Ripp S. Field applications of genetically engineered microorganisms for bioremediation
processes. Curr Opin Biotechnol 2000; 11(3): 286-9.
[http://dx.doi.org/10.1016/S0958-1669(00)00097-5] [PMID: 10851144]

[43]   Okoh AI. Biodegradation alternative in the cleanup of petroleum hydrocarbon pollutants. Biotechnol
Mol Biol Rev 2006; 1(2): 38-50.

[44]   Nhi-Cong LT, Mikolasch A, Awe S, Sheikhany H, Klenk HP, Schauer F. Oxidation of aliphatic,
branched chain, and aromatic hydrocarbons by *Nocardia cyriacigeorgica* isolated from oil-polluted
sand samples collected in the Saudi Arabian Desert. J Basic Microbiol 2010; 50(3): 241-53.
[http://dx.doi.org/10.1002/jobm.200900358] [PMID: 20143352]

[45]   Abha S, Singh CS. Hydrocarbon pollution: effects on living organisms, remediation of contaminated
environments, and effects of heavy metals co-contamination on bioremediation. Introduction to
enhanced oil recovery (EOR) processes and bioremediation of oil-contaminated sites 2012; 23: 318.

[46]   Rojo F. Degradation of alkanes by bacteria. Environ Microbiol 2009; 11(10): 2477-90.
[http://dx.doi.org/10.1111/j.1462-2920.2009.01948.x] [PMID: 19807712]

[47]   Coon MJ. Omega Oxygenases: Nonheme-iron enzymes and P450 cytochromes. Biochem Biophys Res
Commun 2005; 338(1): 378-85.
[http://dx.doi.org/10.1016/j.bbrc.2005.08.169] [PMID: 16165094]

[48] Kotani T, Kawashima Y, Yurimoto H, Kato N, Sakai Y. Gene structure and regulation of alkane monooxygenases in propane-utilizing Mycobacterium sp. TY-6 and Pseudonocardia sp. TY-7. J Biosci Bioeng 2006; 102(3): 184-92.
[http://dx.doi.org/10.1263/jbb.102.184] [PMID: 17046531]

[49] Kniemeyer O, Musat F, Sievert SM, *et al.* Anaerobic oxidation of short-chain hydrocarbons by marine sulphate-reducing bacteria. Nature 2007; 449(7164): 898-901.
[http://dx.doi.org/10.1038/nature06200] [PMID: 17882164]

[50] Rabus R, Wilkes H, Behrends A, *et al.* Anaerobic initial reaction of n-alkanes in a denitrifying bacterium: evidence for (1-methylpentyl)succinate as initial product and for involvement of an organic radical in n-hexane metabolism. J Bacteriol 2001; 183(5): 1707-15.
[http://dx.doi.org/10.1128/JB.183.5.1707-1715.2001] [PMID: 11160102]

[51] Paul D, Pandey G, Pandey J, Jain RK. Accessing microbial diversity for bioremediation and environmental restoration. Trends Biotechnol 2005; 23(3): 135-42.
[http://dx.doi.org/10.1016/j.tibtech.2005.01.001] [PMID: 15734556]

[52] Jain RK, Kapur M, Labana S, *et al.* Microbial diversity: application of microorganisms for the biodegradation of xenobiotics. Curr Sci 2005; 10: 101-12.

[53] Abo-Amer AE. Characterization of a strain of Pseudomonas putida isolated from agricultural soil that degrades cadusafos (an organophosphorus pesticide). World J Microbiol Biotechnol 2012; 28(3): 805-14.
[http://dx.doi.org/10.1007/s11274-011-0873-5] [PMID: 22805799]

[54] Ortiz-Hernández ML, Sánchez-Salinas E, Dantán-González E, Castrejón-Godínez ML. Pesticide biodegradation: mechanisms, genetics and strategies to enhance the process. Biodegradation-life of Science 2013; 10: 251-87.

[55] Senthilkumar S, Anthonisamy A, Arunkumar S, Sivakumari V. Biodegradation of methyl parathion and endosulfan using Pseudomonas aeruginosa and Trichoderma viridae. J Environ Sci Eng 2011; 53(1): 115-22.
[PMID: 22324156]

[56] Singh DK. Biodegradation and bioremediation of pesticide in soil: concept, method and recent developments. Indian J Microbiol 2008; 48(1): 35-40.
[http://dx.doi.org/10.1007/s12088-008-0004-7] [PMID: 23100698]

[57] Satish GP, Ashokrao DM, Arun SK. Microbial degradation of pesticide: A review. Afr J Microbiol Res 2017; 11(24): 992-1012.
[http://dx.doi.org/10.5897/AJMR2016.8402]

[58] Iranzo M, Sainz-Pardo I, Boluda R, Sanchez J, Mormeneo S. The use of microorganisms in environmental remediation. Ann Microbiol 2001; 51(2): 135-44.

[59] Vischetti C, Casucci C, Perucci P. Relationship between changes of soil microbial biomass content and imazamox and benfluralin degradation. Biol Fertil Soils 2002; 35(1): 13-7.
[http://dx.doi.org/10.1007/s00374-001-0433-5]

[60] Huang Y, Xiao L, Li F, *et al.* Microbial degradation of pesticide residues and an emphasis on the degradation of cypermethrin and 3-phenoxy benzoic acid: a review. Molecules 2018; 23(9): 2313.
[http://dx.doi.org/10.3390/molecules23092313] [PMID: 30208572]

[61] Deng W, Lin D, Yao K, *et al.* Characterization of a novel β-cypermethrin-degrading *Aspergillus niger* YAT strain and the biochemical degradation pathway of β-cypermethrin. Appl Microbiol Biotechnol 2015; 99(19): 8187-98.
[http://dx.doi.org/10.1007/s00253-015-6690-2] [PMID: 26022858]

[62] Agarwal N, Solanki VS, Gacem A, *et al.* Bacterial laccases as biocatalysts for the remediation of environmental toxic pollutants: a green and eco-friendly approach—a review. Water 2022; 14(24): 4068.

[http://dx.doi.org/10.3390/w14244068]

[63]   Li CH, Wong YS, Tam NFY. Anaerobic biodegradation of polycyclic aromatic hydrocarbons with amendment of iron(III) in mangrove sediment slurry. Bioresour Technol 2010; 101(21): 8083-92.
[http://dx.doi.org/10.1016/j.biortech.2010.06.005] [PMID: 20594830]

[64]   Perfumo A, Banat IM, Marchant R, Vezzulli L. Thermally enhanced approaches for bioremediation of hydrocarbon-contaminated soils. Chemosphere 2007; 66(1): 179-84.
[http://dx.doi.org/10.1016/j.chemosphere.2006.05.006] [PMID: 16782171]

[65]   Margesin R, Schinner F. Bioremediation (natural attenuation and biostimulation) of diesel-oil contaminated soil in an alpine glacier skiing area. Appl Environ Microbiol 2001; 67(7): 3127-33.
[http://dx.doi.org/10.1128/AEM.67.7.3127-3133.2001] [PMID: 11425732]

[66]   Shan H, Kurtz HD Jr, Freedman DL. Evaluation of strategies for anaerobic bioremediation of high concentrations of halomethanes. Water Res 2010; 44(5): 1317-28.
[http://dx.doi.org/10.1016/j.watres.2009.10.035] [PMID: 19945730]

[67]   Chen M, Shih K, Hu M, *et al.* Biostimulation of indigenous microbial communities for anaerobic transformation of pentachlorophenol in paddy soils of southern China. J Agric Food Chem 2012; 60(12): 2967-75.
[http://dx.doi.org/10.1021/jf204134w] [PMID: 22385283]

[68]   Chen BY, Chen WM, Chang JS. Optimal biostimulation strategy for phenol degradation with indigenous rhizobium *Ralstonia taiwanensis*. J Hazard Mater 2007; 139(2): 232-7.
[http://dx.doi.org/10.1016/j.jhazmat.2006.06.022] [PMID: 16844294]

[69]   Meneghetti LR, Thome A, Schnaid F, Prietto PD, Cavelhão G. Natural attenuation and biostimulation of biodiesel contaminated soils from southern Brazil with different particle sizes. J Enviro Sci Eng B 2012; 1(2): 155-62.

[70]   Coulon F, Pelletier E, Gourhant L, Delille D. Effects of nutrient and temperature on degradation of petroleum hydrocarbons in contaminated sub-Antarctic soil. Chemosphere 2005; 58(10): 1439-48.
[http://dx.doi.org/10.1016/j.chemosphere.2004.10.007] [PMID: 15686763]

[71]   Okolo JC, Amadi EN, Odu CT. Effects of soil treatments containing poultry manure on crude oil degradation in a sandy loam soil. Appl Ecol Environ Res 2005; 3(1): 47-53.
[http://dx.doi.org/10.15666/aeer/0301_047053]

[72]   Chaillan F, Chaîneau CH, Point V, Saliot A, Oudot J. Factors inhibiting bioremediation of soil contaminated with weathered oils and drill cuttings. Environ Pollut 2006; 144(1): 255-65.
[http://dx.doi.org/10.1016/j.envpol.2005.12.016] [PMID: 16487636]

[73]   Leahy JG, Colwell RR. Microbial degradation of hydrocarbons in the environment. Microbiol Rev 1990; 54(3): 305-15.
[http://dx.doi.org/10.1128/mr.54.3.305-315.1990] [PMID: 2215423]

[74]   Abioye OP. Biological remediation of hydrocarbon and heavy metals contaminated soil. Soil Cont 2011; 12(7): 127-42.

[75]   Lendvay JM, Löffler FE, Dollhopf M, *et al.* Bioreactive barriers: a comparison of bioaugmentation and biostimulation for chlorinated solvent remediation. Environ Sci Technol 2003; 37(7): 1422-31.
[http://dx.doi.org/10.1021/es025985u]

[76]   Silva E, Fialho AM, Sá-Correia I, Burns RG, Shaw LJ. Combined bioaugmentation and biostimulation to cleanup soil contaminated with high concentrations of atrazine. Environ Sci Technol 2004; 38(2): 632-7.
[http://dx.doi.org/10.1021/es0300822] [PMID: 14750741]

[77]   Hinchee RE, Fredrickson J, Alleman BC. Bioaugmentation for site remediation. OH (United States): Battelle Press, Columbus 1995.

[78]   Atlas RM, Raymond RL. Stimulated petroleum biodegradation. CRC Crit Rev Microbiol 1977; 5(4):

371-86.
[http://dx.doi.org/10.3109/10408417709102810] [PMID: 334468]

[79]   Lima, D., Viana, P., André, S., Chelinho, S., Costa, C., Ribeiro, R., Sousa, J.P., Fialho, A.M., Viegas, C. A. Evaluating a bioremediation tool for atrazine contaminated soils in open soil microcosms: the effectiveness of bioaugmentation and biostimulation approaches. Chemosphere., 2009, 74(2), 187-192.

[80]   Fulekar MH, Singh A, Bhaduri AM. Genetic engineering strategies for enhancing phytoremediation of heavy metals. Afr J Biotechnol 2009; 8(4).

[81]   Abatenh E, Gizaw B, Tsegaye Z, Wassie M. The role of microorganisms in bioremediation-A review Open J Environ Biol 2017; 2(1): 038-46.
[http://dx.doi.org/10.17352/ojeb.000007]

[82]   Adetunji CO, Anani OA. Recent advances in the application of genetically engineered microorganisms for microbial rejuvenation of contaminated environment. Microorganisms for Sustainability 2021; 27: 303-24.
[http://dx.doi.org/10.1007/978-981-15-7459-7_14]

[83]   Markandey DK. Environmental biotechnology. APH Publishing 2004.

[84]   Wackett LP. Stable isotope probing in biodegradation research. Trends Biotechnol 2004; 22(4): 153-4.
[http://dx.doi.org/10.1016/j.tibtech.2004.01.013] [PMID: 15104106]

[85]   Pereira P, Enguita FJ, Ferreira J, Leitão AL. DNA damage induced by hydroquinone can be prevented by fungal detoxification. Toxicol Rep 2014; 1: 1096-105.
[http://dx.doi.org/10.1016/j.toxrep.2014.10.024] [PMID: 28962321]

[86]   Safiyanu I, Isah AA, Abubakar US, Rita Singh M. Review on comparative study on bioremediation for oil spills using microbes. Res J Pharm Biol Chem Sci 2015; 6: 783-90.

[87]   Safiyanu I, Sani I, Rita SM. Review on bioremediation of oil spills using microbial approach. Int J Eng Sci Res 2015; 3(6): 41-55.

[88]   Bahadure S, Kalia R, Chavan R. Comparative study of bioremediation of hydrocarbon fuels. Int J Biotechnol Bioeng Res 2013; 4(7): 677-86.

[89]   Wolski EA, Barrera V, Castellari C, González JF. Biodegradation of phenol in static cultures by Penicillium chrysogenum ERK1: catalytic abilities and residual phytotoxicity. Rev Argent Microbiol 2012; 44(2): 113-21.
[PMID: 22997771]

[90]   Ability of some soil fungi in biodegradation of petroleum hydrocarbon. J Appl Environ Microbiol 2014; 2(2): 46-52.

[91]   Karigar CS, Rao SS. Role of microbial enzymes in the bioremediation of pollutants: a review. Enzyme Res 2011; 2011: 1-11.
[http://dx.doi.org/10.4061/2011/805187] [PMID: 21912739]

[92]   Singh A, Kumar V, Srivastava JN. Assessment of bioremediation of oil and phenol contents in refinery waste water *via* bacterial consortium. J Pet Environ Biotechnol 2013; 4(3): 1-4.
[http://dx.doi.org/10.4172/2157-7463.1000145]

[93]   Hesham AEL, Khan S, Tao Y, Li D, Zhang Y, Yang M. Biodegradation of high molecular weight PAHs using isolated yeast mixtures: application of meta-genomic methods for community structure analyses. Environ Sci Pollut Res Int 2012; 19(8): 3568-78.
[http://dx.doi.org/10.1007/s11356-012-0919-8] [PMID: 22535224]

[94]   Aranda E, Ullrich R, Hofrichter M. Conversion of polycyclic aromatic hydrocarbons, methyl naphthalenes and dibenzofuran by two fungal peroxygenases. Biodegradation 2010; 21(2): 267-81.
[http://dx.doi.org/10.1007/s10532-009-9299-2] [PMID: 19823936]

[95]   Sivakumar G, Xu J, Thompson RW, Yang Y, Randol-Smith P, Weathers PJ. Integrated green algal technology for bioremediation and biofuel. Bioresour Technol 2012; 107: 1-9.

[http://dx.doi.org/10.1016/j.biortech.2011.12.091] [PMID: 22230775]

[96]   Lin C, Gan L, Chen ZL. Biodegradation of naphthalene by strain *Bacillus fusiformis* (BFN). J Hazard Mater 2010; 182(1-3): 771-7.
[http://dx.doi.org/10.1016/j.jhazmat.2010.06.101] [PMID: 20643503]

[97]   Simarro R, González N, Bautista LF, Molina MC. Assessment of the efficiency of in situ bioremediation techniques in a creosote polluted soil: Change in bacterial community. J Hazard Mater 2013; 262: 158-67.
[http://dx.doi.org/10.1016/j.jhazmat.2013.08.025] [PMID: 24025312]

[98]   Hidayat A, Tachibana S. Biodegradation of aliphatic hydrocarbon in three types of crude oil by Fusarium sp. F 092 under stress with artificial sea water. J Environ Sci Technol 2011; 5(1): 64-73.
[http://dx.doi.org/10.3923/jest.2012.64.73]

[99]   Yadav M, Singh SK, Sharma JK, Yadav KDS. Oxidation of polyaromatic hydrocarbons in systems containing water miscible organic solvents by the lignin peroxidase of *Gleophyllum striatum* MTCC-1117. Environ Technol 2011; 32(11): 1287-94.
[http://dx.doi.org/10.1080/09593330.2010.535177] [PMID: 21970171]

[100] Maliji D, Olama Z, Holail H. Environmental studies on the microbial degradation of oil hydrocarbons and its application in Lebanese oil polluted coastal and marine ecosystem. Int J Curr Microbiol Appl Sci 2013; 2(6): 1-18.

[101] Burghal AA, Abu-Mejdad NMJA, Al-Tamimi WH. Mycodegradation of crude oil by fungal species isolated from petroleum contaminated soil. Int J Innov Res Sci Eng Technol 2016; 5(2): 1517-24.

[102] El-Borai A, Eltayeb K, Mostafa A, El-Assar S. Biodegradation of industrial oil-polluted wastewater in egypt by bacterial consortium immobilized in different types of carriers. Pol J Environ Stud 2016; 25(5): 1901-9.
[http://dx.doi.org/10.15244/pjoes/62301]

[103] Yan J, Niu J, Chen D, Chen Y, Irbis C. Screening of Trametes strains for efficient decolorization of malachite green at high temperatures and ionic concentrations. Int Biodeterior Biodegradation 2014; 87(87): 109-15.
[http://dx.doi.org/10.1016/j.ibiod.2013.11.009]

[104] Shedbalkar U, Jadhav JP. Detoxification of malachite green and textile industrial effluent by Penicillium ochrochloron. Biotechnol Bioprocess Eng; BBE 2011; 16(1): 196-204.
[http://dx.doi.org/10.1007/s12257-010-0069-0]

[105] Kumar S, Chaurasia P, Kumar A. Isolation and characterization of microbial strains from textile industry effluents of Bhilwara, India: analysis with bioremediation. J Chem Pharm Res 2016; 8(4): 143-50.

[106] Adebajo S, Balogun S, Akintokun A. Decolourization of vat dyes by bacterial isolates recovered from local textile mills in Southwest, Nigeria. Microbiol Res J Int 2017; 18(1): 1-8.
[http://dx.doi.org/10.9734/MRJI/2017/29656]

CHAPTER 3

A Study of Green Synthesis of Metal Nanoparticles using Plant Extracts and their Biological and Environmental Applications

Sankara Rao Miditana[1,*] and **Saivenkatesh Korlam**[2]

[1] *Department of Chemistry, Government Degree College, Puttur, Tirupathi, Andhra Pradesh, India*

[2] *Department of Botany, SVA Government Degree College, Srikalahasti, Tirupati, Andhra Pradesh, India*

Abstract: Nanomaterials (NMs)-based technology is a powerful tool in the current scenario because of their size and unique physicochemical properties. Green synthesized NMs are promising in creating new and vital products that are beneficial to the environment, industry, and humans. Due to its simplicity, nontoxicity, and environmentally benign advantages, the synthesis of metal nanoparticles (NPs) using green techniques has received a lot of attention recently. Every day, attention is drawn to recycling waste and putting it to good use. NPs are easily manufactured in a green, energy-free manner using plant extracts that are not intended for human consumption. Metal-based NPs are widely used due to their applications, including medicine, biomedical sciences, biosensing, food, cosmetics, and electronics. NPs produced from novel synthesis techniques using plant extracts have remarkable qualities. In the synthesis of NPs *via* the green approach, many metals such as silver, gold, copper, zinc, manganese, nickel, and magnesium are used due to their unique physical and optical properties. In this chapter, the authors have reported the mechanisms of various green methods of synthesis of NPs, their biological and environmental applications along with their challenges and prospects.

Keywords: Antioxidant, Antimicrobial, Anti-bacterial, Anticancer, Biodegradable, Catalytic activity, Dye degradation, Green synthesis, Nanoparticles.

INTRODUCTION

NMs are materials that have at least one dimension (length, width, or height) in the nanometer scale, typically ranging from 1 to 100 nanometers [1]. They exhibit unique physical and chemical properties that are different from their bulk

---

* Corresponding author Sankara Rao Miditana: Department of Chemistry, Government Degree College, Puttur, Tirupathi, Andhra Pradesh, India; E-mail: sraom90@gmail.com

Neha Agarwal, Vijendra Singh Solanki and Sreekantha B. Jonnalagadda (Eds.)

counterparts due to their small size, high surface area to volume ratio, enhanced catalytic activity, and quantum confinement effects. They have novel electronic, optical, and magnetic properties, making them attractive for a wide range of applications in various fields, including medicine, electronics, energy, and environmental science [2, 3]. To synthesize NPs, two methods are employed, namely the chemical method and the physical method [1]. Chemical synthesis methods include electrochemical techniques, chemical reduction and photochemical reduction [4]. Although most chemical methods succeed in producing pure and well-defined NPs, they are expensive and inefficient and release hazardous wastes into the environment, so more environmentally friendly methods are preferred [5]. On the other hand, physical synthesis methods include condensation, evaporation, and laser ablation.

The traditional methods for synthesizing NPs involve the use of toxic chemicals, high-energy inputs, and complex procedures that are not environmentally friendly and can cause harm to human health [6]. Green synthetic processes are being established as an alternative to physical and chemical processes. Green synthesis has many advantages compared to chemical and physical methods: it is non-toxic, pollution-free, environmentally friendly, economical, and more sustainable [7 - 10]. Environmental issues like agriculture production and catalysis can be approached by green synthesis. Green synthesis of NPs involves the use of natural sources such as plants, microbes, and biopolymers as reducing and stabilizing agents. These methods have advantages such as low cost, biodegradability, and scalability, making them an attractive alternative to traditional synthesis methods. According to Christophe *et al.*, green synthesis *via* plant materials plays a key role in the determination of the size and morphology of prepared NPs [11]. Plant extracts can act as reducing and capping agents. In terms of the size of the NPs formed, the green-synthesized products are larger than those obtained by chemical methods [12]. The green synthetic approach produces an optimum yield of NPs than the physical and chemical techniques.

In recent years, there has been growing interest in developing green synthesis methods for NPs that are sustainable, non-toxic, and eco-friendly [13]. In this context, the study of NPs *via* green synthesis has gained significant attention along with exploring the use of various plant extracts, microorganisms (bacteria, algae, viruses, fungi) and other natural sources (solar mediated and green chemistry synthesis) to synthesize NPs with controlled size, shape, and composition. In the green approach of NP synthesis, metal atoms play a crucial role as they are the building blocks of the NPs. The properties of the metal atoms, such as their size, shape, and surface chemistry, influence the properties of the resulting NPs [14, 15]. Metal atoms are often used as the precursor for NP synthesis. Gold, silver, manganese, magnesium and copper are commonly used metals for the synthesis of metal NPs. Metal atoms can also act as a reducing agent to convert metal ions into NPs. By varying the concentration and type of metal atoms used, it is possible to control the size and shape of NPs. They have a wide range of applications in catalysis, electronic and biomedical fields [16]. In the present study, authors focus on various green methods of synthesis of NPs, which offers a promising avenue for developing sustainable and eco-friendly materials and exploring their potential applications in various fields like drug delivery, biosensing, catalysis, and wastewater treatment. This chapter also reports the various applications of plant-extract mediated green synthesized NPs.

GREEN METHODS OF SYNTHESIS OF METAL NPS

Green synthesis of NPs involves the use of natural sources such as plants, microbes, and biopolymers as reducing and stabilizing agents. Here are some of the commonly used methods, including plant extract-mediated synthesis, microbial-mediated synthesis, solar-mediated synthesis, and green chemistry-based synthesis for the preparation of NPs.

Plant Extract-mediated Synthesis

This method is a green and sustainable approach to the synthesis of NPs. Various parts of the plant, such as leaves, stems, and fruits, have been used for the synthesis of NPs [17 - 21]. The process involves the reduction of metal salts in the presence of plant extracts, which provide the necessary reducing agents and stabilizing agents for the formation of NPs [22]. The detailed procedure is discussed below.

- Select a plant material and prepare an extract by grinding the plant material in a solvent such as water, ethanol, or methanol.
- Heat the extract to a suitable temperature, typically between 60-100°C.
- Add a metal precursor solution to the extract, typically a salt such as silver nitrate, gold chloride, or platinum chloride.
- Allow the reaction to proceed for a certain amount of time, usually from 30 minutes to several hours, while stirring.
- Monitor the formation of NPs by measuring the absorbance of the solution using UV-Vis spectroscopy. The absorption peak indicates the formation of NPs. The scheme of preparation of NPs by plant extracts is shown in Fig. (1).

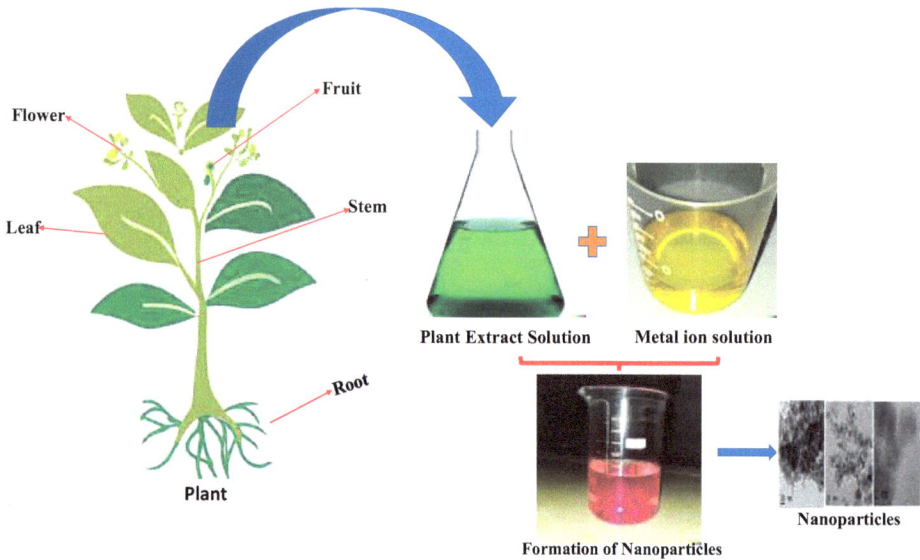

Fig. (1). Synthesis of NPs using extracts from various parts of a plant.

Microbial-mediated Synthesis

Microbial-mediated synthesis of NPs is another green and sustainable approach to the synthesis of NPs. The process involves the use of microorganisms such as bacteria, fungi, algae, viruses and yeast as reducing and stabilizing agents [23 - 27]. They produce enzymes and other biomolecules that can act as reducing and stabilizing agents [23]. Individual procedures for the preparation of NPs by microbial-mediated synthesis are reported here.

Green Synthesis using Bacteria

Green synthesis of NPs by bacteria involves the use of microorganisms such as bacteria to reduce metal ions into their corresponding NPs. This method is eco-friendly, cost-effective, and has minimal environmental impacts [28]. The procedure for green synthesis of NPs by bacteria can be summarized into the following steps [29].

Bacterial culture: The first step in the green synthesis of NPs by bacteria is to culture the bacteria in a nutrient-rich medium. The bacterial strain used should be selected based on its ability to reduce metal ions to their corresponding NP.

Metal ion solution: The metal ion solution is prepared by dissolving the metal salt in deionized water. The concentration of the metal ion solution is optimized based on the desired size and concentration of the NPs.

Incubation: The bacterial culture is incubated with the metal ion solution for a specific duration under optimized conditions such as temperature, pH, and aeration. During incubation, the bacteria reduce the metal ions to their corresponding NPs.

Purification: After the incubation period, the bacterial culture is centrifuged to separate the NPs from the bacterial cells. The NPs are then washed several times with deionized water to remove any impurities or excess reactants.

Green Synthesis using Algae

Green synthesis of NPs using algae is an eco-friendly approach that utilizes the reducing and stabilizing properties of algae to form NPs. The following is a general procedure for the green synthesis of NPs using algae [30].

Algae selection: The first step is to select a suitable species of algae that can produce the desired NPs. The algae are then cultured under controlled conditions in a laboratory or greenhouse.

Algae cultivation: The algae are cultured in a suitable growth medium under controlled conditions to promote their growth and biomass production.

Harvesting of algae: The algae are harvested and washed to remove impurities and excess growth medium.

Preparation of the extract: The harvested algae are then washed and dried, and their cell walls are disrupted using techniques such as sonication or homogenization. The resulting extract is then filtered to remove any debris.

Synthesis of NPs: The extract is then mixed with a suitable reducing agent, such as sodium borohydride, and heated to a specific temperature for a specific period. This results in the formation of NPS.

Green Synthesis using Fungi

Green synthesis of NPs using fungi involves the use of fungal extracts or whole cells to reduce metal ions into their corresponding NPs. The procedure can be summarized into the following steps [31, 32].

Fungal culture: The first step in the green synthesis of NPs using fungi is to culture the fungi in a nutrient-rich medium. The fungal strain used should be selected based on its ability to reduce metal ions to their corresponding NPs.

Filtration of fungal extract: If fungal extract is used, the extract is filtered to remove any impurities or debris.

Metal ion solution: The metal ion solution is prepared by dissolving the metal salt in deionized water. The concentration of the metal ion solution is optimized based on the desired size and concentration of the NPs.

Incubation: The fungal extract or whole cells are added to the metal ion solution and incubated for a specific duration under optimized conditions such as temperature, pH, and aeration. During incubation, the fungi reduce the metal ions to their corresponding NPs.

Purification: After the incubation period, the solution is centrifuged to separate the NPs from the fungal cells or extract. The NPs are then washed several times with deionized water to remove any impurities or excess reactants.

Green Synthesis using Viruses

Green synthesis of NPs using viruses involves the use of viral vectors to synthesize NPs. The procedure can be summarized into the following steps [33 - 35].

Viral vector selection: The first step in the green synthesis of NPs using viruses is to select the appropriate viral vector. The viral vector should be able to infect the host cell and express the genes responsible for NP synthesis.

Plasmid construction: The gene responsible for NP synthesis is cloned into the viral vector, and the resulting plasmid is transformed into a bacterial host for amplification.

Host cell transfection: The plasmid is then transfected into the host cell, which can be either prokaryotic or eukaryotic.

NP synthesis: The host cell expresses the genes responsible for NP synthesis, leading to the formation of NPs within the cell.

Purification: After NP synthesis, the cells are lysed, and the NPs are purified using various techniques such as centrifugation, chromatography, or filtration.

Green Chemistry-based Synthesis

Green chemistry-based synthesis of NPs is an environmentally friendly and sustainable approach that involves the use of natural, renewable, and non-toxic materials as reducing and stabilizing agents [36 - 40]. This method involves the use of non-toxic and environmentally friendly chemicals such as ascorbic acid and sodium borohydride as reducing agents for the synthesis of NPs [36]. The general procedure for the preparation of NPs by green chemistry-based synthesis is reported below:

- Dissolve the metal precursor in a suitable solvent such as water or ethanol.
- Add natural reducing and stabilizing agents, such as plant extracts, proteins, or carbohydrates, to the metal precursor solution.
- Incubate the mixture under suitable conditions, such as temperature, pH, and duration, for the reduction of metal ions to NPs.
- Collect the NP suspension and purify it by centrifugation or filtration.
- Characterize the synthesized NPs using various analytical techniques.

Solar-mediated Synthesis

In this method, solar radiation is used as a source of energy for the synthesis of NPs [41 - 43]. This method is environmentally friendly and does not require any external energy sources [43]. Solar-mediated synthesis of NPs is an eco-friendly and sustainable approach that involves the use of solar energy for the reduction of metal ions to NPs [44]. The general procedure for the preparation of NPs by solar-mediated synthesis is reported below.

- Dissolve the metal precursor in a suitable solvent such as water or ethanol.
- Add natural reducing and stabilizing agents, such as plant extracts, proteins, or carbohydrates, to the metal precursor solution.
- Place the reaction mixture in a suitable container and expose it to sunlight or artificial light sources such as UV lamps or LEDs.
- Incubate the mixture under suitable conditions, such as temperature, pH, and duration, for the reduction of metal ions to nanoparticles.
- Collect the NP suspension and purify it by centrifugation or filtration.
- Characterize the synthesized NPs using various analytical techniques.

Among all green methods, plant-extract-mediated synthesis of NPs is an emerging field of research that has gained significant attention in recent years. It involves the use of plant extracts as reducing agents for the synthesis of NPs. Plant-extract mediated synthesis of NPs is an important area of research that offers several advantages over conventional synthesis methods, including eco-friendliness, cost-effectiveness, biocompatibility, and versatility.

MECHANISM OF METAL NP SYNTHESIS USING PLANT EXTRACTS

Metal NP (MNP) synthesis using plant extracts contains a variety of active reducing agents, such as alkaloids, phenols, terpenoids, quinines, amides, flavonoids, proteins, and alcohols [45]. Reducing active ingredients like flavonoids and phenols can convert some metal cations into MNPs which also act as stabilizers to prevent MNP aggregation, and as a result, they play a crucial role in the green synthesis of MNPs [46, 47].

In addition to reducing agents, stabilizing agents are essential in the synthesis of NPs from plant extracts to prevent their aggregation and ensure their stability. Some common stabilizing agents that can be found in plant extracts are proteins and amino acids, polysaccharides, polyphenols, and lipids [48]. It is important to note that the stabilizing agents present in plant extracts can vary depending on the plant species and the extraction method used. Additionally, the concentration and type of stabilizing agent required for NP synthesis can vary depending on the type and size of NPs being synthesized, and the desired properties of the final product. Functional groups present in the plant extract, which are reacting with precursors, are shown in Table **1**.

Table 1. **Main functional groups available in plant extracts that react with precursors.**

Functional Group	Formula
Hydroxyl group	R-OH
Carbonyl group	R-CO-R
Amine group	R-NH2
Methoxy group	R-O-CH3

It is known that the synthesis of MNPs utilizing plant extracts may be roughly divided into three stages: the reduction phase, the growth phase, and the termination phase. Metal ions are converted to zero-valent metal atoms during the reduction stage by electron transfer by the reducing photoactive agent. The zero-valent metal atoms then increase in size by aggregating into nanometallic particles in the development stage that can have several shapes, including linear, rod-shaped, triangular, hexagonal, or cubic. To keep MNPs stable throughout the termination stage, phytoactive components with antioxidant characteristics are enhanced around MNPs [49, 50].

According to Sajadi *et al.*, flavonolignans from *Silybum marianum* L. seed extracts were used to decrease copper ions to Cu NPs and alter their structure from keto to enol [51]. Using nanocellulose, Yu *et al.* generated AgNPs by attracting positively charged silver ions to the negatively charged carboxyl and hydroxyl groups on the material [52]. The hydroxyl groups reduced the silver ions at 80 °C and then oxidized to form aldehyde groups. To prepare AgNPs, Jigyasa and Rajput utilized polyphenols (rutin/curcumin) [53]. They hypothesized that rutin/curcumin may have reduced silver ions through carbonyl or phenolic hydroxyl groups to produce AgNPs. Employing starch as a starting material, Wang *et al.* reported that the numerous hydroxyl and carboxyl groups on the linear chain of starch reacted with the silver ions to reduce them to AgNPs [46]. In addition, it has been found that the produced AgNPs were stable because of their passivated surface and negatively charged carboxyl group around them. Plant extracts have both reducing and antioxidant characteristics, so they prepare MNPs by reducing them while stabilizing them to prevent oxidation. According to Veisi *et al.*, phenolic, flavonoid, and tannin substances bind to the surface of AgNPs to maintain their stability through electronic interactions [54]. Nasrollahzadeh *et al.* noticed the existence of active components with antioxidant characteristics in plant extracts that reduce copper ions and concurrently protect the stability of Cu NPs [55].

APPLICATIONS OF PLANT EXTRACT -MEDIATED GREEN SYNTHESIZED MNPS

Biological Applications

Researchers are intrigued by nanotechnology due to the microscopic size and substantial surface-to-volume ratio of NPs. These characteristics lead to notable chemical and physical alterations in their properties. As a consequence of these unique attributes, NPs find diverse applications across various areas, including biological and environmental fields. Moreover, green-synthesized NPs have a wide range of potential applications in these fields. This chapter focuses on recent findings from biological studies of MNPs using plant extracts, and these findings are presented in Table **2**.

Table 2. Biological applications of green synthesized NPs.

Metals used for NPs	Plant Extracts used	Size (nm)	Applications	Refs.
Ag	*Catharanthus roseus*/Flower	30	Antioxidant, Antimicrobial and Photocatalytic activity	[56]
Ag	*Eugenia roxburghii.DC*/Leaves	25-50	Antimicrobial	[57]
Ag	*Croton sparsiflorus morong*/Leaf	22-52	Antibacterial and Antifungal	[58]
Ag	*Grewia flaviscences*/Leaf	50-70	Antimicrobial	[59]

(Table 2) cont.....

Metals used for NPs	Plant Extracts used	Size (nm)	Applications	Refs.
Ag	*Delphinium denudatum*/Root	< 85	Antibacterial and Mosquito activity	[60]
Ag	*Anisomeles Indica*/Leaf	50-100	Mosquitocidal potential against Malaria, dengue and Japanese encephalitis vectors	[61]
Ag	*Solanum nigrum L*/Leaves and Green berry	50-100	Mosquito Larvicidal and Antimicrobial activity	[62]
Ag	*Mangifera indica*/ Flower	10-20	Antibacterial activity	[63]
Ag	*Salvia hispanica L*/ Seeds	1-27	Antibacterial activity	[64]
Cu	*Jatropha cura*/Leaves	10	CT-DNA binding and photocatalytic activity	[65]
Cu	*Ageratum houstoniaum*/ Leaf	80	Photocatalytic activity and antibacterial activities	[66]
Cu	*Cinnamomum Zelanicum*/Medicinal plant	19.55-69.70	Antioxidant and anti-human lung carcinoma	[67]
Cu	*Neem*/ Flower	5	Antibacterial activity	[68]
Cu	*Green coffee bean*	5-8	Degradation of dyes	[69]
Cu	*Frunus nepolensis*/Fruit	35-50	Anticancer activity	[70]
Cu	*Prunus mahalab L*	20-30	Biological activities	[71]
Mn	*Eucalyptus robusta and corymbia citriodora*/Leaf	16.61-23.68 and 21.77-28.63	Agriculture applications	[72]
Mn	*Cinnamomum verum*/Stem	50-100	Photocatalytic and antimicrobial activity	[73]
Mn	*Pumpkin*/Seeds	50	Anticancer efficiency	[74]
Mn	*Brassica oleraceae*/Cabbage leaves	-	Antibacterial activity	[75]
Mn	*Ctenolepis garcinia (Burm.f) C.B clark*	57-69	-	[76]
Fe, Mn	*Cannabis sativa*/Leaf	20-80	Photocatalytic removal of dyes	[77]
Mg	*Rosa floribunda charisma*/ Flower	35.25-55.14	Antioxidant, antiaging and antibiofilm activities	[78]
Mg, Ag	*Hydrangea paniculate*/Flower	-	Healthcare applications	[79]
Au	*Citrullus lanatus rind*/ Fruit	20-140	Proteasome inhibitory activity, antibacterial and antioxidant potential	[80]
Au	*Atriplex halimus and Chenopodium amperosidies*	2-10	Anticancer, antioxidant and catalytic efficiencies	[81]

(Table 2) cont.....

Metals used for NPs	Plant Extracts used	Size (nm)	Applications	Refs.
Au	*Bauhinia purpurea/* Leaf	-	Anticancer, antimicrobial antioxidant and catalytic activities	[82]
Au	*Jatrapha integerrima Jacq./*Flower	28-43	Antibacterial activity	[83]
Au	*Vitex negundo/*Leaf	20-70	Antioxidant and antibacterial activity	[84]
Au	*Curuma psedomontana*	39	Antimicrobial, antioxidant and anti-inflammatory	[85]
Au	*Onion peel*	25-70	Biological activities	[86]
Au	*Carambola/*Fruit	10-20	Anticancer and antioxidant	[87]
Au	*Corchorus olitorius/* Leaf	37-50	Antiproliferative effect in cancer cells	[88]
Au	*Nerium oleander*	10-100	Anticancer activity 0n MCF-7 cells and catalytic activity	[89]

Environmental Applications

Continuous monitoring of environmental pollution stands as a crucial and fundamental necessity in the realm of environmental pollution control. The primary challenge lies in the effective elimination of pollutants, such as those found in air, soil, and water, which pose both short and long-term threats to the environment and human health. Tackling pollution across such a broad spectrum proves challenging and necessitates a focus on eliminating contaminants at their source. Various methods have been implemented globally to address air, water, and soil pollution, with nanotechnology emerging as a transformative solution.

The application of nanotechnology facilitates the reduction of air, water, and soil pollution. Moreover, recent research highlights the green-synthesised MNPs' fascinating environmental applications, a rapidly expanding field in nanotechnology. Through sustainable and eco-friendly synthesis methods employing plant extracts, researchers have harnessed MNPs for diverse environmental purposes. These NPs exhibit exceptional efficacy in water remediation, showcasing their potential as powerful nanocatalysts for pollutant degradation. Here, we present environmental application findings in Table **3**, emphasizing the mitigation of water contamination through the utilization of MNPs.

Table 3. Environmental applications of green synthesized NPs.

The Metal used for NP synthesis	Plant Extract used	Size nm	Reported Dye	Degradation %	Degradation Time	Refs.
Ag	*Cestrum nocturnum* L. /Leaf	50-90	1. 4-Nitrophenol 2. Congo red 3. 4-Nitroaniline 4. Methylene blue	90 90 78-79 78-79	8 min 15 min 8 min 18 min	[90]
Ag	*Morinda tinctoria*/ Leaf	79-96	Methylene Blue	95	72 hrs	[91]
Ag	*Cichorium intybus*/ Medicinal plant	15-30	Brilliant Blue R	50	5 min	[92]
Ag	*Salvia officinalis*	40	Congo Red	82	50 min	[93]
Au and Ag	*Mussaenda glabrata*	-	1. Rhodamine B 2. Methyl orange	- -	5 and 9 min 4 and 7 min	[94]
Ag	*Cicer arietinum*/Leaf	88.8	1. Methylene blue 2. Congo red	- -	16 min 14 min	[95]
Ag	*Cryptocarya alba*/Leaf	16.1	Methylene Blue	100	<1 min	[96]
Ag	*Carica papaya*/ Leaf	10-70	Yellow 3RS and Blue CP (Multi)	90	30 min	[97]
Ag@CdO	*Citrus limen*/Seed	7-50	Brilliant green	96	90 min	[98]
Cu	*Andrographis paniculata.*	68	1. Methyl red 2. Eosin Dye	92 95	48 hrs 48 hrs	[99]
CuO	*Seriphidium oliverianum*/ Leaf	12	1. Methyl green 2. Methyl orange	65 65	60 min 60 min	[100]
Fe	*Catharanthus roseus*	-	Methyl Orange	50	2 hrs	[101]
Fe	*Shorea robusta*	54-80	Congo Red	96	15 min	[102]
Fe	*Chlorophytum comosum*/Leaf	< 100	Methyl Orange	77	6 hrs	[103]
NiO	*Tribulus terrestris*/Medicinal plant	60-90	Congo red	92	20 min	[104]

ADVANTAGES OF PLANT EXTRACT MEDIATED SYNTHESIS

Plant-extract mediated synthesis of NPs is a promising eco-friendly and cost-effective method, but it has some limitations and challenges that need to be addressed. To maximize the benefits of this method, more research is needed to improve reproducibility, control particle size and shape, and develop standardized protocols [105 - 107].

Three toxic chemicals used for the synthesis of NPs (precursor, reducing agent and capping or stabilizing agent) may have major harmful impacts on both human beings and the environment. As a result, a completely new environmentally conscious methodology is to be implemented for the synthesis of NPs. When humans come in contact with these NPs, even minute amounts of hazardous residues prevent safe applications [108]. Plant extract-mediated synthesis has the following advantages [109 - 111]:

- Affordability, eco-friendliness, sustainability, and simplicity of operation.
- Non-toxicity, homogeneity and stability in the synthesized NPs.
- Biocompatibility and adaptability for clinical applications of the NPs.
- Energy economy and ease of synthesis procedure.
- Utilization of secondary metabolites from plants as both reductants and stabilizers.

DRAWBACKS OF PLANT EXTRACT-MEDIATED SYNTHESIS

Plant extract-mediated synthesis of NPs has several advantages, but it also has a few limitations. Batch-to-batch variability, complexity in standardization, and potential contamination of plant extracts can affect the NPs' size, shape, and stability. Standardization can be challenging due to the complexity of plant extracts, and purification can be cumbersome. Limited control over size and shape is also a challenge. The presence of bioactive compounds in plant extracts complicates the formation of NPs. Scaling up the process for industrial applications can be challenging and longer synthesis times may be necessary for industries with high efficiency and productivity. However, certain constraints associated with this method must be acknowledged and addressed, which include [112, 113]:

- Complexity and diversity of phytochemicals in plant systems.
- Challenges associated with bio-reduction reactions.
- Issues related to homogeneity in synthesis.
- Difficulties in scaling up the synthesis process.
- Concerns regarding reproducibility of results.
- Accessibility to raw materials for consistent synthesis.
- Considerations of product stability over time.

CONCLUSION AND FUTURE OUTLOOK

The traditional synthesis of NPs produces potentially poisonous substances, but the plant-mediated synthesis of NPs holds the potential to bring about a transformative shift in NP production methodologies. The generation of NPs using extracts of different species, or green method, has emerged as an important front in nanotechnology. In this chapter, a variety of green methods like plant extract-mediated, microbial-mediated, green-chemistry-based and solar-mediated synthesis were discussed. Being environmentally friendly, cost-effective, biocompatible, tunable, and easily scalable, the plant extract-mediated method is recommended among all methods of synthesis. These properties make it an attractive approach for the production of NPs for various applications. The green methods reported in this chapter have shown more stability than many conventional methods. However, there is a need to develop more advanced methods of green synthesis of NPs using plants and other natural resources that could control the desired properties of NPs and address several constraints and toxicities associated with their use. Simply said, NPs make our lives comfortable and simple, therefore, nanotechnology has a promising future because it is used in a variety of scientific fields.

REFERENCES

[1]   Baig N, Kammakakam I, Falath W. Nanomaterials: a review of synthesis methods, properties, recent progress, and challenges. Mater Adv 2021; 2(6): 1821-71.
        [http://dx.doi.org/10.1039/D0MA00807A]

[2]   Khan I, Saeed K, Khan I. Nanoparticles: Properties, applications and toxicities Jarabjc 2019; 12(7): 908-31.

[3]   Sonia T, Debora R. Applications of nanomaterials Extremo Applic Nanotech. Chapter 5. Cham: Springer, 2016; pp. 163-93.

[4]   Salem SS, Fouda A. Green synthesis of metallic nanoparticles and their prospective biotechnological applications: an overview. Biol Trace Elem Res 2021; 199(1): 344-70.
        [http://dx.doi.org/10.1007/s12011-020-02138-3] [PMID: 32377944]

[5]   Philip D, Unni C, Aromal SA, Vidhu VK. *Murraya Koenigii* leaf-assisted rapid green synthesis of silver and gold nanoparticles. Spectrochim Acta A Mol Biomol Spectrosc 2011; 78(2): 899-904.
        [http://dx.doi.org/10.1016/j.saa.2010.12.060] [PMID: 21215687]

[6]   Khan N, Ali S, Latif S, Mehmood A. Biological synthesis of nanoparticles and their applications in sustainable agriculture production. Nat Sci 2022; 14(6): 226-34.
        [http://dx.doi.org/10.4236/ns.2022.146022]

[7]   Devi HS, Boda MA, Shah MA, Parveen S, Wani AH. Green synthesis of iron oxide nanoparticles using *Platanus orientalis* leaf extract for antifungal activity. Green Processing and Synthesis 2019; 8(1): 38-45.
        [http://dx.doi.org/10.1515/gps-2017-0145]

[8]   Alsammarraie FK, Wang W, Zhou P, Mustapha A, Lin M. Green synthesis of silver nanoparticles using turmeric extracts and investigation of their antibacterial activities. Jcolsurb 2018; 171: 398-405.

[9]   Kataria N, Garg VK. Green synthesis of Fe_3O_4 nanoparticles loaded sawdust carbon for cadmium (II) removal from water: Regeneration and mechanism. Chemosphere 2018; 208: 818-28.
        [http://dx.doi.org/10.1016/j.chemosphere.2018.06.022] [PMID: 29906756]

[10]   Nasrollahzadeh M, Mohammad Sajadi S. Green synthesis of copper nanoparticles using *Ginkgo biloba L.* leaf extract and their catalytic activity for the Huisgen [3 + 2] cycloaddition of azides and alkynes at room temperature. J Colloid Interface Sci 2015; 457: 141-7.
        [http://dx.doi.org/10.1016/j.jcis.2015.07.004] [PMID: 26164245]

[11] Hano C, Abbasi BH. Plant-based green synthesis of nanoparticles: production, characterization and applications. Biomolecules 2021; 12(1): 31.
[http://dx.doi.org/10.3390/biom12010031] [PMID: 35053179]

[12] Ying S, Guan Z, Ofoegbu PC, Clubb P, Rico C. He F *et al*. Green synthesis of nanoparticles: Current developments and limitations, J. Env. Tech Inno 2022; 26: 102336.

[13] Singh NB, Jain P, De A, Tomar R. Green Synthesis and Applications of Nanomaterials. Curr Pharm Biotechnol 2021; 22(13): 1705-47.
[http://dx.doi.org/10.2174/1389201022666210412142734] [PMID: 33845733]

[14] Li H, Wang Y. Metal nanocrystals: synthetic methods and potential applications. Chem Soc Rev 2014; 43(5): 1543-74.
[http://dx.doi.org/10.1039/C3CS60296F] [PMID: 24356335]

[15] Kim SK, Choi HJ, Lee CH. Metal nanoparticles and their applications. Chem Soc Rev 2010; 39(7): 2674-86.

[16] Huang J, Li W, Xie Y. Controlled synthesis of noble metal nanoparticles: an overview of recent advances. Syn. App Nano 2012; 8(2): 167-75.

[17] Sathishkumar M, Sneha K, Won SW, Cho CW, Kim S, Yun YS. Cinnamon zeylanicum bark extract and powder mediated green synthesis of nano-crystalline silver particles and its bactericidal activity. Colloids Surf B Biointerfaces 2009; 73(2): 332-8.
[http://dx.doi.org/10.1016/j.colsurfb.2009.06.005] [PMID: 19576733]

[18] Ahmed S, Ahmad M, Swami BL, Ikram S. A review on plants extract mediated synthesis of silver nanoparticles for antimicrobial applications: A green expertise. J Adv Res 2016; 7(1): 17-28.
[http://dx.doi.org/10.1016/j.jare.2015.02.007] [PMID: 26843966]

[19] Vijayaraghavan K, Nalini SP, Prakash NU. Biogenic synthesis of multi-applicative silver nanoparticles by using *Ziziphus jujuba* leaf extract. Spectrochim Acta A Mol Biomol Spectrosc 2012; 97: 1-5.

[20] MubarakAli D, Thajuddin N, Jeganathan K, Gunasekaran M. Plant extract mediated synthesis of silver and gold nanoparticles and its antibacterial activity against clinically isolated pathogens. Colloids Surf B Biointerfaces 2011; 85(2): 360-5.
[http://dx.doi.org/10.1016/j.colsurfb.2011.03.009]

[21] Shankar S, Ahmad A, Pasricha R, Sastry M. Bioreduction of chloroaurate ions by geranium leaves and its endophytic fungus yields gold nanoparticles of different shapes. J Mater Chem A Mater Energy Sustain 2003; 13(7): 1822-6.

[22] Gopinath V, MubarakAli D, Priyadarshini S, Priyadharsshini NM, Thajuddin N, Velusamy P. Biosynthesis of silver nanoparticles from Tribulus terrestris and its antimicrobial activity: A novel biological approach. Colloids Surf B Biointerfaces 2012; 96: 69-74.
[http://dx.doi.org/10.1016/j.colsurfb.2012.03.023] [PMID: 22521683]

[23] Singh R, Shedbalkar UU, Wadhwani SA. Microbial-mediated synthesis of nanoparticles: current trends and future prospects. J Cluster Sci 2021; 32(2): 347-64.

[24] Das RK, Goudar VS, Nadagouda MN, Vootla SK. Microbial-mediated synthesis of nanoparticles: a review on recent developments and future prospects. J Environ Chem Eng 2021; 9(3): 105259.

[25] Kannan S, Krishnamoorthy K, Thangavelu KP. Microbial-mediated synthesis of nanoparticles and their applications: a review. J Environ Manage 2021; 292: 112826.

[26] Asghari S, Sharafi S. Microbial-mediated synthesis of nanoparticles: a comprehensive review. J Microbiol Biotechnol Food Sci 2021; 11(4): 934-45.

[27] Shahverdi AR, Fakhimi A, Shahverdi HR. Microbial-mediated synthesis of metallic nanoparticles: challenges and opportunities. Nano Lett 2021; 13(3): 159-66.

[28] Kalimuthu K, Suresh Babu R, Venkataraman D, Bilal M, Gurunathan S. Biosynthesis of silver

nanocrystals by Bacillus licheniformis. Colloids Surf B Biointerfaces 2008; 65(1): 150-3.
[http://dx.doi.org/10.1016/j.colsurfb.2008.02.018] [PMID: 18406112]

[29]   Singaravelu G, Arockiamary JS, Kumar VG. Extracellular synthesis of silver nanoparticles by a marine alga, *Sargassum wightii* Grevilli and their antibacterial effects. J Nanosci Nanotechnol 2007; 7(7): 2445-9.

[30]   Nadagouda MN, Speth TV, Varma R, Lowry GV. Synthesis of silver and gold nanoparticles using antioxidant extracts of algae (Desmococcus spp. and Dioschloris spp.). Green Chem 2012; 14(4): 1039-45.

[31]   Jain N, Bhargava A, Majumdar S. Fungus-mediated synthesis of silver nanoparticles and their immobilization in the mycelial matrix: A novel biological approach to nanoparticle synthesis. Nano-Micro Lett 2011; 3(3): 132-9.

[32]   Ahmad A, Mukherjee P, Senapati S, *et al.* Extracellular biosynthesis of silver nanoparticles using the fungus *Fusarium oxysporum.* Colloids Surf B Biointerfaces 2003; 28(4): 313-8.
[http://dx.doi.org/10.1016/S0927-7765(02)00174-1]

[33]   Koudelka KJ, Pitek AS, Manchester M, Steinmetz NF. Virus-based nanoparticles as versatile nanomachines. Annu Rev Virol 2015; 2(1): 379-401.
[http://dx.doi.org/10.1146/annurev-virology-100114-055141] [PMID: 26958921]

[34]   Wang Q, Chen B, Wang D. HuY, Zhou X. Viral nanoparticles as versatile nanomaterials for medicine: Opportunities and challenges. J Mater Chem B Mater Biol Med 2018; 6(43): 6846-60.
[PMID: 32254577]

[35]   Zhang X, Sun H, Yu H. Green synthesis of nanoparticles by virus-mediated gene expression. Crit Rev Biotechnol 2018; 38(2): 198-213.

[36]   Das SK, Das AR. Green chemistry-based synthesis of nanoparticles and their biomedical applications: a review. J Drug Deliv Sci Technol 2021; 66: 102882.

[37]   Srivastava S, Kulkarni S, Choudhury AR. Green synthesis of nanoparticles: a review. Green Chem Lett Rev 2021; 14(1): 58-74.

[38]   Rathod VK, Yadav AK. Green chemistry-based synthesis of metallic nanoparticles: a review. J Mater Sci Mater Electron 2021; 32(5): 4645-67.

[39]   Narayanankutty A, Divya P. Green synthesis of nanoparticles: an eco-friendly and sustainable approach. Green Chemistry and Sustainable Technology. Springer 2021; pp. 307-32.

[40]   Gangula A, Podila R, Karanam L. Green synthesis of nanoparticles and their applications: a review. Curr Opin Chem Eng 2021; 32: 102-9.

[41]   Sahoo SK, Paria S. Solar-mediated synthesis of metallic nanoparticles: a review. Mater Sci Eng 2021; 266: 115085.

[42]   Vijayakumar S, Vinothkumar K, Shankar R. Solar-mediated synthesis of nanoparticles: a sustainable approach. J Environ Chem Eng 2021; 9(1): 104823.

[43]   Pahari B, Sahoo SK. Solar-mediated synthesis of nanoparticles: a sustainable approach for catalysis and sensing applications. Nano Adv 2021; 3(8): 2183-93.

[44]   Zhang Y, Yang Z, Wang X. Solar-mediated synthesis of metal nanoparticles for sustainable applications. Nanomaterials 2021; 11(8): 2074.
[PMID: 34443905]

[45]   Bao Y, He J, Song K, Guo J, Zhou X, Liu S. Plant-extract-mediated synthesis of metal nanoparticles. J Chem 2021; 6562687

[46]   Wang X, Yuan L, Deng H, Zhang Z. Structural characterization and stability study of green synthesized starch stabilized silver nanoparticles loaded with isoorientin. Food Chem 2021; 338: 127807.

[http://dx.doi.org/10.1016/j.foodchem.2020.127807] [PMID: 32818865]

[47]   Naseer A, Ali A, Ali S, *et al.* Biogenic and eco-benign synthesis of platinum nanoparticles (Pt NPs) using plants aqueous extracts and biological derivatives: environmental, biological and catalytic applications. J Mater Res Technol 2020; 9(4): 9093-107.
[http://dx.doi.org/10.1016/j.jmrt.2020.06.013]

[48]   Küünal S, Rauwel P, Rauwel E. Plant extract mediated synthesis of nanoparticles. Micro and Nano Technol, Emerg Applica of Nanopar and Archi Nanostr. Elsevier 2018; pp. 411-46.

[49]   Dinesh GK, Pramod M, Chakma S. Sonochemical synthesis of amphoteric Cu^0-Nanoparticles using Hibiscus rosa-sinensis extract and their applications for degradation of 5-fluorouracil and lovastatin drugs. J Hazard Mater 2020; 399: 123035-48.
[http://dx.doi.org/10.1016/j.jhazmat.2020.123035] [PMID: 32512280]

[50]   Nasrollahzadeh M, Ghorbannezhad F, Issaabadi Z, Sajadi SM. Recent developments in the biosynthesis of Cu based recyclable nanocatalysts using plant extracts and their application in the chemical reactions. Chem Rec 2019; 19(2-3): 601-43.
[http://dx.doi.org/10.1002/tcr.201800069] [PMID: 30230690]

[51]   Sajadi SM, Nasrollahzadeh M, Maham M. Aqueous extract from seeds of *Silybum marianum* L. as a green material for preparation of the Cu/Fe_3O_4 nanoparticles: A magnetically recoverable and reusable catalyst for the reduction of nitroarenes. J Colloid Interface Sci 2016; 469: 93-8.
[http://dx.doi.org/10.1016/j.jcis.2016.02.009] [PMID: 26874271]

[52]   Yu Z, Hu C, Guan L, Zhang W, Gu J. Green synthesis of cellulose nanofibrils decorated with Ag nanoparticles and their application in colorimetric detection of L-cysteine. ACS Sustain Chem& Eng 2020; 8(33): 12713-21.
[http://dx.doi.org/10.1021/acssuschemeng.0c04842]

[53]   Jigyasa , Rajput JK. Bio-polyphenols promoted green synthesis of silver nanoparticles for facile and ultra-sensitive colorimetric detection of melamine in milk. Biosens Bioelectron 2018; 120: 153-9.
[http://dx.doi.org/10.1016/j.bios.2018.08.054] [PMID: 30173011]

[54]   Veisi H, Kavian M, Hekmati M, Hemmati S. Biosynthesis of the silver nanoparticles on the graphene oxide's surface using *Pistacia atlantica* leaves extract and its antibacterial activity against some human pathogens. Polyhedron 2019; 161: 338-45.
[http://dx.doi.org/10.1016/j.poly.2019.01.034]

[55]   Nasrollahzadeh M, Sajjadi M, Sajadi SM. Green synthesis of Cu/ zirconium silicate nanocomposite by using *Rubia tinctorum* leaf extract and its application in the preparation of *N* -benzyl- *N* -arylcyan amides. Appl Organomet Chem 2019; 33(2): e4705.
[http://dx.doi.org/10.1002/aoc.4705]

[56]   Kandiah M, Chandrasekaran KN. Green Synthesis of Silver Nanoparticles Using *Catharanthus roseus* Flower Extracts and the Determination of Their Antioxidant, Antimicrobial, and Photocatalytic Activity. J Nanotechnol 2021; 2021: 1-18.
[http://dx.doi.org/10.1155/2021/5512786]

[57]   Giri AK, Jena B, Biswal B, *et al.* Green synthesis and characterization of silver nanoparticles using *Eugenia roxburghii* DC. extract and activity against biofilm-producing bacteria. Sci Rep 2022; 12(1): 8383.
[http://dx.doi.org/10.1038/s41598-022-12484-y] [PMID: 35589849]

[58]   Kathiravan V, Ravi S, Ashokkumar S, Velmurugan S, Elumalai K, Khatiwada CP. Green synthesis of silver nanoparticles using *Croton sparsiflorus* morong leaf extract and their antibacterial and antifungal activities. Spectrochim Acta A Mol Biomol Spectrosc 2015; 139: 200-5.
[http://dx.doi.org/10.1016/j.saa.2014.12.022] [PMID: 25561298]

[59]   Sana SS, Badineni VR, Arla SK, Naidu Boya VK. Eco-friendly synthesis of silver nanoparticles using leaf extract of *Grewia flaviscences* and study of their antimicrobial activity. Mater Lett 2015; 145: 347-50.

[http://dx.doi.org/10.1016/j.matlet.2015.01.096]

[60]   Suresh G, Gunasekar PH, Kokila D, *et al.* Green synthesis of silver nanoparticles using *Delphinium denudatum* root extract exhibits antibacterial and mosquito larvicidal activities. Spectrochim Acta A Mol Biomol Spectrosc 2014; 127: 61-6.
[http://dx.doi.org/10.1016/j.saa.2014.02.030] [PMID: 24632157]

[61]   Govindarajan M, Rajeswary M, Veerakumar K, Muthukumaran U, Hoti SL, Benelli G. Green synthesis and characterization of silver nanoparticles fabricated using *Anisomeles indica*: Mosquitocidal potential against malaria, dengue and Japanese encephalitis vectors. Exp Parasitol 2016; 161: 40-7.
[http://dx.doi.org/10.1016/j.exppara.2015.12.011] [PMID: 26708933]

[62]   Anjali R, Anupam G, Goutam C. Mosquito larvicidal and antimicrobial activity of synthesized nano-crystalline silver particles using leaves and green berry extract of *Solanum nigrum* L. (Solanaceae: Solanales). Jactatropica 2013; 128(3): 613-22.

[63]   Ameen F, Srinivasan P, Selvankumar T, *et al.* Phytosynthesis of silver nanoparticles using *Mangifera indica* flower extract as bioreductant and their broad-spectrum antibacterial activity. Bioorg Chem 2019; 88: 102970.
[http://dx.doi.org/10.1016/j.bioorg.2019.102970] [PMID: 31174009]

[64]   Hernández-Morales L, Espinoza-Gómez H, Flores-López LZ, *et al.* Study of the green synthesis of silver nanoparticles using a natural extract of dark or white *Salvia hispanica* L. seeds and their antibacterial application. Appl Surf Sci 2019; 489: 952-61.
[http://dx.doi.org/10.1016/j.apsusc.2019.06.031]

[65]   Ghosh MK, Sahu S, Gupta I, Ghorai TK. Green synthesis of copper nanoparticles from an extract of *Jatropha curcas* leaves: characterization, optical properties, CT-DNA binding and photocatalytic activity. RSC Advances 2020; 10(37): 22027-35.
[http://dx.doi.org/10.1039/D0RA03186K] [PMID: 35516624]

[66]   Chandraker SK, Lal M, Ghosh MK, Tiwari V, Ghorai TK, Shukla R. Green synthesis of copper nanoparticles using leaf extract of *Ageratum houstonianum Mill.* and study of their photocatalytic and antibacterial activities. Nano Express 2020; 1(1): 010033.
[http://dx.doi.org/10.1088/2632-959X/ab8e99]

[67]   Liu H, Wang G, Liu J, *et al.* Green synthesis of copper nanoparticles using *Cinnamomum zelanicum* extract and its applications as a highly efficient antioxidant and anti-human lung carcinoma. J Exp Nanosci 2021; 16(1): 410-23.
[http://dx.doi.org/10.1080/17458080.2021.1991577]

[68]   Gopalakrishnan V, Muniraj S. Neem flower extract assisted green synthesis of copper nanoparticles – Optimisation, characterisation and anti-bacterial study. Mater Today Proc 2021; 36: 832-6.
[http://dx.doi.org/10.1016/j.matpr.2020.07.013]

[69]   Wang G, Zhao K, Gao C, *et al.* Green synthesis of copper nanoparticles using green coffee bean and their applications for efficient reduction of organic dyes. J Environ Chem Eng 2021; 9(4): 105331.
[http://dx.doi.org/10.1016/j.jece.2021.105331]

[70]   Biresaw SS, Taneja P. Copper nanoparticles green synthesis and characterization as anticancer potential in breast cancer cells (MCF7) derived from *Prunus nepalensis* phytochemicals. Mater Today Proc 2022; 49(8): 3501-9.
[http://dx.doi.org/10.1016/j.matpr.2021.07.149]

[71]   Dashtizadeh Z, Jookar Kashi F, Ashrafi M. Phytosynthesis of copper nanoparticles using *Prunus mahaleb* L. and its biological activity. Mater Today Commun 2021; 27: 102456.
[http://dx.doi.org/10.1016/j.mtcomm.2021.102456]

[72]   Joao G, Janaina S, Lusitâneo MD, Natan P, Cintia S, Humberto R. green synthesis of manganese based nanoparticles mediated by eucalyptus robusta and *Corymbia citriodora* for agricultural applications. Jcolsurfa 2021; 636: 128180.

[73] Kamran U, Bhatti HN, Iqbal M, Jamil S, Zahid M. Biogenic synthesis, characterization and investigation of photocatalytic and antimicrobial activity of manganese nanoparticles synthesized from *Cinnamomum verum* bark extract. J Mol Struct 2019; 1179: 532-9.
[http://dx.doi.org/10.1016/j.molstruc.2018.11.006]

[74] Alafaleq NO, Torki A. Biogenic synthesis of Cu-Mn bimetallic nanoparticles using pumpkin seeds extract and their characterization and anticancer efficacy. Nano mat 2023; 13(7): 1201.

[75] Sharmila A, Santu A. Biosynthesis of Manganese Nanoparticles (MnNPs) from *Brassica oleraceae* (Cabbage leaves) and its Antibacterial Activity. Asian J Phys Chem Sci 2021; 9: 1-11.

[76] Pau JL, Iniya U C. Photocatalytic removal of dyes from aqueous medium by Fe,Mn and Fe-Mn nanoparticles synthesized using *cannabis sativa* leaf extract. Water 2022; 14: 3535.

[77] Naz S, Kalsoom R, Ali F, *et al.* Photocatalytic removal of dyes from aqueous medium by fe, mn and fe-mn nanoparticles synthesized using cannabis sativa leaf extract. Water 2022; 14(21): 3535.
[http://dx.doi.org/10.3390/w14213535]

[78] Younis I, El-Hawary SS, Eldahshan O. Eldahshan, Abdel-Aziz M.M, Ali Z. Green synthesis of magnesium nanoparticles mediated from *Rosa foribunda* charisma extract and its antioxidant, antiaging and antibioflm activities. Sci Rep 2021; 11: 16868.
[http://dx.doi.org/10.1038/s41598-021-96377-6] [PMID: 34413416]

[79] Karunakaran G, Jagathambal M, Venkatesh M, *et al. Hydrangea paniculata* flower extract-mediated green synthesis of MgNPs and AgNPs for health care applications. Powder Technol 2017; 305: 488-94.
[http://dx.doi.org/10.1016/j.powtec.2016.10.034]

[80] Patra JK, Baek KH. Novel green synthesis of gold nanoparticles using *Citrullus lanatus* rind and investigation of proteasome inhibitory activity, antibacterial, and antioxidant potential. Int J Nanomedicine 2015; 10: 7253-64.
[PMID: 26664116]

[81] Hosny M, Fawzy M, Abdelfatah AM, Fawzy EE, Eltaweil AS. Comparative study on the potentialities of two halophytic species in the green synthesis of gold nanoparticles and their anticancer, antioxidant and catalytic efficiencies. Adv Powder Technol 2021; 32(9): 3220-33.
[http://dx.doi.org/10.1016/j.apt.2021.07.008]

[82] Vijayan R, Joseph S, Mathew B. Anticancer, antimicrobial, antioxidant, and catalytic activities of green-synthesized silver and gold nanoparticles using *Bauhinia purpurea* leaf extract. Bioprocess Biosyst Eng 2019; 42(2): 305-19.
[http://dx.doi.org/10.1007/s00449-018-2035-8] [PMID: 30421171]

[83] Suriyakala G, Sathiyaraj S, Babujanarthanam R, *et al.* Green synthesis of gold nanoparticles using *Jatropha integerrima Jacq.* flower extract and their antibacterial activity. J King Saud Univ Sci 2022; 34(3): 101830.
[http://dx.doi.org/10.1016/j.jksus.2022.101830]

[84] Veena S, Devasena T, Sathak SSM, Yasasve M, Vishal LA. Green synthesis of gold nanoparticles from *Vitex negundo* leaf extract: characterization and *in vitro* evaluation of antioxidant–antibacterial activity. J Cluster Sci 2019; 30(6): 1591-7.
[http://dx.doi.org/10.1007/s10876-019-01601-z]

[85] Muniyappan N, Pandeeswaran M, Amalraj A. Green synthesis of gold nanoparticles using *Curcuma pseudomontana* isolated curcumin: Its characterization, antimicrobial, antioxidant and anti-inflammatory activities. Environmental Chemistry and Ecotoxicology 2021; 3: 117-24.
[http://dx.doi.org/10.1016/j.enceco.2021.01.002]

[86] Patra JK, Kwon Y, Baek KH. Green biosynthesis of gold nanoparticles by onion peel extract: Synthesis, characterization and biological activities. Adv Powder Technol 2016; 27(5): 2204-13.
[http://dx.doi.org/10.1016/j.apt.2016.08.005]

[87] Babu AB, Bhavani Anagani KD, Sravanthi P. Green synthesis of gold nanoparticles using *Carambola* fruit extract and evaluation of their antioxidant and anticancer efficiency. Asian J Chem 2023; 35(3): 739-47.
[http://dx.doi.org/10.14233/ajchem.2023.26988]

[88] Ismail E, Saqer A, Assirey E, Naqvi A, Okasha R. Successful green synthesis of gold nanoparticles using a *Corchorus olitorius* extract and their antiproliferative effect in cancer cells. Int J Mol Sci 2018; 19(9): 2612.
[http://dx.doi.org/10.3390/ijms19092612] [PMID: 30177647]

[89] Barai AC, Paul K, Dey A, *et al.* Green synthesis of *Nerium oleander*-conjugated gold nanoparticles and study of its *in vitro* anticancer activity on MCF-7 cell lines and catalytic activity. Nano Converg 2018; 5(1): 10.
[http://dx.doi.org/10.1186/s40580-018-0142-5] [PMID: 29682442]

[90] Kumar P, Dixit J, Singh AK, *et al.* Tiwari Efficient catalytic degradation of selected toxic dyes by green biosynthesized silver nanoparticles using aqueous leaf extract of *Cestrum nocturnum* L. Nanomaterials 2022; 12(21): 3851.
[http://dx.doi.org/10.3390/nano12213851] [PMID: 36364627]

[91] Vanaja M, Paulkumar K, Baburaja M, *et al.* Degradation of methylene blue using biologically synthesized silver nanoparticles. Bioinorg Chem Appl 2014; 2014: 1-8.
[http://dx.doi.org/10.1155/2014/742346] [PMID: 24772055]

[92] Sidorowicz A, Szymański T, Rybka JD. Photodegradation of biohazardous dye brilliant blue R using organometallic silver nanoparticles synthesized through a green chemistry method. Biology 2021; 10(8): 784.
[http://dx.doi.org/10.3390/biology10080784] [PMID: 34440016]

[93] Royji Albeladi SS, Malik MA, Al-thabaiti SA. Facile biofabrication of silver nanoparticles using *Salvia officinalis* leaf extract and its catalytic activity towards Congo red dye degradation. J Mater Res Technol 2020; 9(5): 10031-44.
[http://dx.doi.org/10.1016/j.jmrt.2020.06.074]

[94] Francis S, Joseph S, Koshy EP, Mathew B. Green synthesis and characterization of gold and silver nanoparticles using *Mussaenda glabrata* leaf extract and their environmental applications to dye degradation. Environ Sci Pollut Res Int 2017; 24(21): 17347-57.
[http://dx.doi.org/10.1007/s11356-017-9329-2] [PMID: 28589274]

[95] Arya G, Sharma N, Ahmed J, *et al.* Degradation of anthropogenic pollutant and organic dyes by biosynthesized silver nano-catalyst from *Cicer arietinum* leaves. J Photochem Photobiol B 2017; 174: 90-6.
[http://dx.doi.org/10.1016/j.jphotobiol.2017.07.019] [PMID: 28756157]

[96] Recio-Sánchez G, Tighe-Neira R, Alvarado C, *et al.* Assessing the effectiveness of green synthetized silver nanoparticles with *Cryptocarya alba* extracts for remotion of the organic pollutant methylene blue dye. Environ Sci Pollut Res Int 2019; 26(15): 15115-23.
[http://dx.doi.org/10.1007/s11356-019-04934-4] [PMID: 30919197]

[97] Jain A, Ahmad F, Gola D, *et al.* Multi dye degradation and antibacterial potential of Papaya leaf derived silver nanoparticles. Environ Nanotechnol Monit Manag 2020; 14: 100337.
[http://dx.doi.org/10.1016/j.enmm.2020.100337]

[98] Mahmoud ME, Amira MF, Seleim SM, Nabil GM, Abouelanwar ME. Multifunctionalized graphene oxide@nanopolyaniline@zirconium silicate nanocomposite for rapid microwable removal of dyes. J Nanostructure Chem 2021; 11(4): 645-62.
[http://dx.doi.org/10.1007/s40097-021-00390-0]

[99] Rajeshkumar S, Vanaja M, Kalirajan A. Degradation of toxic dye using phytomediated copper nanoparticles and its free-radical scavenging potential and antimicrobial activity against environmental pathogens. Bioinorg Chem Appl 2021; 2021: 1-10.

[http://dx.doi.org/10.1155/2021/1222908] [PMID: 34899884]

[100] Aroob S, Carabineiro SAC, Taj MB, *et al.* Green synthesis and photocatalytic dye degradation activity of CuO nanoparticles. Catalysts 2023; 13(3): 502.
[http://dx.doi.org/10.3390/catal13030502]

[101] Roy A, Singh V, Sharma S, *et al.* Antibacterial and dye degradation activity of green synthesized iron nanoparticles. J Nanomater 2022; 2022: 1-6.
[http://dx.doi.org/10.1155/2022/3636481]

[102] Jha AK, Chakraborty S. Photocatalytic degradation of Congo Red under UV irradiation by zero valent iron nano particles (nZVI) synthesized using *Shorea robusta* (Sal) leaf extract. Water Sci Technol 2020; 82(11): 2491-502.
[http://dx.doi.org/10.2166/wst.2020.517] [PMID: 33339802]

[103] Shaker Ardakani L, Alimardani V, Tamaddon AM, Amani AM, Taghizadeh S. Green synthesis of iron-based nanoparticles using *Chlorophytum comosum* leaf extract: methyl orange dye degradation and antimicrobial properties. Heliyon 2021; 7(2): e06159.
[http://dx.doi.org/10.1016/j.heliyon.2021.e06159] [PMID: 33644459]

[104] Khan ZUH, Khan A, Shah NS, *et al.* Photocatalytic and biomedical investigation of green synthesized NiONPs: Toxicities and degradation pathways of Congo red dye. Surf Interfaces 2021; 23: 100944.
[http://dx.doi.org/10.1016/j.surfin.2021.100944]

[105] Miditana SR, Tirukkovalluri SR, Raju IM. Synthesis and antibacterial activity of transition metal (Ni/Mn) co-doped TiO_2 nanophotocatalyst on different pathogens under visible light irradiation. Nanosystems: Physics, Chemistry, Mathematics 2022; 13(1): 104-14.
[http://dx.doi.org/10.17586/2220-8054-2022-13-1-104-114]

[106] Miditana SR, Tirukkovalluri S, Imandi M. A., B., A, R. Review on the synthesis of doped TiO_2 nanomaterials by Sol-gel method and description of experimental techniques. J Water Enviro Nanotechnol 2022; 7(2): 218-29.

[107] Patil MP, Kim GD. Eco-friendly approach for nanoparticles synthesis and mechanism behind antibacterial activity of silver and anticancer activity of gold nanoparticles. Appl Microbiol Biotechnol 2017; 101(1): 79-92.
[http://dx.doi.org/10.1007/s00253-016-8012-8] [PMID: 27915376]

[108] Sau TK, Rogach AL. Nonspherical noble metal nanoparticles: colloid-chemical synthesis and morphology control. Adv Mater 2010; 22(16): 1781-804.
[http://dx.doi.org/10.1002/adma.200901271] [PMID: 20512953]

[109] Ferdous Z, Nemmar A. Health impact of silver nanoparticles: A review of the biodistribution and toxicity following various routes of exposure. Int J Mol Sci. 2020;21(7):2375.
[http://dx.doi.org/10.3390/ijms21072375]

[110] Vanlalveni C, Lallianrawna S, Biswas A, Selvaraj M, Changmai B, Rokhum SL. Green synthesis of silver nanoparticles using plant extracts and their antimicrobial activities: a review of recent literature. RSC Advances 2021; 11(5): 2804-37.
[http://dx.doi.org/10.1039/D0RA09941D] [PMID: 35424248]

[111] Agarwal N, Solanki VS, Pare B, Singh N, Jonnalagadda SB. Current trends in nanocatalysis for green chemistry and its applications- a mini-review. Curr Opin Green Sustain Chem 2023; 41: 100788.
[http://dx.doi.org/10.1016/j.cogsc.2023.100788]

[112] Ankamwar B, Kirtiwar S, Shukla AC. Plant-mediated green synthesis of nanoparticles. Advances in Pharmaceutical Biotechnology. Springer 2020; pp. 221-34.

[113] Huq MA, Ashrafudoulla M, Rahman MM, Balusamy SR, Akter S. Green synthesis and potential antibacterial applications of bioactive silver nanoparticles: A review. Polymers 2022; 14(4): 742.
[http://dx.doi.org/10.3390/polym14040742] [PMID: 35215655]

CHAPTER 4

Current Trends in Green Bioremediation of Environmental Organic Pollutants

Amrit K. Mitra[1,*]

[1] *Department of Chemistry, Government General Degree College, Singur, Hooghly, West Bengal, India*

Abstract: A biological process termed bioremediation transforms waste into a form that can be used and reused by other microbes. Recent research has revealed that xenobiotic pollution and other associated refractory substances pose a serious threat to both human health and the environment. Many contaminants, including heavy metals, polychlorinated biphenyls, plastics and different agrochemicals, are prevalent in the environment because of their toxicity and inability to biodegrade. The key objective of bioremediation is the degradation of pollutants and their transformation into less harmful forms. Depending on several variables like cost, kind and concentration of the contaminant and other considerations, *ex-situ* or *in-situ* bioremediation may be used. Bioremediation can be done with the help of microorganisms that can withstand all circumstances due to their metabolic potential. Microbes have a tremendous nutritional capacity, making them useful in the bioremediation of environmental contaminants. With the complete and coordinated activity of microorganisms, bioremediation is significantly involved with the decomposition, expulsion, immobility or decontamination of different chemical pollutants and physically dangerous chemicals from the natural atmosphere. Enzymatic processes and other techniques, including bioventing, bioaugmentation, biostimulation, biopiles and bioattenuation, are widely used throughout the world based on their characteristics, benefits and drawbacks. This chapter aims to portray the current advancements in green bioremediation methods, how various contaminants are broken down by microorganisms and what the future holds for bioremediation in terms of lowering global pollution levels.

Keywords: Biodegradation, Bioattenuation, Biostimulation, Bioaugmentation, Bioventing, Biopiles, Bioremediation, Contaminants, Environment, Microorganisms, Monitoring and stimulation, Pollutants, Sustainable technologies.

---

[*] **Corresponding author Amrit K. Mitra**: Department of Chemistry, Government General Degree College, Singur, Hooghly, West Bengal, India; Tel: +91-33-2630-0126; +91-9432164011; Fax: +91-33-2630-0126; E-mail: ambrosia12june@gmail.com

Neha Agarwal, Vijendra Singh Solanki and Sreekantha B. Jonnalagadda (Eds.)

INTRODUCTION

Microorganisms are regarded as the first living things to have evolved. They can adjust themselves to a wide range of difficult environmental conditions.

They are distributed across the biosphere due to their outstanding metabolic capability and ease of growth in a variety of environmental circumstances. From extreme environmental circumstances like frozen settings, acidic lakes, and bottoms of deep oceans to the small intestines of animals, they play a significant role in regulating biogeochemical cycles. The global biogeochemical cycle is governed by microorganisms which are also in charge of carbon and nitrogen fixation and methane and sulfur metabolism [1].

Bioremediation is a well-organized activity of microorganisms where a variety of metabolic enzymes are produced, which can be used to remove contaminants in a green manner [2]. This can be done either by directly destroying the pollutants or by converting them into less toxic intermediates. The process is continued depending on the specific capacity of microbes to transform hazardous contaminants to produce biomass and generate energy [3, 4]. Microorganisms provide a suitable platform for bioremediation of plastics, heavy metals, hydrocarbons, greenhouse gases *etc*. [2 - 6]. As a result, bioremediation employs low-cost and less technical approaches that can be performed on-site frequently. However, because of the narrow spectrum of pollutants onto which it is effective, the lengthy time frames needed and the inappropriateness of the achievable residual contamination levels, it may not always be suitable [7]. Even though the procedures used in bioremediation are not technically sophisticated, considerable expertise may be required to develop and implement them successfully [3]. The principal agents of bioremediation are bacteria, archaea and fungi. The terms 'bioremediation' and 'biodegradation' are increasingly interchangeable [3]. Various sites across the world, particularly in Europe, have tried bioremediation with varied degrees of effectiveness. Unfortunately, there is a lack of knowledge and understanding of the principles, methods, benefits and drawbacks of bioremediation, particularly among site owners and regulators [8 - 10].

This book chapter has been written to discuss the current trends in bioremediation to address environmental threats. This is a significant research area because microorganisms are environmentally friendly and have the potential to produce valuable genetic material that can be used effectively [3]. This chapter will also provide a practical view of the bioremediation processes, the benefits and drawbacks of the method and the considerations to be taken into account when dealing with a proposal for bioremediation.

PRINCIPLES OF BIOREMEDIATION

Anthropogenic activities cause the release of huge amounts of pollutants into the environment every year. These releases can be intentional and strictly controlled in some situations (such as industrial emissions) or unintentional (such as chemical or oil spills). Many of these substances persist in both terrestrial and aquatic habitats and are harmful. The accumulation of these harmful substances above the allowable amounts causes the poisoning of land and water resources [11 - 13]. The bioremediation process reduces the organic wastes to a benign state or levels below concentration limits set by regulatory authorities under controlled settings. A pollutant's biodegradation is frequently the consequence of the activity of numerous organisms. Bioaugmentation occurs when microorganisms are brought to a contaminated site to aid in degradation [14].

Microbial enzymes attack the contaminants and transform them into toxic compounds because bioremediation is only successful in those sites that support microbial activity and growth. But, some pollutants, like high aromatic hydrocarbons or chlorinated organic compounds, are immune to microbial attack. Bioremediation techniques can remediate some pollutants on-site, minimizing exposure hazards for workers. They are often more cost-effective than conventional procedures like cremation and are operated in aerobic circumstances. Operating a system under anaerobic conditions may allow microbial organisms to break down resistant compounds [15]. A representative diagram related to the principles of bioremediation is in Fig. (**1**).

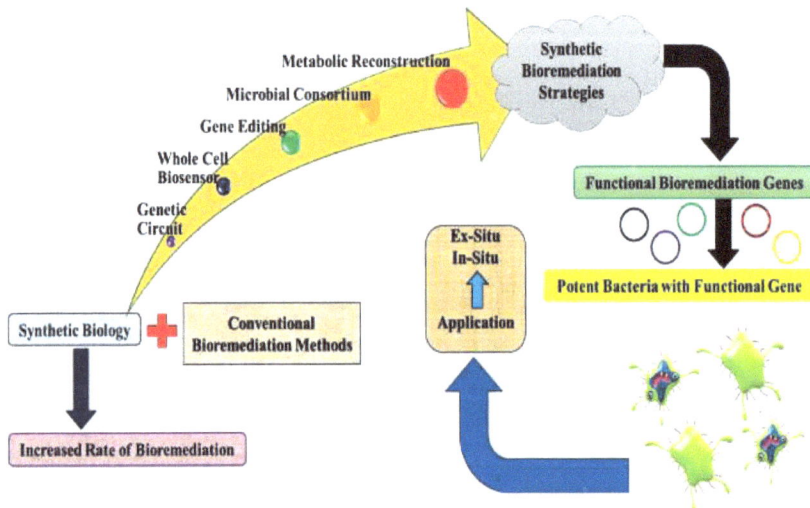

Fig. (1). Representative diagram related to the principles of bioremediation.

FACTORS AFFECTING BIOREMEDIATION

Microorganisms act as biocatalysts that fasten the biochemical reactions and destroy the intended contaminant with the help of their enzymes [3, 4]. Microorganisms can combat pollution when they have access to a variety of material substances that can assist them in producing energy and nutrients to grow additional cells. The physicochemical properties of the environment, the chemical makeup, pollutant concentration, and their accessibility to microorganisms play a role in determining the effectiveness of bioremediation [9]. Controlling and optimizing bioremediation is a complex system due to several factors, including the role of a microbial population, the accessibility of contaminants to the microbes, and different environmental factors such as temperature, pH, soil type, presence of oxygen and other nutrients [3].

Biological Factors

Bacteriophages and protozoa can prey on one another, or soil microorganisms may compete with one another for carbon sources, all of which can affect the breakdown of organic compounds [3]. The contaminant concentration and the concentration of the catalyst are directly related to the rate of biodegradation. The rate of breakdown of contaminants can be changed by the cells' production of particular enzymes. To have an affinity for the contamination and to be available, enzymes must also participate in the metabolism process. The three main biological aspects are population size, composition, and interaction (competition, predation and succession) [16].

Environmental Factors

The metabolic potential of the microorganisms and the physicochemical qualities of the pollutants dictate the potential interactions during the bioremediation process. The activity of microorganisms is influenced by different factors such as solubility in water, pH, temperature, redox potential, moisture, availability of nutrients, and oxygen content [16]. A pH range of 6.5 to 8.5 is typically ideal for biodegradation in the majority of environments and ecosystems. Moisture has a great influence on the metabolic rate of pollutants because it influences the availability and kind of soluble materials, osmotic pressure and pH of terrestrial and aquatic systems [17, 18].

Availability of Nutrients

The amount of nutrients modifies the critical nutritional balance for microbial growth and affects the efficacy and pace of biodegradation. By maximizing the bacterial C: N: P ratio, particularly the delivery of important nutrients like N and

P, the efficiency of biodegradation can be increased. Microorganisms need different nutrients like carbon, nitrogen and phosphorus to exist and carry out their microbiological functions [3, 19]. By adding the right amount of nutrients, it is possible to increase the metabolic activity of microorganisms and, consequently, the rate of biodegradation [19, 20]. These nutrients are present in small amounts in the natural environment, but they are nevertheless available [21].

Temperature

Temperature has a great influence on biochemical reactions, and many of them get doubled with every 10°C rise in temperature. But the cells degenerate at a particular temperature. The sluggish rate of biochemical reactions of the breakdown of natural oil in the Arctic region forces the microbes to remove the spilt oil in the frigid environment. The subzero water temperature closes the transport channels inside the microbial cells. This could even cause the cytoplasm to freeze, rendering the metabolic activity of the oleophilic microorganisms inert. The microbial enzymes can be active in the degradation pathway only at a certain temperature. Their metabolic rate is different at every temperature because for the degradation of different pollutants, different temperatures are required, and physiological characteristics of bacteria are also influenced by temperature change [22].

Concentration of Oxygen

The concentration of oxygen decides whether a system is aerobic or anaerobic. Since oxygen gas is required by almost all living things, biological degradation occurs both in aerobic and anaerobic environments. Different microorganisms have varying needs for oxygen and the rate of biodegradation depends on the oxygen requirement. Most of the time, the availability of oxygen can improve the degradation of hydrocarbons [21, 23].

Effect of pH

The pH also has a great influence on the metabolic activities of microbes. The pH of the soil can be used to forecast microbial development. Metabolic activities are significantly impacted by even slight pH changes. The pH level above and below the optimal range produces undesired effects [24]. It is feasible to adjust the pH by adding lime if the soil has an excessive amount of acid.

Metal Ions

Small amounts of metals are necessary for bacteria and fungi. However, large amounts of metals hinder cell metabolism. Metal compounds affect the rate of deterioration both directly and indirectly [3, 9].

TYPES OF BIOREMEDIATIONS

Bioremediation is a preferred substitute for chemical and physical methods for rejuvenation at polluted sites. There are essentially two ways of waste treatment: (a) *in situ* bioremediation and (b) *ex situ* bioremediation.

In Situ Bioremediation

These methods are cheaper and cause less disruption because they provide treatment where it is needed without excavating toxins. Several methods, including bioventing, biosparging, bioaugmentation, bioslurping, biofilters, and biostimulation, can help the bioremediation process [25 - 27]. A representative diagram related to the techniques of bioremediation is in Fig. (2). Different methods of *in situ* bioremediation are explained below:

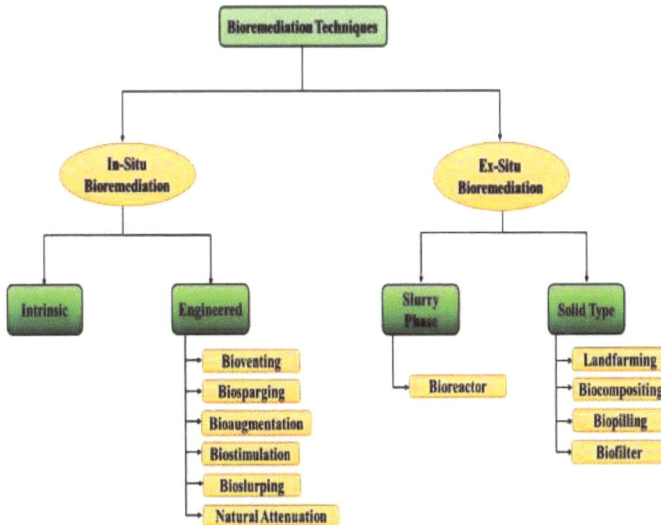

Fig. (2). Representative diagram related to the techniques of bioremediation.

Bioventing

It encourages the aerobic breakdown process. Delivering oxygen into an unsaturated zone improves the natural ability of local microorganisms to break down the organic pollutants adsorbed into soil. Air is directly injected into the contaminated area through vertical and horizontal wells. In this method, only the necessary amount of air is used to promote deterioration. Additionally, it reduces the release of pollutants into the environment and their volatilization [28 - 31].

Biosparging

It involves aerating groundwater under pressure to raise oxygen concentrations, which in turn encourages native bacteria to degrade toxins more quickly. This method is very similar to air sparging in the viewpoint that volatile substances are separated by desorption and volatilization from the saturated zone. The degradation process is accelerated by pressurized air because volatile chemicals flow upward toward the unsaturated zone [32, 33].

Bioaugmentation

According to Andréolli *et al.*, bioaugmentation is the technique of introducing a particular mix of naturally existing or genetically modified microbial strains with enhanced capabilities in contaminated places to fasten the natural degradation process. This method is used for the remediation of soil and groundwater, which are contaminated with chlorinated ethenes and trichloroethylene [34 - 38].

Bioslurping

This method is used to treat soil and groundwater that is contaminated with volatile and semi-volatile organic chemicals. Bioslurping has the combined advantage of the effectiveness of vacuum-enhanced pumping, soil vapor extraction and bioventing [39 - 41].

Biofilters

By immobilizing microorganisms on a solid base, biofilters are used to digest pollutants found in air emissions. Through biofilters a humid, polluted air stream is sent over a porous packed bed that naturally immobilizes a mixed culture of organisms. This way, air pollutants are destroyed because they are adsorbed onto microbial biofilms. Sulfur gases, hydrogen sulfide, nitrous oxide, and dimethyl sulfides *etc.*, are removed using biofilters [42 - 44]. Recent research on biological waste air treatment using biofilters has focused on fundamental concerns such as microbial dynamics, microscopical characterization of the process culture and oxygen and nutrient limits.

Biostimulation

The addition of various aqueous solutions with nutrients to the contaminated site promotes biodegradation capabilities and the expansion of native microbial communities. This method is intended for the bioremediation of hydrocarbon, volatile organic compounds, pesticides, herbicides and other chemical-containing soil and groundwater [45].

In situ procedures do not require excavation or transportation because the treatment is delivered directly to the soil. This method is quite easy to implement and has the important benefit of preventing contamination from spreading, which might happen during transport. The addition of air, nutrients and/or microorganisms (either native or foreign) requires only minimal technological equipment and is done *in situ* on the polluted soil. The main drawbacks of this method are its slow kinetics, which requires prolonged treatment times to complete the biodegradation process, and the depth of soil that oxygen can reach is typically limited to the superficial layer. The operational plan is to monitor the level of pollution over time which can be rather difficult depending on the properties of the soil. In addition, *in situ* processes are uncontrolled, so it is difficult to anticipate the extent of remediation of the contaminated site. As a result, *in situ* bioremediation approaches are economical but not always appropriate, particularly *in situations* where there is significant pollution and/or where quick intervention measures are needed [46, 47].

Ex-situ Bioremediation

Ex-situ bioremediation means removing the contaminated material from the region to be treated. *Ex situ* technology can provide a useful option in certain situations. The treatment of excavated soil is a common element of a wide range of technologies that fall within the category of *ex-situ* but are distinguished by varying degrees of sophistication. The disadvantages include greater expense involved and the potential of contamination spreading during excavation and transportation [48, 49].

Composting

Composting is a traditional process for treating solid wastes and sewage sludge, although it has only relatively recently been utilized for soil bioremediation [50, 51]. To promote the growth of bacterial populations capable of degrading the pollutants in the soil *via* co-metabolic pathways, the contaminated soil is mixed with nonhazardous organic amendments, typically other solid wastes (such as manure and agricultural wastes) suitable for composting applications. The procedure, which is carried out in an aerobic environment, makes use of the heat produced by the biodegradation reactions, which raise the ambient temperature by a significant amount (about 50–60 °C) [52, 53].

Bioreactors

Microbial bioreactors provide a controlled environment to manage important process parameters to optimize the bioremediation process. Their use in remediation is particularly alluring. Another benefit is the flexibility in bioreactor design to suit the application or the reactor's intended function. However, bioremediation through bioreactors requires pollutant displacement, which can involve excavation for soils and sediments, transportation and potential confinement of the contaminated media, increasing the cost of the procedure [54 - 56].

Biopiles

Excavated soil that contains hydrocarbons that are remediable by aerobic means can be treated in 'biopiles'. During the biodegradation process, biopiles are employed to lower petroleum pollutant concentrations in the excavated soils. A system of pipework and pumps that either pushes air into the biopile under positive pressure or pulls air through the pile under negative pressure is used to supply air to the biopile system during this process. Microbial respiration increases microbial activity, which, in turn, leads to a high level of breakdown of adsorbed petroleum pollutants [57 - 59].

MICROORGANISMS AND POLLUTANTS

Bacteria are utilized for the biodegradation of diverse natural and synthetic chemicals for the reduction of hazards. There are a number of bacterial bioremediation potentials that are helpful both economically and environmentally. By utilizing microorganisms' innate metabolic capacity to transform, degrade or accumulate toxic compounds like hydrocarbons, pharmaceuticals, radionuclides and toxic metals, bioremediation and biotransformation methods have been used [60]. The purpose of bioremediation is to provide nutrients and other compounds that induce microorganisms to eliminate pollutants. The current bioremediation technologies rely on microorganisms endemic to contaminated places, encouraging them to operate by providing optimal quantities of nutrients and other substances required for their metabolism. However, scientists are still looking for ways to add non-native bacteria and genetically modified microbes designed to degrade the pollutants of concern at specific places of contaminated areas [61]. Microorganisms obtain energy *via* catalyzing energy-producing chemical processes which require the breakdown of chemical reactions. To create more cells, the energy gained from these electron exchanges is subsequently used, along with carbon obtained from the contaminants [62].

It is fascinating to understand how gram-positive and gram-negative bacteria respond to hazardous organic and inorganic toxins in the environment. Gram-positive bacteria (Actinomyces, Mycobacterium and Staphylococci) have a diverse metabolic repertoire and survive in environments contaminated with toxic metals and aromatic chemicals. Oil pollution can be reduced during bioremediation by utilizing Actinomycetes [5]. Actinobacteria have significant bioremediation ability as they can degrade heavy metals, chemical substances and pesticides in high concentrations. On the other hand, mycobacteria are good candidates for use in bioremediation of soils contaminated with organic pollutants because they are rich in G+C-containing genera and exhibit a wide range of properties. Mycobacterium is typically found in environmental conditions such as soil contaminated with PAHs. Furthermore, the presence of fatty acid-saturated cell walls has increased this bacterium's resilience to toxic and harsh environmental conditions, allowing it to survive in polluted ecosystems. Since they can be found in a range of surroundings, including water, non-tuberculous mycobacteria can be used to clean up contamination from soil and water. The existence of pathogenic biofilms, such as those of *Staphylococcus aureus* and *Pseudomonas aeruginosa*, is required for the establishment of a biofilm-mediated bioremediation process [63].

Pseudomonas fluorescens, *Pseudomonas putida*, *Alcaligenes faecalis* and *Enterobacter cloacae* are the examples of gram-negative bacteria. *Pseudomonas fluorescens* promotes pollutant degradation. Other bacteria of the same species have quantified biosurfactants, similar to how *Pseudomonas putida* and *Pseudomonas aeruginosa* increased the availability of the pollutant. *Pseudomonas putida* has been shown to serve as a soil inoculant to treat naphthalene-contaminated soils. It may convert styrene oil into the biodegradable compound PHA. This technique is most commonly applied to marine oil spills, in which oil is spilt into the sea or nearby seas. *Pseudomonas putida* is a rod-shaped bacterium, also known as 'oil guzzlers'. Pseudomonas has a great potential for the breakdown of hydrocarbons due to the variety of their metabolic processes and their resistance to chemical remediation treatments found at pollution sites. *Pseudomonas putida* promotes plant growth and has a commensal relationship with plants. Several key gene functions that promote nutrient mobilization, disease protection and effective niche colonization are involved in the interaction. *Alcaligenes faecalis* strains can be used in the environmental business to biodegrade organic pollutants and industrial wastewater because they produce enzymes that break down organic toxins. Using the immobilized *Alcaligenes faecalis* strain WT14, removal of high concentrations of nitrate is possible. An opportunistic pathogen called *Enterobacter cloacae* has been widely exploited in the domains of bioremediation and bioprotection. It is also known as plant growth-promoting rhizobacteria (PGPR) [64 - 70].

MULTIPLE APPLICATIONS OF GREEN BIOREMEDIATION

Removal of Heavy Metals

One of the most significant environmental issues is heavy metal contamination which is brought on by different anthropogenic and natural processes. The removal of these harmful heavy metals from the environment has been suggested using several physical and chemical techniques. In terms of restrictions, cost-effectiveness and the production of dangerous compounds, they are the least successful. Since microorganisms do not create any by-products and are extremely effective even at low metal concentrations, they represent a solution to these issues. Heavy metals may only be converted from one to another oxidation state or from one organic compound to the other but cannot be eliminated because the nuclear structure of the element is non-degradable. Microorganisms can protect themselves from heavy metal toxicity through a variety of methods, including absorption, adsorption, methylation, oxidation and reduction. Since methylated substances are highly volatile, microbial methylation is important for the bioremediation of heavy metals. Mercury, Hg (II), for instance, can be biomethylated to gaseous methyl mercury by multiple species of bacteria such as *Pseudomonas aeruginosa* and *Brevibacterium iodinium* [71 - 73].

Canstein *et al.* described consortia of marine bacteria efficiently removing mercury in a bioreactor by a disturbance-independent method [74]. The possible breakdown of phenol and heavy metals by a novel combination of genetic mechanisms in bacteria was also described. According to reports of the marine bacterium *Enterobacter cloaceae*, bacteria also can chelate heavy metals, thereby removing them from contaminated environments by the release of exopolysaccharides [75]. In addition, it has been discovered that some purple nonsulfur bacterial isolates, such as *Rhodobium marinum* and *Rhodobacter sphaeroides,* can biosorb or biotransform heavy metals from contaminated environments. Bacteria are chosen because of their tolerance and biosorption of heavy metals to assess pollution. However, for a limited number of bacteria, the genetic processes of bioremediation toward hazardous metals are lowered [5]. Table **1** showcases the list of microorganisms that can serve for utilizing heavy metals.

Table 1. Microorganisms that serve for utilizing heavy metals.

Microorganisms	Compounds	Refs.
Saccharomyces cerevisiae	Heavy metals, mercury, lead and nickel-based compounds	[76, 77]
Cunninghamella elegans	Heavy metal-based compounds	[78]

(Table 1) cont.....

Microorganisms	Compounds	Refs.
Pseudomonas aeruginosa	Compounds containing Fe^{2+}, Pb^{2+}, Mn^{2+} and Cu^{2+} ions	[79]
Lysinibacillus sphaericus CBAM5	Compounds containing Co^{2+}, Cu^{2+}, Pb^{2+} and Cr^{3+} ions	[80]
Microbacterium pofundi strain Shh49T	Compounds containing Fe^{2+} ions	
Aspergillus versicolr, a. fumigatus, Paecilomyces sp., Paecilomyces sp., Terichoderma sp., Microsporum sp., Cladosporium sp.,	Compounds containing Cd^{2+} ions	[81 - 83]
Bacillus safensis (JX123862) strain (PB-5 and RSA-4) Geobacter spp.	Compounds containing Fe^{3+} ions	[84]
Pseudomonas aeruginosa, Aeromonas sp.	Compounds containing Cu^{2+}, Ni^{2+}, Cr^{3+} ions	[85]
Aerococcus sp., Rhodopseudomonas palustris	Compounds containing Pb^{2+}, Cr^{3+}, Cd^{2+} ions	[86, 87]

Degradation of Polyaromatic Hydrocarbons (PAHs)

PAHs are a serious environmental threat due to their pervasiveness, toxicity, and carcinogenicity. However, it has been observed that certain marine bacteria have the ability to bio-remediate the same by producing CO_2 and metabolic intermediates throughout the metabolism process, acquiring energy and carbon for cell growth [5]. Latha *et al.* demonstrated how to introduce a catabolic plasmid of *Pseudomonas putida* bearing a genotype for hydrocarbon breakdown into a marine bacterium, which improved the bacteria's bioremediation capacity. To improve the biodegradation of PAHs in a marine environment, certain unique marine bacterial species, such as *Cycloclasticus spirillensus, Lutibacterium anuloederans* and *Neptunomonas naphthovorans*, have also been used [88]. *Burkholderia cepacia, Comamonas testosteroni, Bacillus cereus, Pseudomonas paucimobilis, Moraxella sp., Corynebacterium renale, Pseudomonas putida, Cyclotrophicus sp., Pseudomonas fluorescens, Streptomyces sp., Mycobacterium sp., Brevundimonas vesicularis, Achromobacter denitrificans, Rhodococcus sp. and Vibrio sp.* have been isolated from marine resources, which are well capable of mineralizing naphthalene. It has been discovered that bacterial isolates like *Sphingomonas paucimobilis* EPA505 use fluoranthene as their exclusive carbon source [5, 89].

Petroleum and Diesel Biodegradation

Crude oil is the most significant organic contaminant in the environment [90]. The oil-eating bacteria that are present in the environment and use it as a source of carbon and energy can break down these organic pollutants. *Nocardia,*

Micrococcus, Planococcus, Acinetobacter, Marinococcus, Methylobacterium and Rhodococcus are a few of the significant bacterial genera that may break down oil. Deppe *et al.* established a consortium for the considerable breakdown of crude oil and its constituent parts using Arctic bacteria such as *Pseudomonas, Marinobacter, Agreia, Psychrobacter, Pseudoalteromonas and Shewanella.* Bioaugmented and biostimulated marine bacterium products have recently been reported for oil remediation in marine environments [91].

Some microbes are capable of degrading the hydrocarbon structure in diesel [92]. Because of their flexibility and resilience, different species such as *Pseudomonas sp., Bacillus sp. and Acinetobacter sp.* are considered to have superior degrading capability [93, 94]. These species have unique metabolism to keep them alive by altering various metabolic and enzymatic pathways [95]. According to Bhuvaneswar *et al.*, the following bacteria were isolated from diesel-contaminated environments: *Micrococcus* sp., *Actinomycetes* sp., *Flavobacterium* sp., *Pseudomonas* sp., *Achromobacter* sp., *Acetobacter* sp., *etc.* These bacteria worked together in synergistic ways to degrade the hydrocarbons present in diesel and use it as a source of carbon [96].

The majority of the aliphatic hydrocarbons in diesel are alkanes (*n*-alkanes, iso-alkanes and cycloalkane) and saturated chain hydrocarbons (also referred to as paraffin). Since they make up the majority of diesel, a bacterium's capacity to biodegrade alkanes will influence how well it can break down diesel as a whole [97]. Alkanes are typically biodegraded as a result of aerobic processes. Aerobic bacteria with the potential to degrade alkanes go through specific degradation routes, as shown in Scheme **1** [92, 98].

With the help of the enzymes monooxygenase and dioxygenase, aliphatic molecules begin to degrade. The initial degradation step involves attaching an oxygen atom to the terminal or sub-terminal carbon under aerobic conditions. Through convergent routes, this mechanism will transform aliphatic molecules into a few core intermediates, such as primary and secondary alcohol. Alcohol dehydrogenase then aids in the conversion of the alcohol into aldehyde. The following process involves the alcohol dehydrogenase enzyme reacting with aldehydes to create fatty acids. Following the β-oxidation process, the fatty acids interact with coenzyme A (CoA) to create acetyl CoA. The carboxylation reaction is another name for the entire conversion. This process produces acetyl CoA, which can enter the central metabolic pathway, such as the Krebs cycle. Before entering into the bacteria's main metabolic system, *n*-alkanes and *iso*-alkanes travel *via* similar pathways [92, 98].

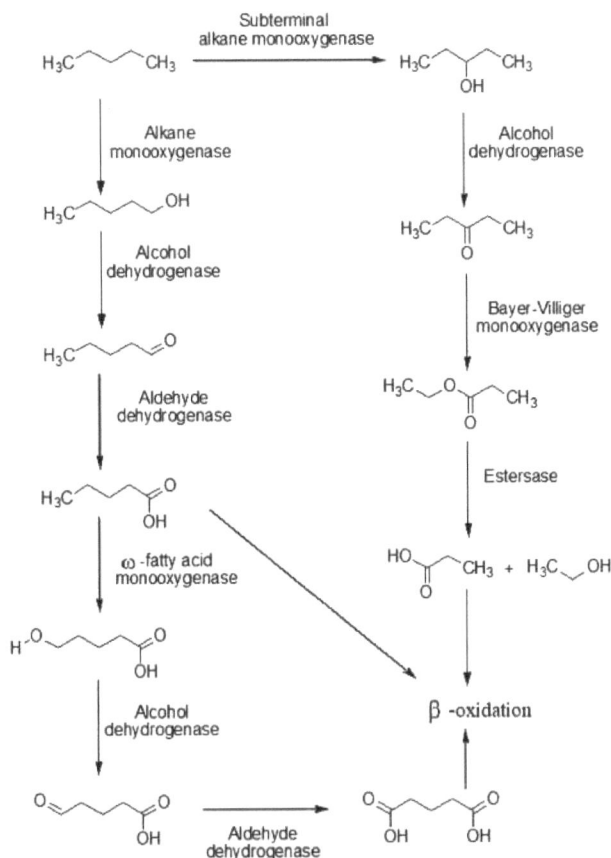

Scheme (1). Biodegradation mechanism of alkane hydrocarbons [92, 98].

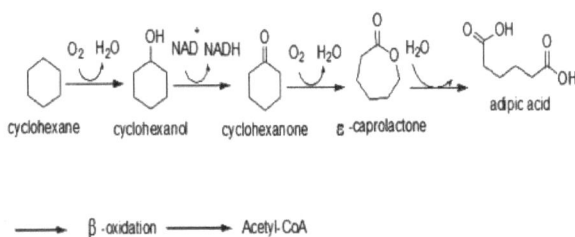

Scheme (2). Biodegradation mechanism of cycloalkane [92].

The biodegradation of cycloalkanes follows a slightly different pathway, as depicted in Scheme **2** [92, 99, 100]. It is understood that only a small number of bacterial species may use cycloalkanes as the only carbon source for their metabolism. The co-metabolism of a bacterial mixed culture is a more typical process for the breakdown of cycloalkanes.

When molecular oxygen is present, the enzyme known as cyclohexane monooxygenase attacks the alkyl side chain, resulting in the breakdown of cycloalkanes. As a result of this reaction, cyclohexanol is produced, which is oxidized by cyclohexanol dehydrogenase to produce cyclohexanone. Caprolactone is produced from cyclohexanone in the presence of cyclohexanone monooxygenase. Caprolactone hydrolase produces adipic acid as an intermediate compound, which is subsequently subjected to the β-oxidation process and produces acetyl-CoA. The bacteria may employ acetyl CoA as a component in the Krebs cycle [92]. The alkynes present in diesel follow a similar pathway, although it is extremely uncommon for a single bacterial species to degrade an alkyne. *Aquincola tertiaricarbonis* L108 and *Methylibium petroleiphilum* PM1 have demonstrated their ability to do so by facilitating the cleavage of the triple bond [101]. The presence of water, H^+ and Hg^{2+} ions starts the breakdown of the alkynes because strong acids and Hg^{2+} are required for the hydration of alkynes. Alkyne hydratase acts as a catalyst to speed up this reaction, which yields an intermediate product of enol tautomer that quickly transforms into keto-tautomer. Under the severe conditions of a permanganate oxidizing agent, the carboxylase catalyst oxidizes the keto form to generate carboxylic acid. To create acetyl-CoA, the generated carboxylic acid is subjected to oxidation [92, 101]. Complete hydrocarbon breakdown yields ATP for bacterial metabolism, followed by CO_2 and H_2O as by-products of the aerobic metabolic processes.

The second important ingredient in diesel is aromatic hydrocarbons. In contrast to alkanes, cycloalkanes, or alkynes, aromatic hydrocarbons degrade *via* more intricate pathways, as shown in Scheme **3**. Due to the involvement of multiple enzymes, the degradation of aromatic hydrocarbons is more complicated than that of aliphatic hydrocarbons. The dioxygenase enzyme joins two oxygen atoms with the benzene ring as the initial step in the breakdown of aromatic hydrocarbons. By catalyzing the union of oxygen with the substrate molecule, the enzyme causes the benzene to undergo hydroxylation, yielding the less stable product *cis*-benzene dihydro-diol. The *cis*-diol dehydrogenase enzyme will dehydrogenate this substance, resulting in catechol. The dioxygenase enzyme combines to break the catechol ring, which results in the termination of the aromatic ring by the addition of oxygen molecules. The aldehydes and carboxylic acids produced from this pathway will go through the same procedure as that of aliphatic hydrocarbon because the reaction's result is a linear hydrocarbon. The β-oxidation mechanism in this pathway leads to the formation of Acetyl-CoA, which subsequently enters the Krebs cycle, which is bacteria's main metabolic system [92, 102]. Table **2** shows the list of microorganisms that can be used for oil bioremediation, and Table **3** displays the interactions of hydrocarbons with a wide array of microorganisms.

Table 2. Important microorganisms for oil bioremediation.

Microorganisms	Compound	References
Fusarium sp., Alcaligenes odorans, Corynebacterium propinquum, Pseudomonas aeru	Oil	[103, 104]
B. brevis, P. aeruginosa KH6, B. locheniformis and B. sphaericus, Bacillus cereus, Citrobacter koseri and Serratia ficaria, Candida krusei and Saccharomyces cerevisiae	Crude oil	[105 - 107]
Psedomonas aeriginosa, P. putida, Athobacter sp and Bacillus sp., Bacillus cereus A, Pseudomonas cepacia, Bacillus cereus, Citrobacter koseri and Serratia ficaria	Diesel oil	[108, 109]

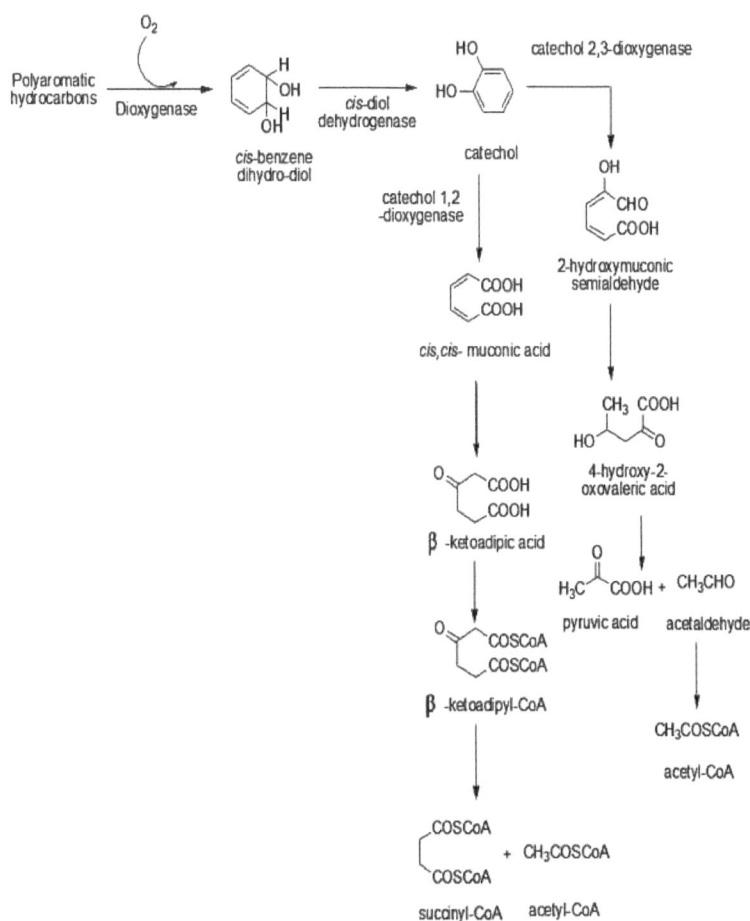

Scheme (3). Biodegradation mechanism of aromatic hydrocarbon [92, 102].

Table 3. Interaction of hydrocarbons with microorganisms.

Microorganisms	Compound	References
Penicillium chrysogenum, Pseudomonas putida	Aromatic hydrocarbons (monocyclic), benzene, toluene, ethyl benzene, phenolic compounds	[110, 111]
P. alcaligenes and P. putida P. veronii, Achromobacter, Flavobacterium, Acinetobacter	polycyclic aromatic hydrocarbons, toluene	[112, 113]
Phanerochaete chrysosporium	Biphenyl and triphenylmethane	[114]
A. fumigatus, F. saloni and P. funiculosum, Tyromyces palustris, Gloeophyllum trabeum, Trametes versicolor	Hydrocarbon	[115]
Coprinellus radians	PAHs, methylnaphthalenes and dibenzofurans	[116]
Alcaligenes odorans, Bacillus subtilis, Corynebacterium propinquum,	Phenol	[103]
Candida viswanathii	Phenanthrene, benzopyrene	[117]
Cyanobacteria, green algae and diatoms and Bacillus licheniformis	Naphthalene	[118, 119]
Acinobacter sp., Pseudomonas sp., Ralstonia sp. And Microbacterium sp,	Aromatic hydrocarbons	
Gleophyllum striatum	Striatum pyrene, 9-methylanthracene, dibenzothisphene, and Lignin peroxidase	[120]

Biodegradation of Dyes

Rapid industrialization and urbanization result in massive amounts of trash being discharged into the environment, which, in turn, increases pollution. The majority of colored effluents are dyes released from the textile, dyestuff and dyeing industries. Owing to their color, biorecalcitrance and possible toxicity to humans and animals, azo dyes represent a significant group of dyes that raise environmental concerns [121, 122]. Due to their high BOD, COD, heat, color pH and metal content, effluents from the textile and dye industries are extremely difficult to treat by physical and chemical procedures. The main goal of bioremediation is to increase the native species' capacity for natural degradation. However, azo dyes are xenobiotics, and their natural breakdown is quite challenging. Therefore, the only procedure that may safely degrade textile azo colors is aerobic treatment. Microbial species of bacteria, fungi, algae and actinomycetes are capable of removing azo dye *via* biotransformation, biodegradation, or liberalization [123, 124].

Microperoxidases are enzymes that can be produced by some bacteria, including *Bacillus sterothermophilis*. Using *Pseudomonas* bacteria, a degradative pathway for sulfonated azo dyes was discovered [125]. Peroxidase, which is similarly produced by *Flavibacterium*, can destroy azo dyes in an aerobic environment [126]. Without investigating the individual microbial populations, microbial consortiums are employed as 'black boxes' for environmental cleanup. Microbial consortiums can react to a wide range of contaminants due to their intricacy [127]. Table **4** portrays the list of microorganisms that are involved in the bioremediation of dyes.

Table 4. Microorganisms involved in the bioremediation of dyes.

Microorganism	Compound	Refs.
B. subtilis strain NAP1, NAP2, NAP4	Oil-based paints	[3, 121]
Phanerochaete chrysosporium, Penicillium ochrochloron, Myrothecium rorodum IM 6482	Industrial dyes	[3]
Bacillus sp (ETL -2012), Pseudomonas aeruginosa, Bacillus pumilus HKG212	Textile Dye (Remazol Black B), Sulfonated di-azo dye Reactive Red HE8B, RNB dye [3]	[3]
Exiguobacterium indicum, Micrococcus luteus, Listeria denitrificans and Nocardia atlantica, Bacillus cereus and Acinetobacter baumanii	Textile Azo Dyes and Azo dyes effluents	[3]
Bacillus firmus, Staphylococcus aureus and Klebsiella oxytoca	Textile effluents and vat dyes	[3]

Biodegradation of Pesticides

The beginning of the industrial revolution and the production of different pesticides have undoubtedly increased the yield of our agricultural products and protected the bulk of our crops from pests. Although pesticides currently play a significant role in increasing yield and benefit our farmers. The economical use of pesticides in agricultural fields is a big concern due to the rise in soil pollution. Numerous pollutants (pesticides as one of the most dangerous pollutants) represent a substantial risk to human health and the natural ecology. Physical or chemical treatments that are currently accessible are either expensive or insufficient. Pesticide detoxification is made possible by bioremediation, which is affordable, effective and environmentally friendly [128].

Pesticide-degrading bacteria are found in the genera *Flavobacterium, Arthobacter, Aztobacter, Burkholderia* and *Pseudomonas* [3, 128]. Pesticide degradation has recently been linked to the bacterium *Raoultella* species. The parent ingredient of the pesticide must be completely oxidized to produce carbon dioxide and water. Pesticide-degrading microflora should be added externally to the soil when the natural microbial community cannot handle pesticides. Table **5** depicts the list of microorganisms that are responsible for the degradation of pesticides.

Table 5. Microorganisms involved in the bioremediation of pesticides.

Microorganisms	Compound	Refs.
Bacillus, Staphylococcus	Endosulfan	[129]
Enterobacter	Chlorpyrifos	[130]
Pseudomonas putida, Acinetobacter sp., Arthrobacter sp.	Ridomil MZ 68 MG, Fitoraz WP 76, Decis 2.5 EC, malation	[130]
Acenetobacter sp., Pseudomonas sp., Enterobacter sp. And Photobacterium sp.	Chlorpyrifos and methyl parathion	[130]

GENETICALLY ENGINEERED MICROORGANISMS

A wide range of contaminants can be degraded using microorganisms' extensive metabolic and physiological adaptability. Microbial systems are now used in bioremediation projects, most notably in the treatment of organic polluted soils and streams. In some cases, native populations are unsuitable for use in polluted site remediation because natural microbial communities destroy chemicals at a sluggish rate to restore the polluted site. Hence, genetically engineered microorganisms (GEMs) are being examined for *in situ* bioremediation of contaminated ecosystems [131, 132].

Recombinant DNA technology is used to describe this kind of scientific technique. By developing GEMs, genetic engineering has enhanced the utilization and eradication of dangerous, undesirable wastes in laboratory settings. GEMs with increased degradation capabilities encompass a wide range of chemical contaminants that have shown promising applications in the bioremediation of soil, groundwater, and activated sludge. The generation of novel strains with desirable features through the designing of pathways and the adjustment of enzyme specificity and substrate affinity have been the key components of using GEMs in bioremediation. A few strategies are needed to be carried out in GEMs:

a. Development, monitoring and controlling of bioprocess
b. Construction and regulation of pathways
c. The applications of bio reporter-based sensors for chemical sensing, toxicity reduction and endpoint analysis.

A few examples are provided below to demonstrate the diversity of advancements that have been made, the methodologies that have been employed and the species that have been utilized in such applications. Aromatic compounds are a focus of bioremediation studies utilizing engineered bacteria because they represent a particularly varied collection of pollutants in soil and water. In Comamonas testosteroni strain VP44, Hrywna *et al.* cloned and expressed the *ohb* operon from *P. aeruginosa* and the *fcb* operon from *Arthrobacter globiformis* (capable of encoding enzymes that can metabolize chlorobenzoic acids) [133]. The genes that transform chlorinated biphenyls into *ortho-* and *para-* chlorobenzoic acids are naturally present in the host strain. Thus, a strain capable of fully mineralizing monochlorobiphenyls was created by transforming the host with plasmids carrying the *ohb* and *fcb* operons.

According to Monti *et al.*, *Pseudomonas fluorescens* ATCC 17400 was modified with the genes encoding the 2,4-dintirotoluene (2,4-DNT) breakdown pathway from *Burkholderia sp.* strain DNT [134]. When 2,4 DNT was given to the recombinant strain as the only supply of nitrogen, the compound was entirely broken down, and the cell utilized the carbon, which resulted in co-metabolism. The capacity of the recombinant strain to break down 2,4-DNT (at low temperatures) and its lower toxicity to different plants under specific circumstances made it superior to *Burkholderia sp.* Again, Wu *et al.* reported that *Comamonas sp.* strains CNB1 genes, which encode a novel reductive pathway for 4-chloronitrobenzene and nitrobenzene, have also been cloned and expressed in *E. coli* [135].

At waste sites where radionuclides and organic pollutants coexist, another class of contaminants is heavy metals, including radionuclides. Due to the radiation from these radionuclides' toxicity to the majority of bacteria, bioremediation of the organics in such sites presents challenges. However, owing to its strong resilience to acute exposures to ionizing radiation, *Deinococcus radiodurans* is an advantageous host for genetic engineering techniques in which mixed waste exists. According to Lange *et al.*, it has been modified with toluene dioxygenase genes from *Pseudomonas putida* F1 [136]. In a highly irradiated environment, the modified strain was demonstrated to efficiently oxidize toluene, 3,4-dichloro-1-butene and trichloroethylene. Renninger *et al.* used engineered *Pseudomonas aeruginosa* to successfully bioremediate uranium *via* the uranyl group, and they did this by over-expressing *P. aeruginosa* polyphosphate kinase. In this instance, the *tac-lac* promoter was used to clone the endogenous genes encoding polyphosphate synthesis and degradation into a plasmid with a wide host range [137]. In comparison to the parent strain, the modified strain accumulates 100 times more polyphosphate. Large amounts of phosphate are secreted when the polyphosphate is broken down; these phosphate molecules then combine with the uranyl group and precipitate at the cell membrane. Additionally, the ability to bioremediate heavy metals has been genetically engineered into E. coli and employed as a host. Crameri *et al.* used the *Staphylococcus aureus* arsenate resistance operon and DNA shuffling techniques to construct an arsenate detoxification pathway. The recombinant *E. coli* that resulted showed improved arsenate resistance and was capable of detoxifying it *via* reduction [138].

Mercury detoxification was proven in another application using genetically altered *E. coli* carrying the sequence encoding the Hg(II) binding domain of the MerR protein from *Shigella flexneri*. In comparison to the parental cells lacking Hg(II) binding domain, the recombinant cells produced a significant amount of Hg(II) binding domain on the cell surface. It is possible to generate engineered strains with enhanced bioremediation properties by transferring genes from one species to another. Oxygenases incorporate monooxygenases or dioxygenases O atoms into the substrate that is being oxidized to catalyze the reduction of O_2. Thereby, they are capable of initiating the mineralization of a variety of organic contaminants. This benefit of oxidative breakdown of pollutants can be further enhanced by protein engineering of oxygenases. Laccase plays a role in the biodegradation of polyaromatic hydrocarbons as well as the degradation of lignin [139]. Additionally, genetic engineering has been utilized to enhance the efficiency of enzymes that break down non-aromatic molecules. For instance, the catabolic phase that comes after the ring opening of aromatic and chloroaromatic compounds is catalyzed by muconate and chloromuconate cycloisomerases, respectively. The activity of *P. putida* muconate cycloisomerase against numerous substituted (halogenated as well as methylated) muconates was improved by site-directed mutagenesis to resemble chloromuconate isomerases [132]. Again, the catabolic genes from *Rhodococcus erythropolis* were cloned into two *Pseudomonas* strains by Gallardo *et al.* [140]. The enzymes responsible for desulfurizing dibenzothiophene, a representative of the sulfur-containing compounds found in fossil fuels, are encoded by these genes [138]. These enzymes are valuable because they can eliminate sulfur without reducing the amount of fuel. Therefore, their strategy aims to eliminate the contamination's source before it even has a chance to enter the environment [138, 140].

CONCLUSION AND FUTURE PERSPECTIVES

Biodegradation is a very promising method for resolving a polluted environment through microbial activity. It is, however, restricted to biodegradable substances. There are concerns that the by-products of biodegradation could be more toxic than the original compound. Several biological processes are extremely specific. The availability of competent microbial populations, optimum environmental conditions and appropriate quantity of nutrients are crucial requirements for success. Bioremediation requires more time as compared to alternative treatment methods including excavation and soil removal. To resolve this problem, GEMs can be used to speed up the recovery of polluted waste sites and to increase degradation. In laboratory investigations, it has become a common practice to utilize genetic engineering to create microbes with the ability to degrade particular pollutants.

Yet, the use of GEMs for the remediation of contaminated places is still debatable and highly limited. This is attributed to the concern that GEMs are not 'natural' and would linger in the ecosystem, thereby changing the ecology. Concerns over GEMs' stability and the transfer of modified DNA from them are critical due to the possible consequences of their release for bioremediation. The use of GEMs in the field may bring promising results for the remediation of polluted sites in the future. However, substantial research will probably be necessary to ascertain the true risks and advantages associated with such sites.

ACKNOWLEDGEMENT

The author would like to acknowledge the financial assistance provided by the Department of Science & Technology and Biotechnology, Government of West Bengal, India (Memo No. 860(Sanc.)/STBT-11012(25)/5/2019-ST SEC dated 03/11/2023).

REFERENCES

[1]   Tang CY, Fu QS, Criddle CS, Leckie JO. Effect of flux (transmembrane pressure) and membrane properties on fouling and rejection of reverse osmosis and nanofiltration membranes treating perfluorooctane sulfonate containing wastewater. Environ Sci Technol 2007; 41(6): 2008-14.
       [http://dx.doi.org/10.1021/es062052f] [PMID: 17410798]

[2]   Strong PJ, Burgess JE. Treatment methods for wine-related and distillery wastewaters: a review. Bioremediat J 2008; 12(2): 70-87.
       [http://dx.doi.org/10.1080/10889860802060063]

[3]   Abatenh E, Gizaw B, Tsegaye Z, Wassie M. The role of microorganisms in bioremediation-A review Open JEnviron Bio 2017; 2(1): 038-46.

[4]   El Fantroussi S, Agathos SN. Is bioaugmentation a feasible strategy for pollutant removal and site remediation? Curr Opin Microbiol 2005; 8(3): 268-75.
       [http://dx.doi.org/10.1016/j.mib.2005.04.011] [PMID: 15939349]

[5]   Das S, Dash HR. Microbial bioremediation: A potential tool for restoration of contaminated areas. Micro Biodegr and Biorem. Elsevier 2014; 1: pp. 1-21.

[6]   Angelim AL, Costa SP, Farias BCS, Aquino LF, Melo VMM. An innovative bioremediation strategy using a bacterial consortium entrapped in chitosan beads. J Environ Manage 2013; 127: 10-7.
       [http://dx.doi.org/10.1016/j.jenvman.2013.04.014] [PMID: 23659866]

[7]   Kensa VM. Bioremediation-an overview. I Control Pollution 2011; 27(2): 161-8.

[8]   Demnerová K, Mackova M, Speváková V, et al. Two approaches to biological decontamination of groundwater and soil polluted by aromatics-characterization of microbial populations. Int Microbiol 2005; 8(3): 205-11.
       [PMID: 16200499]

[9]   Sarao LK, Kaur S. Chapter-3 Microbial Bioremediation. N Vist Micro Sci. 2021; p. 47.

[10] Bala S, Garg D, Thirumalesh BV, *et al.* Recent strategies for bioremediation of emerging pollutants: a review for a green and sustainable environment. Toxics 2022; 10(8): 484.
[http://dx.doi.org/10.3390/toxics10080484] [PMID: 36006163]

[11] Singh A, Ward OP. Biotechnology and bioremediation—an overview. Biodegradation and bioremediation. 2004; pp. 1-7.

[12] Kumar A, Bisht BS, Joshi VD, Dhewa T. Review on bioremediation of polluted environment: a management tool. Int J Environ Sci 2011; 1(6): 1079-93.

[13] Ławniczak Ł, Woźniak-Karczewska M, Loibner AP, Heipieper HJ, Chrzanowski Ł. Microbial degradation of hydrocarbons—basic principles for bioremediation: a review. Molecules 2020; 25(4): 856.
[http://dx.doi.org/10.3390/molecules25040856] [PMID: 32075198]

[14] Eweis JB, Ergas SJ, Chang DP, Schroeder ED. Bioremediation principles. McGraw-Hill Book Company Europe 1998.

[15] Azubuike CC, Chikere CB, Okpokwasili GC. Bioremediation techniques–classification based on site of application: principles, advantages, limitations and prospects. World J Microbiol Biotechnol 2016; 32(11): 180.
[http://dx.doi.org/10.1007/s11274-016-2137-x] [PMID: 27638318]

[16] Boopathy R. Factors limiting bioremediation technologies. Bioresour Technol 2000; 74(1): 63-7.
[http://dx.doi.org/10.1016/S0960-8524(99)00144-3]

[17] Vidali M. Bioremediation. An overview. Pure Appl Chem 2001; 73(7): 1163-72.
[http://dx.doi.org/10.1351/pac200173071163]

[18] Omokhagbor Adams G, Tawari Fufeyin P, Eruke Okoro S, Ehinomen I. Bioremediation, biostimulation and bioaugmention: a review. Int J Environ Bioremediat Biodegrad 2020; 3(1): 28-39.
[http://dx.doi.org/10.12691/ijebb-3-1-5]

[19] Couto N, Fritt-Rasmussen J, Jensen PE, Højrup M, Rodrigo AP, Ribeiro AB. Suitability of oil bioremediation in an Artic soil using surplus heating from an incineration facility. Environ Sci Pollut Res Int 2014; 21(9): 6221-7.
[http://dx.doi.org/10.1007/s11356-013-2466-3] [PMID: 24488519]

[20] Thavasi R, Jayalakshmi S, Banat IM. Application of biosurfactant produced from peanut oil cake by *Lactobacillus delbrueckii* in biodegradation of crude oil. Bioresour Technol 2011; 102(3): 3366-72.
[http://dx.doi.org/10.1016/j.biortech.2010.11.071] [PMID: 21144745]

[21] Macaulay BM. Understanding the behaviour of oil-degrading micro-organisms to enhance the microbial remediation of spilled petroleum. Appl Ecol Environ Res 2015; 13(1): 247-62.

[22] Das N, Chandran P. Microbial degradation of petroleum hydrocarbon contaminants: an overview Biotechnol Res Int 2011; 2011: 941810.
[http://dx.doi.org/10.4061/2011/941810]

[23] Asira EE. Factors that determine bioremediation of organic compounds in the soil. Academic Journal of Interdisciplinary Studies 2013; 2(13): 125.
[http://dx.doi.org/10.5901/ajis.2013.v2n13p125]

[24] Wang Q, Zhang S, Li Y, Klassen W. Potential approaches to improving biodegradation of hydrocarbons for bioremediation of crude oil pollution. J Environ Prot 2011; 2(1): 47-55.
[http://dx.doi.org/10.4236/jep.2011.21005]

[25] Jørgensen KS. *In situ* bioremediation. Adv Appl Microbiol 2007; 61: 285-305.
[http://dx.doi.org/10.1016/S0065-2164(06)61008-3] [PMID: 17448793]

[26] Pandey J, Chauhan A, Jain RK. Integrative approaches for assessing the ecological sustainability of *in situ* bioremediation. FEMS Microbiol Rev 2009; 33(2): 324-75.
[http://dx.doi.org/10.1111/j.1574-6976.2008.00133.x] [PMID: 19178567]

[27]   Farhadian M, Vachelard C, Duchez D, Larroche C. *In situ* bioremediation of monoaromatic pollutants in groundwater: A review. Bioresour Technol 2008; 99(13): 5296-308.
[http://dx.doi.org/10.1016/j.biortech.2007.10.025] [PMID: 18054222]

[28]   Hoeppel RE, Hinchee RE, Arthur MF. Bioventing soils contaminated with petroleum hydrocarbons. J Ind Microbiol Biotechnol 1991; 8(3): 141-6.

[29]   Dupont RR. Fundamentals of bioventing applied to fuel contaminated sites. Environ Prog 1993; 12(1): 45-53.
[http://dx.doi.org/10.1002/ep.670120109]

[30]   Lee TH, Byun IG, Kim YO, Hwang IS, Park TJ. Monitoring biodegradation of diesel fuel in bioventing processes using *in situ* respiration rate. Water Sci Technol 2006; 53(4-5): 263-72.
[http://dx.doi.org/10.2166/wst.2006.131] [PMID: 16722077]

[31]   Agarry S, Latinwo G. Biodegradation of diesel oil in soil and its enhancement by application of bioventing and amendment with brewery waste effluents as biostimulation-bioaugmentation agents. J Ecol Eng 2015; 16(2): 82-91.
[http://dx.doi.org/10.12911/22998993/1861]

[32]   Kao CM, Chen CY, Chen SC, Chien HY, Chen YL. Application of *in situ* biosparging to remediate a petroleum-hydrocarbon spill site: Field and microbial evaluation. Chemosphere 2008; 70(8): 1492-9.
[http://dx.doi.org/10.1016/j.chemosphere.2007.08.029] [PMID: 17950413]

[33]   Johnson PC, Johnson RL, Bruce CL, Leeson A. Advances in *in situ* air sparging/biosparging. Bioremediat J 2001; 5(4): 251-66.
[http://dx.doi.org/10.1080/20018891079311]

[34]   Gentry T, Rensing C, Pepper I. New approaches for bioaugmentation as a remediation technology. Crit Rev Environ Sci Technol 2004; 34(5): 447-94.
[http://dx.doi.org/10.1080/10643380490452362]

[35]   Herrero M, Stuckey DC. Bioaugmentation and its application in wastewater treatment: A review. Chemosphere 2015; 140: 119-28.
[http://dx.doi.org/10.1016/j.chemosphere.2014.10.033] [PMID: 25454204]

[36]   Tyagi M, da Fonseca MMR, de Carvalho CCCR. Bioaugmentation and biostimulation strategies to improve the effectiveness of bioremediation processes. Biodegradation 2011; 22(2): 231-41.
[http://dx.doi.org/10.1007/s10532-010-9394-4] [PMID: 20680666]

[37]   Thompson IP, Van Der Gast CJ, Ciric L, Singer AC. Bioaugmentation for bioremediation: the challenge of strain selection. Environ Microbiol 2005; 7(7): 909-15.
[http://dx.doi.org/10.1111/j.1462-2920.2005.00804.x] [PMID: 15946288]

[38]   Singer AC, van der Gast CJ, Thompson IP. Perspectives and vision for strain selection in bioaugmentation. Trends Biotechnol 2005; 23(2): 74-7.
[http://dx.doi.org/10.1016/j.tibtech.2004.12.012] [PMID: 15661343]

[39]   Kim S, Krajmalnik-Brown R, Kim JO, Chung J. Remediation of petroleum hydrocarbon-contaminated sites by DNA diagnosis-based bioslurping technology. Sci Total Environ 2014; 497-498: 250-9.
[http://dx.doi.org/10.1016/j.scitotenv.2014.08.002] [PMID: 25129160]

[40]   Keet BA. Bioslurping state of the art. Battelle Press, Columbus, OH (United States); 1995; 31.

[41]   Connolly MD, Gibbs BM, Keet B. Bioslurping applied to a gasoline and diesel spill in fractured rock. Battelle Press, Columbus, OH (United States); 1995; 31..

[42]   Deshusses MA. Biological waste air treatment in biofilters. Curr Opin Biotechnol 1997; 8(3): 335-9.
[http://dx.doi.org/10.1016/S0958-1669(97)80013-4] [PMID: 9206016]

[43]   Lith C, Leson G, Michelsen R. Evaluating design options for biofilters. J Air Waste Manag Assoc 1997; 47(1): 37-48.
[http://dx.doi.org/10.1080/10473289.1997.10464410]

[44] Pujol R, Hamon M, Kandel X, Lemmel H. Biofilters: flexible, reliable biological reactors. Water Sci Technol 1994; 29(10-11): 33-8.
[http://dx.doi.org/10.2166/wst.1994.0742]

[45] Hazen TC. Biostimulation Lawrence Berkeley National Lab 2009; 1.

[46] Bento FM, Camargo FAO, Okeke BC, Frankenberger WT. Comparative bioremediation of soils contaminated with diesel oil by natural attenuation, biostimulation and bioaugmentation. Bioresour Technol 2005; 96(9): 1049-55.
[http://dx.doi.org/10.1016/j.biortech.2004.09.008] [PMID: 15668201]

[47] Arjoon A, Olaniran AO, Pillay B. Enhanced 1,2-dichloroethane degradation in heavy metal co-contaminated wastewater undergoing biostimulation and bioaugmentation. Chemosphere 2013; 93(9): 1826-34.
[http://dx.doi.org/10.1016/j.chemosphere.2013.06.034] [PMID: 23835411]

[48] Tomei MC, Daugulis AJ. *Ex situ* bioremediation of contaminated soils: an overview of conventional and innovative technologies. Crit Rev Environ Sci Technol 2013; 43(20): 2107-39.
[http://dx.doi.org/10.1080/10643389.2012.672056]

[49] Lin TC, Pan PT, Cheng SS. *Ex situ* bioremediation of oil-contaminated soil. J Hazard Mater 2010; 176(1-3): 27-34.
[http://dx.doi.org/10.1016/j.jhazmat.2009.10.080] [PMID: 20053499]

[50] Debertoldi M, Vallini G, Pera A. The biology of composting: A review. Waste Manag Res 1983; 1(2): 157-76.
[http://dx.doi.org/10.1016/0734-242X(83)90055-1]

[51] Imbeah M. Composting piggery waste: A review. Bioresour Technol 1998; 63(3): 197-203.
[http://dx.doi.org/10.1016/S0960-8524(97)00165-X]

[52] Stentiford EI. Composting control: principles and practice The science of composting 1996; 49-59.

[53] Cooperband L. The art and science of composting. Center for Integrated agricultural systems. 2002; 29..

[54] Martin I, Wendt D, Heberer M. The role of bioreactors in tissue engineering. Trends Biotechnol 2004; 22(2): 80-6.
[http://dx.doi.org/10.1016/j.tibtech.2003.12.001] [PMID: 14757042]

[55] Cooney CL. Bioreactors: Design and Operation. Science 1983; 219(4585): 728-33.
[http://dx.doi.org/10.1126/science.219.4585.728] [PMID: 17814034]

[56] Betts JI, Baganz F. Miniature bioreactors: current practices and future opportunities. Microb Cell Fact 2006; 5(1): 21.
[http://dx.doi.org/10.1186/1475-2859-5-21] [PMID: 16725043]

[57] Wang X, Wang Q, Wang S, Li F, Guo G. Effect of biostimulation on community level physiological profiles of microorganisms in field-scale biopiles composed of aged oil sludge. Bioresour Technol 2012; 111: 308-15.
[http://dx.doi.org/10.1016/j.biortech.2012.01.158] [PMID: 22357295]

[58] Gomez F, Sartaj M. Optimization of field scale biopiles for bioremediation of petroleum hydrocarbon contaminated soil at low temperature conditions by response surface methodology (RSM). Int Biodeterior Biodegradation 2014; 89: 103-9.
[http://dx.doi.org/10.1016/j.ibiod.2014.01.010]

[59] Delille D, Duval A, Pelletier E. Highly efficient pilot biopiles for on-site fertilization treatment of diesel oil-contaminated sub-Antarctic soil. Cold Reg Sci Technol 2008; 54(1): 7-18.
[http://dx.doi.org/10.1016/j.coldregions.2007.09.003]

[60] Karigar CS, Rao SS. Role of microbial enzymes in the bioremediation of pollutants: a review. Enzyme Res 2011; 2011: 1-11.

[http://dx.doi.org/10.4061/2011/805187] [PMID: 21912739]

[61]   Balba MT, Al-Awadhi N, Al-Daher R. Bioremediation of oil-contaminated soil: microbiological methods for feasibility assessment and field evaluation. J Microbiol Methods 1998; 32(2): 155-64.
[http://dx.doi.org/10.1016/S0167-7012(98)00020-7]

[62]   Brim H, McFarlan SC, Fredrickson JK, *et al.* Engineering *Deinococcus radiodurans* for metal remediation in radioactive mixed waste environments. Nat Biotechnol 2000; 18(1): 85-90.
[http://dx.doi.org/10.1038/71986] [PMID: 10625398]

[63]   Buck JD. Nonstaining (KOH) method for determination of gram reactions of marine bacteria. Appl Environ Microbiol 1982; 44(4): 992-3.
[http://dx.doi.org/10.1128/aem.44.4.992-993.1982] [PMID: 6184019]

[64]   Chung WK, King GM. Isolation, characterization, and polyaromatic hydrocarbon degradation potential of aerobic bacteria from marine macrofaunal burrow sediments and description of *Lutibacterium anuloederans* gen. nov., sp. nov., and Cycloclasticus spirillensus sp. nov. Appl Environ Microbiol 2001; 67(12): 5585-92.
[http://dx.doi.org/10.1128/AEM.67.12.5585-5592.2001] [PMID: 11722910]

[65]   De J, Ramaiah N, Vardanyan L. Detoxification of toxic heavy metals by marine bacteria highly resistant to mercury. Mar Biotechnol 2008; 10(4): 471-7.
[http://dx.doi.org/10.1007/s10126-008-9083-z] [PMID: 18288535]

[66]   Hua X, Wu Z, Zhang H, *et al.* Degradation of hexadecane by Enterobacter cloacae strain TU that secretes an exopolysaccharide as a bioemulsifier. Chemosphere 2010; 80(8): 951-6.
[http://dx.doi.org/10.1016/j.chemosphere.2010.05.002] [PMID: 20537678]

[67]   Lin C, Gan L, Chen ZL. Biodegradation of naphthalene by strain *Bacillus fusiformis* (BFN). J Hazard Mater 2010; 182(1-3): 771-7.
[http://dx.doi.org/10.1016/j.jhazmat.2010.06.101] [PMID: 20643503]

[68]   Banerjee G, Pandey S, Ray AK, Kumar R. Bioremediation of heavy metals by a novel bacterial strain Enterobacter cloacae and its antioxidant enzyme activity, flocculant production, and protein expression in presence of lead, cadmium, and nickel. Water Air Soil Pollut 2015; 226(4): 91.
[http://dx.doi.org/10.1007/s11270-015-2359-9]

[69]   Chen L, Chen L, Pan D, *et al.* Heterotrophic nitrification and related functional gene expression characteristics of *Alcaligenes faecalis* SDU20 with the potential use in swine wastewater treatment. Bioprocess Biosyst Eng 2021; 44(10): 2035-50.
[http://dx.doi.org/10.1007/s00449-021-02581-z] [PMID: 33978835]

[70]   Yadav S, Singh K, Chandra R. Plant growth–promoting rhizobacteria (PGPR) and bioremediation of industrial waste. Micro Sustain Devel Biorem. CRC press 2019; 13: p. (207)241.

[71]   Diels L, De Smet M, Hooyberghs L, Corbisier P. Heavy metals bioremediation of soil. Mol Biotechnol 1999; 12(2): 149-58.
[http://dx.doi.org/10.1385/MB:12:2:149] [PMID: 10596372]

[72]   Ahmed T, Liaqat I, Murtaza R, Rasheed A. Bioremediation approaches for E-waste management: A step toward sustainable environment Elect Waste Poll: Environ Occur and Treat Techno 2019; 267-90.

[73]   Zhang, W., Chen, L., Liu, D. Characterization of a marine-isolated mercury-resistant *Pseudomonas putida* strain SP1 and its potential application in marine mercury reduction. Applied Microbiology and Biotechnology., 2012, 93, 1305-1314

[74]   von Canstein H, Kelly S, Li Y, Wagner-Döbler I. Species diversity improves the efficiency of mercury-reducing biofilms under changing environmental conditions. Appl Environ Microbiol 2002; 68(6): 2829-37.
[http://dx.doi.org/10.1128/AEM.68.6.2829-2837.2002] [PMID: 12039739]

[75]   Iyer A, Mody K, Jha B. Biosorption of heavy metals by a marine bacterium. Mar Pollut Bull 2005; 50(3): 340-3.

[http://dx.doi.org/10.1016/j.marpolbul.2004.11.012] [PMID: 15757698]

[76]   Chen C, Wang JL. Characteristics of Zn^{2+} biosorption by *Saccharomyces cerevisiae*. Biomed Environ Sci 2007; 20(6): 478-82.
        [PMID: 18348406]

[77]   Talos K, Pager C, Tonk S, *et al.* Cadmium biosorption on native *Saccharomyces cerevisiae* cells in aqueous suspension. Acta Univ Sapientiae Agric Environ 2009; 1: 20-30.

[78]   Tigini V, Prigione V, Giansanti P, Mangiavillano A, Pannocchia A, Varese GC. Fungal biosorption, an innovative treatment for the decolourisation and detoxification of textile effluents. Water 2010; 2(3): 550-65.
        [http://dx.doi.org/10.3390/w2030550]

[79]   Paranthaman SR, Karthikeyan B. Bioremediation of heavy metal in paper mill effluent using Pseudomonas spp. Int J Microbiol 2015; 1(4): 1-5.

[80]   Peña-Montenegro TD, Lozano L, Dussán J. Genome sequence and description of the mosquitocidal and heavy metal tolerant strain *Lysinibacillus sphaericus* CBAM5. Stand Genomic Sci 2015; 10(1): 2.
        [http://dx.doi.org/10.1186/1944-3277-10-2] [PMID: 25685257]

[81]   Wu YH, Zhou P, Cheng H, Wang CS, Wu M, Xu XW. Draft genome sequence of Microbacterium profundi Shh49T, an Actinobacterium isolated from deep-sea sediment of a polymetallic nodule environment. Genome Announc 2015; 3(3): 00642-15.
        [http://dx.doi.org/10.1128/genomeA.00642-15] [PMID: 26067975]

[82]   Mohammadian Fazli M, Soleimani N, Mehrasbi M, Darabian S, Mohammadi J, Ramazani A. Highly cadmium tolerant fungi: their tolerance and removal potential. J Environ Health Sci Eng 2015; 13(1): 19.
        [http://dx.doi.org/10.1186/s40201-015-0176-0] [PMID: 25806110]

[83]   Rajesh P, Athiappan M, Paul R, Raj KD. Bioremediation of cadmium by *Bacillus safensis* (JX126862), a marine bacterium isolated from mangrove sediments. Int J Curr Microbiol Appl Sci 2014; 3(12): 326-35.

[84]   Magnuson TS, Hodges-Myerson AL, Lovley DR. Characterization of a membrane-bound NADH-dependent Fe $^{3+}$ reductase from the dissimilatory Fe $^{3+}$ -reducing bacterium *Geobacter sulfurreducens*. FEMS Microbiol Lett 2000; 185(2): 205-11.
        [http://dx.doi.org/10.1111/j.1574-6968.2000.tb09063.x] [PMID: 10754249]

[85]   Zhao XQ, Wang RC, Lu XC, Lu JJ, Li J, Hu H. Tolerance and biosorption of heavy metals by *Cupriavidus metallidurans* strain XXKD-1 isolated from a subsurface laneway in the Qixiashan Pb-Zn sulfide minery in Eastern China. Geomicrobiol J 2012; 29(3): 274-86.
        [http://dx.doi.org/10.1080/01490451.2011.619637]

[86]   Sinha SN, Biswas M, Paul D, Rahaman S. Biodegradation potential of bacterial isolates from tannery effluent with special reference to hexavalent chromium. Biotechnology Bioinformatics and Bioengineering 2011; 1(3): 381-6.

[87]   Sinha SN, Paul D. Heavy metal tolerance and accumulation by bacterial strains isolated from waste water. J Chem Biol Phys Sci 2013; 4(1): 812.

[88]   Latha K, Lalithakumari D. Transfer and expression of a hydrocarbon-degrading plasmid pHCL from *Pseudomonas putida* to marine bacteria. World J Microb Biotech 2001; 17: 523-8.

[89]   Andrade LL, Leite DCA, Ferreira EM, *et al.* Microbial diversity and anaerobic hydrocarbon degradation potential in an oil-contaminated mangrove sediment. BMC Microbiol 2012; 12(1): 186.
        [http://dx.doi.org/10.1186/1471-2180-12-186] [PMID: 22935169]

[90]   McKew BA, Coulon F, Osborn AM, Timmis KN, McGenity TJ. Determining the identity and roles of oil-metabolizing marine bacteria from the Thames estuary, UK. Environ Microbiol 2007; 9(1): 165-76.
        [http://dx.doi.org/10.1111/j.1462-2920.2006.01125.x] [PMID: 17227421]

[91]   Deppe U, Richnow HH, Michaelis W, Antranikian G. Degradation of crude oil by an arctic microbial consortium. Extremophiles 2005; 9(6): 461-70.
[http://dx.doi.org/10.1007/s00792-005-0463-2] [PMID: 15999222]

[92]   Imron MF, Kurniawan SB, Ismail NI, Abdullah SRS. Future challenges in diesel biodegradation by bacteria isolates: A review. J Clean Prod 2020; 251: 119716.
[http://dx.doi.org/10.1016/j.jclepro.2019.119716]

[93]   Chen Y, Lin J, Chen Z. Remediation of water contaminated with diesel oil using a coupled process: Biological degradation followed by heterogeneous Fenton-like oxidation. Chemosphere 2017; 183: 286-93.
[http://dx.doi.org/10.1016/j.chemosphere.2017.05.120] [PMID: 28551205]

[94]   Purwanti IF, Abdullah SRS, Hamzah A, *et al.* Biodegradation of diesel by bacteria isolated from *Scirpus mucronatus* rhizosphere in diesel-contaminated sand. Adv Sci Lett 2015; 21(2): 140-3.
[http://dx.doi.org/10.1166/asl.2015.5843]

[95]   Wu B, Lan T, Lu D, Liu Z. Ecological and enzymatic responses to petroleum contamination. Environ Sci Process Impacts 2014; 16(6): 1501-9.
[http://dx.doi.org/10.1039/C3EM00731F] [PMID: 24765642]

[96]   Bhuvaneswar C, Swathi G, Bhaskar BV, Munichandrababu T, Rajendra W. Effective synergetic biodegradation of diesel oil by bacteria. Int J Environ Biol 2012; 2(4): 195-9.

[97]   Whale GF, Dawick J, Hughes CB, Lyon D, Boogaard PJ. Toxicological and ecotoxicological properties of gas-to-liquid (GTL) products. 2. Ecotoxicology. Crit Rev Toxicol 2018; 48(4): 273-96.
[http://dx.doi.org/10.1080/10408444.2017.1408567] [PMID: 29309204]

[98]   Singh SN, Kumari B, Mishra S. Microbial degradation of alkanes Micro Degra Xenob. 2012; pp. 439-69.
[http://dx.doi.org/10.1007/978-3-642-23789-8_17]

[99]   Presentato A, Cappelletti M, Sansone A, *et al.* Aerobic growth of *Rhodococcus aetherivorans* BCP1 using selected naphthenic acids as the sole carbon and energy sources. Front Microbiol 2018; 9: 672.
[http://dx.doi.org/10.3389/fmicb.2018.00672] [PMID: 29706937]

[100]   Salamanca D, Engesser KH. Isolation and characterization of two novel strains capable of using cyclohexane as carbon source. Environ Sci Pollut Res Int 2014; 21(22): 12757-66.
[http://dx.doi.org/10.1007/s11356-014-3206-z] [PMID: 24969427]

[101]   Schuster J, Schäfer F, Hübler N, *et al.* Bacterial degradation of tert-amyl alcohol proceeds *via* hemiterpene 2-methyl-3-buten-2-ol by employing the tertiary alcohol desaturase function of the Rieske nonheme mononuclear iron oxygenase MdpJ. J Bacteriol 2012; 194(5): 972-81.
[http://dx.doi.org/10.1128/JB.06384-11] [PMID: 22194447]

[102]   Kothari V, Panchal M, Srivastava N. Presence of catechol metabolizing enzymes in *Virgibacillus salarius*. J Enviro Conserv Res 2013; 1(2): 29-36.
[http://dx.doi.org/10.12966/jecr.08.03.2013]

[103]   Singh A, Kumar V, Srivastava JN. Assessment of bioremediation of oil and phenol contents in refinery waste water *via* bacterial consortium. J Pet Environ Biotechnol 2013; 4(3): 1-4.
[http://dx.doi.org/10.4172/2157-7463.1000145]

[104]   Hidayat A, Tachibana S. Biodegradation of aliphatic hydrocarbon in three types of crude oil by Fusarium sp. F 092 under stress with artificial sea water. J Environ Sci Technol 2011; 5(1): 64-73.
[http://dx.doi.org/10.3923/jest.2012.64.73]

[105]   Maliji D, Olama Z, Holail H. Environmental studies on the microbial degradation of oil hydrocarbons and its application in Lebanese oil polluted coastal and marine ecosystem. Int J Curr Microbiol Appl Sci 2013; 2(6): 1-8.

[106]   Burghal AA, Abu-Mejdad NM, Al-Tamimi WH. Mycodegradation of crude oil by fungal species

isolated from petroleum contaminated soil. Int J Innov Res Sci Eng Technol 2016; 5(2): 1517-24.

[107] El-Borai A, Eltayeb K, Mostafa A, El-Assar S. Biodegradation of industrial oil-polluted wastewater in egypt by bacterial consortium immobilized in different types of carriers. Pol J Environ Stud 2016; 25(5): 1901-9.
[http://dx.doi.org/10.15244/pjoes/62301]

[108] Sukumar S, Nirmala P. Screening of diesel oil degrading bacteria from petroleum hydrocarbon contaminated soil. Int J Adv Res Biol Sci 2016; 3(8): 18-22.

[109] Fagbemi OK, Sanusi AI. Effectiveness of augmented consortia of *Bacillus coagulans, Citrobacter koseri and Serratia ficaria* in the degradation of diesel polluted soil supplemented with pig dung. Afr J Microbiol Res 2016; 10(39): 1637-44.
[http://dx.doi.org/10.5897/AJMR2016.8249]

[110] Pereira P, Enguita FJ, Ferreira J, Leitão AL. DNA damage induced by hydroquinone can be prevented by fungal detoxification. Toxicol Rep 2014; 1: 1096-105.
[http://dx.doi.org/10.1016/j.toxrep.2014.10.024] [PMID: 28962321]

[111] Abdulsalam S, Adefila SS, Bugaje IM, Ibrahim S. Bioremediation of soil contaminated with used motor oil in a closed system. J Bioremediat Biodegrad 2012; 3(12): 1-7.

[112] Safiyanu I, Isah AA, Abubakar US, Rita Singh M. Review on comparative study on bioremediation for oil spills using microbes. Res J Pharm Biol Chem Sci 2015; 6: 783-90.

[113] Safiyanu I, Sani I, Rita SM. Review on bioremediation of oil spills using microbial approach. Int J Eng Sci Res 2015; 3(6): 41-55.

[114] Wolski EA, Barrera V, Castellari C, González JF. Biodegradation of phenol in static cultures by Penicillium chrysogenum ERK1: catalytic abilities and residual phytotoxicity. Rev Argent Microbiol 2012; 44(2): 113-21.
[PMID: 22997771]

[115] AI- Jawhari IFH. Ability of some soil fungi in biodegradation of petroleum hydrocarbon. J Appl Environ Microbiol 2014; 2(2): 46-52.

[116] Aranda E, Ullrich R, Hofrichter M. Conversion of polycyclic aromatic hydrocarbons, methyl naphthalenes and dibenzofuran by two fungal peroxygenases. Biodegradation 2010; 21(2): 267-81.
[http://dx.doi.org/10.1007/s10532-009-9299-2] [PMID: 19823936]

[117] Hesham AEL, Khan S, Tao Y, Li D, Zhang Y, Yang M. Biodegradation of high molecular weight PAHs using isolated yeast mixtures: application of meta-genomic methods for community structure analyses. Environ Sci Pollut Res Int 2012; 19(8): 3568-78.
[http://dx.doi.org/10.1007/s11356-012-0919-8] [PMID: 22535224]

[118] Sivakumar G, Xu J, Thompson RW, Yang Y, Randol-Smith P, Weathers PJ. Integrated green algal technology for bioremediation and biofuel. Bioresour Technol 2012; 107: 1-9.
[http://dx.doi.org/10.1016/j.biortech.2011.12.091] [PMID: 22230775]

[119] Fang, C. R., Yao, J., Zheng, Y. G., Jiang, C. J., Hu, L. F., Wu, Y. Y., Shen, D. S. Dibutyl phthalate degradation by Enterobacter sp. T5 isolated from municipal solid waste in landfill bioreactor. Int Biodeter Biodegrad 2010, 64(6), 442-446.
[http://dx.doi.org/10.1016/j.ibiod.2010.04.010]

[120] Yadav M, Singh SK, Sharma JK, Yadav KDS. Oxidation of polyaromatic hydrocarbons in systems containing water miscible organic solvents by the lignin peroxidase of *Gleophyllum striatum* MTCC-1117. Environ Technol 2011; 32(11): 1287-94.
[http://dx.doi.org/10.1080/09593330.2010.535177] [PMID: 21970171]

[121] Senan RC, Abraham TE. Bioremediation of textile azo dyes by aerobic bacterial consortium. Biodegradation 2004; 15(4): 275-80.
[http://dx.doi.org/10.1023/B:BIOD.0000043000.18427.0a] [PMID: 15473556]

[122] Levine WG. Metabolism of azo dyes: implication for detoxication and activation. Drug Metab Rev 1991; 23(3-4): 253-309.
[http://dx.doi.org/10.3109/03602539109029761] [PMID: 1935573]

[123] Banat IM, Nigam P, Singh D, Marchant R. Microbial decolorization of textile-dyecontaining effluents: A review. Bioresour Technol 1996; 58(3): 217-27.
[http://dx.doi.org/10.1016/S0960-8524(96)00113-7]

[124] Chung KT, Stevens SE Jr. Degradation azo dyes by environmental microorganisms and helminths. Environ Toxicol Chem Int J 1993; 12(11): 2121-32.

[125] Kulla HG, Klausener F, Meyer U, Lüdeke B, Leisinger T. Interference of aromatic sulfo groups in the microbial degradation of the azo dyes Orange I and Orange II. Arch Microbiol 1983; 135: 1-7.
[http://dx.doi.org/10.1007/BF00419473]

[126] Cao W, Mahadevan B, Crawford DL, Crawford RL. Characterization of an extracellular azo dye-oxidizing peroxidase from *Flavobacterium* sp. ATCC 39723. Enzyme Microb Technol 1993; 15(10): 810-7.
[http://dx.doi.org/10.1016/0141-0229(93)90091-F]

[127] Watanabe K, Baker PW. Environmentally relevant microorganisms. J Biosci Bioeng 2000; 89(1): 1-11.
[http://dx.doi.org/10.1016/S1389-1723(00)88043-3] [PMID: 16232691]

[128] Uqab B, Mudasir S, Nazir R. Review on bioremediation of pesticides. J Bioremediat Biodegrad 2016; 7(343): 2.

[129] Mohamed AT, El Hussein AA, El Siddig MA, Osman AG. Degradation of oxyfluorfen herbicide by soil microorganisms biodegradation of herbicides. Biotechnology 2011; 10(3): 274-9.
[http://dx.doi.org/10.3923/biotech.2011.274.279]

[130] Niti C, Sunita S, Kamlesh K, Rakesh K. Bioremediation: An emerging technology for remediation of pesticides. Res J Chem Environ 2013; 17: 4.

[131] Pérez M, Rueda OD, Bangeppagari M, *et al.* Evaluation of various pesticides-degrading pure bacterial cultures isolated from pesticide-contaminated soils in Ecuador. Afr J Biotechnol 2016; 15(40): 2224-33.
[http://dx.doi.org/10.5897/AJB2016.15418]

[132] Urgun-Demirtas M, Stark B, Pagilla K. Use of genetically engineered microorganisms (GEMs) for the bioremediation of contaminants. Crit Rev Biotechnol 2006; 26(3): 145-64.
[http://dx.doi.org/10.1080/07388550600842794] [PMID: 16923532]

[133] Hrywna Y, Tsoi TV, Maltseva OV, Quensen JF III, Tiedje JM. Construction and characterization of two recombinant bacteria that grow on ortho- and para-substituted chlorobiphenyls. Appl Environ Microbiol 1999; 65(5): 2163-9.
[http://dx.doi.org/10.1128/AEM.65.5.2163-2169.1999] [PMID: 10224015]

[134] Monti MR, Smania AM, Fabro G, Alvarez ME, Argaraña CE. Engineering Pseudomonas fluorescens for biodegradation of 2,4-dinitrotoluene. Appl Environ Microbiol 2005; 71(12): 8864-72.
[http://dx.doi.org/10.1128/AEM.71.12.8864-8872.2005] [PMID: 16332883]

[135] Wu J, Jiang C, Wang B, Ma Y, Liu Z, Liu S. Novel partial reductive pathway for 4-chloronitrobenzene and nitrobenzene degradation in *Comamonas* sp. strain CNB-1. Appl Environ Microbiol 2006; 72(3): 1759-65.
[http://dx.doi.org/10.1128/AEM.72.3.1759-1765.2006] [PMID: 16517619]

[136] Lange CC, Wackett LP, Minton KW, Daly MJ. Engineering a recombinant *Deinococcus radiodurans* for organopollutant degradation in radioactive mixed waste environments. Nat Biotechnol 1998; 16(10): 929-33.
[http://dx.doi.org/10.1038/nbt1098-929] [PMID: 9788348]

[137] Renninger N, Knopp R, Nitsche H, Clark DS, Keasling JD. Uranyl precipitation by *Pseudomonas aeruginosa via* controlled polyphosphate metabolism. Appl Environ Microbiol 2004; 70(12): 7404-12.
[http://dx.doi.org/10.1128/AEM.70.12.7404-7412.2004] [PMID: 15574942]

[138] Crameri A, Dawes G, Rodriguez E Jr, Silver S, Stemmer WPC. Molecular evolution of an arsenate detoxification pathway by DNA shuffling. Nat Biotechnol 1997; 15(5): 436-8.
[http://dx.doi.org/10.1038/nbt0597-436] [PMID: 9131621]

[139] Agarwal N, Solanki VS, Gacem A, *et al.* Bacterial laccases as biocatalysts for the remediation of environmental toxic pollutants: A green and eco-friendly approach- A review. Water 2022; 14(24): 4068.
[http://dx.doi.org/10.3390/w14244068]

[140] Gallardo ME, Ferrández A, De Lorenzo V, García JL, Díaz E. Designing recombinant Pseudomonas strains to enhance biodesulfurization. J Bacteriol 1997; 179(22): 7156-60.
[http://dx.doi.org/10.1128/jb.179.22.7156-7160.1997] [PMID: 9371464]

CHAPTER 5

Carbon Dots and their Environmental Applications

N. Maurya[1], S. Mishra[2,*] and M.K. Gupta[3]

[1] *Department of Chemistry, Kamla Rai College, Gopalganj, Jai Prakash University, Chapra, India*

[2] *Department of Chemistry, Jai Prakash University, Chapra, India*

[3] *Department of Chemistry, H. R. College, Jai Prakash University, Amnour, Chapra, India*

Abstract: CDs (Carbon dots) are a new group of zero-dimensional luminescent nanomaterials. They have drawn a lot of attention due to their excellent properties, for example, easy preparation, strong optical properties, low toxicity, significant biocompatibility, low cost, facile functionalization, tunable porous structures and high specific surface area. CDs have found many applications in the fields of bioimaging, sensing, catalysis, optoelectronics and energy conversion. Recently, CDs have demonstrated promising applications in the control of environmental pollution and remediation. CDs have been applied for environmental pollutants sensing, adsorption of contaminants, membrane-based separation, photocatalytic degradation of pollutants, and antimicrobial coatings for protection. In this chapter, we have discussed the classification of CDs, synthesis, properties and applications of CDs in environmental pollution control and environmental protection measures.

Keywords: Antimicrobial, Air pollution, Contaminant adsorption, Carbon dots, Environment, Hydrothermal, Membrane separation, Pollutants sensing, Photocatalytic, Water pollution.

INTRODUCTION

Global development and urbanization are leading to an alarming increase in environmental issues. Concerns are being raised worldwide, particularly, about the harmful and toxic pollutants resulting as a consequence. Over the past few decades, several efforts have been made worldwide to reduce environmental pollution and safeguard human health [1]. Among them, the use of advanced materials is one of the best ways to safeguard the environment. Nanomaterials, particularly those with carbon, metal, and metal-organic framework bases, have developed rapidly and has been successfully utilized to treat and protect the environment [2 - 4].

---

* **Corresponding author S. Mishra:** Department of Chemistry, Jai Prakash University, Chapra, India; E-mail: shachicdri2013@gmail.com

Numerous environmental applications, including membrane-based separation, energy conversion and storage, pollutant sensing, adsorptive removal of contaminants, and catalytic degradation, have made extensive use of carbon-based nanomaterials, such as pristine graphene, graphene oxide, carbon nanotubes (CNTs), CDs, *etc.* [5 - 9]. Among carbon-based nanomaterials, CDs have attracted more attention due to their unique qualities, such as excellent and tunable photoluminescence (PL) efficiency, easy preparation, low toxicity, notable biocompatibility, tiny cost, facile functionalization, substantial pore sizes, and high specific surface areas [8, 9]. The typical CDs are defined as zero-dimensional fluorescent carbon nanoparticles (NPs) with a size below 20 nm, consisting of a graphitic and turbostratic sp^2/sp^3 carbon core with special functional groups (*e.g.*, -COOH, -OH, and -NH$_2$) on the surface [10, 11]. Unique properties of CDs have also been utilized for optoelectronic devices, solar cells, biomedical applications, and environmental applications such as adsorption and elimination of contaminants, photocatalytic degradation of pollutants, creating water treatment membranes, and serving as antimicrobial materials [8 - 17].

Major developments in CD synthesis, their unique properties and applications in several fields, such as sensing, bioanalysis, bioimaging, energy conversion, environmental control and remediation, *etc.*, have been reviewed and reported from time to time by researchers [8 - 17]. Recently, Long *et al.* reported the recent developments and different environmental applications of carbon dots [8]. Hebbar et. al. summarized various methods of CD synthesis, characterization techniques and their environmental applications [9]. In this outlook, firstly, we will discuss the classification of CDs, recent developments in synthetic measures of CDs apropos to the environment and their properties. Then, the current progress in the applications of CDs for the adsorption of contaminants, sensing of environmental pollutants, membrane-based separation, photocatalytic degradation of pollutants, and antimicrobial coatings for protection have been discussed in detail.

CLASSIFICATION OF CDS

CDs are a generic group of carbon NPs made up of distinct and almost spherical NPs. They were initially discovered through single-walled carbon nanotube purification in 2004 [18]. Initially, these carbon NPs were termed "carbon quantum dots (CQDs)" by Sun *et al.*, who anticipated a method to produce CDs by chemical modification and surface passivation to boost fluorescence emission [19]. In 2016, Cayuela *et al.* recommended CD sub-categories, viz. CQDs, graphene quantum dots (GQDs), and carbon nanodots (CNDs) based on specific carbon core structures, surface groups, and properties [11]. GQDs are π-conjugated single sheets or multiple layers of small graphene fragments with supremacy of sp^2 carbon and chemical groups connected within the interlayer defect or on the edge/surface. They are anisotropic with typical dimensions of less than 20 nm in width, exhibit a quantum confinement effect, and are mostly synthesized by the top-down approach. CQDs are usually quasi-spherical, with a mixture of sp^3 and sp^2 hybridized carbon possessing multi-layered graphitic structures and chemical groups lying on the surface. They also exhibit a quantum confinement effect. CNDs are amorphous and have high carbonization and polymer features without obvious crystallinity and quantum confinement [20, 21]. Some of the researchers have classified CDs into four categories, namely CPDs, CQDs, GQDs and CNDs (Fig. **1**) [22, 23]. CPDs have polymer frame-carbon cluster hybrid structures resulting from non-conjugated linear polymers or molecules. However, few studies have pointed out that polymer dots are completely different entities based on their structural features and methods of preparation [24, 25].

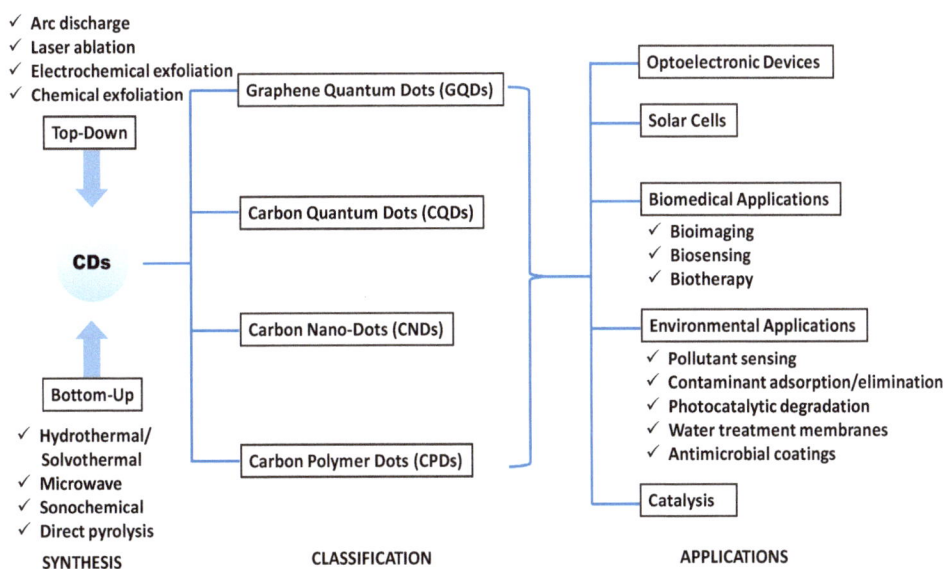

Fig. (1). Synthesis, classification and applications of carbon dots.

SYNTHESIS OF CDS

Xu and colleagues serendipitously discovered CDs through the purification of single-walled carbon nanotubes (SWCNTs) from arc-discharged soot through gel electrophoresis in 2004 [18]. Since then, many synthetic techniques have been developed to synthesize CDs. Broadly, there are two types of synthetic approaches: "top-down" and "bottom-up" techniques, as mentioned in Fig. (**1**).

Top-down Synthesis

In this approach, carbonaceous materials, for example, carbon powders, carbon fibers, graphene and even graphite rods, are cut or broken down using laser ablation, electrochemical oxidation, chemical oxidation and ultrasonic techniques. Sun and colleagues pioneered the CD synthesis by laser ablation technique with argon as a carrier gas in the occurrence of water vapor. CDs obtained upon acid oxidative treatment followed by surface passivation showed bright fluorescence [19]. Nguyen *et al.* obtained CDs from graphite powders in polyethylene glycol (PEG200N) solution *via* femtosecond laser ablation [26]. The first electrochemical technique to create CQDs was described by Zhou's group as growing multi-walled carbon nanotubes on a carbon paper, followed by its insertion into an electrochemical cell containing degassed ACN with supporting electrolyte (0.1 M tetrabutylammonium perchlorate) [27]. Ming and colleagues used water as an electrolyte and readily available low-cost graphite as a carbon source to synthesize CDs of high purity [28]. Recently, Li *et al.* reported quick and large-scale production of CDs by NaCl-assisted electrochemical exfoliation of graphite rods [29]. However, the top-down method often employs a non-selective cutting procedure, which makes it challenging to obtain fine control over the dimensions and characteristics of carbon-based dot products. Also, it involves costly materials, harsh reaction conditions and time-consuming reactions.

Bottom-up Synthesis

This synthetic approach involves step-by-step carbonization of miniature molecule precursors by microwave-assisted, pyrolysis, hydrothermal processes or sonochemical synthesis. The microwave and hydrothermal methods are the most widely used synthetic methods of CDs for environmental applications. They utilize highly abundant and economical precursors, straightforward synthesis, eco-friendly procedures, and are scalable. Hydrothermal synthesis is often referred to as "green synthesis" [30]. Organic acids, juice or waste peels, carbohydrates as well as plastic wastes have been used as carbon sources [31, 32]. Typically, a hydrothermal autoclave is needed for the reaction at high temperatures and high pressures. In microwave-assisted synthesis, electromagnetic waves between 300 MHz and 300 GHz frequencies are irradiated, which disrupts the chemical bonds in raw materials, leading to the production of CDs [33]. Recently, Singh et. al. reported blue fluorescent CDs from human hair using the hydrothermal and microwave methods. The size and quantum yield of CDs obtained *via* the hydrothermal approach was 11 nm and 38% as compared to 78 nm and 17% from the microwave method. Overall, it has been observed that even though microwave synthesis can synthesize CDs within minutes with a scale-up potential, low quantum yields are obtained [34].

PROPERTIES OF CDS

Due to a wide range of raw materials and synthetic techniques employed for the development of CDs, a variety of physical and chemical characteristics are displayed by them. The optical properties of carbon dots are one of the most interesting features. CDs typically exhibit broad absorption spectra in the range of 230 - 320 nm (UV-region) with the extension of the tail into the visible range, stemming from a variety of n– π* (C=O) and π– π* (C=C) transitions in the core and the particles lying on th surface. The size, shape, composition, and surface functional groups of π-domains have the greatest impact on the absorption properties of CDs [35]. The fascinating photoluminescent (PL) emission characteristics of CDs have garnered a lot of interest. The PL mechanism of CDs is a combination of different mechanisms and is influenced by several variables, including surface state, quantum size effect, molecular and carbon-core states, surface passivation/functionalization, and conjugation (Fig. **2**) [36 - 39]. Most of the reported CDs exhibit excitation-dependent fluorescence emission characteristics due to the polydispersity in particle size and have shown red-shift with the increase in excitation wavelength [40]. However, some CDs exhibit excitation-independent emission spectrums that arise mainly on uniform surface states. For instance, Liu *et al.* reported developing new CDs with potassium persulfate and acrylamide as a precursor. Interestingly, the fluorescence emission peak at 434 nm barely altered when the excitation wavelength was modified within the range of 280-390 nm, demonstrating the independence of CDs from excitation [41]. Rangel *et al.* reported CDs from canola oil as a carbon source. The prepared CDs showed excitation wavelength-dependent and independent emission bands in UV (327 nm) and the visible region (633 nm), respectively [42]. Baragau *et al.* synthesized blue-luminescent N-doped carbon quantum dots (NCQDs) from citric acid in the presence of ammonia *via* a Continuous Hydrothermal Flow Synthesis (CHFS). It exhibited excitation-independent optical properties and was used for sensing Cr(VI) in aqueous solutions [43].

ENVIRONMENTAL AND OTHER DOMAINS OF APPLICATIONS OF CDS

Sensing of Environmental Pollutants

Due to increasing commercialization, population, ecological degradation, and dangerous domestic/industrial unregulated discharges, toxic emergent effluents in the form of heavy metals, herbicide and pesticide residues in farming, unabsorbed antibiotics and other toxic compounds (pyridines, hydrogen peroxide or hydrazine's, *etc.*) are a serious worry that reduces the quality of water [44 - 48]. Fluorescence detection technique has proved to be advantageous over standard pollution analytical approaches due to its speed, simplicity, and efficiency. CD-based sensors have emerged as one of the most promising materials for the recognition of environmental pollutants. Mainly, two methodologies are used for the development of CD-based fluorescent sensors for environmental contaminants sensing. Firstly, direct contact between analytes and CDs causes a change in the CD fluorescence signal. Secondly, the functionalization of CDs with recognition molecules is also utilized to detect the target [8]. Applications of the developed sensors are based on the principle of interactions between CDs and environmental pollutants, which lead to either quenching of fluorescence or enhancement of fluorescence. Most of the reported CD-based sensors work on the principle of quenching upon interaction with the analyte. The dynamic quenching effect, Foster Resonance Energy Transfer (FRET), Inner Filter Effect (IFE), Photoinduced Electron Transfer (PET), Surface Energy Transfer (SET) and Static Quenching Effect (SQE) are the basic quenching-based mechanisms of fluorescent CDs in environmental monitoring [8, 9].

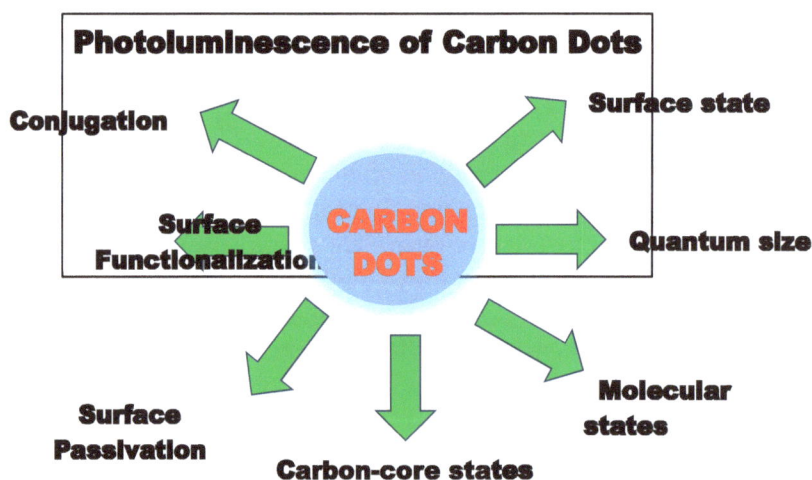

Fig. (2). Photoluminescence mechanism of carbon dots.

CDs for Heavy Metal ions Sensing in Water

Heavy metal pollution can have a multitude of harmful consequences on human health as well as on the environment, even at very low concentrations [49]. CDs have become one of the major promising materials for the sensing of heavy metal ions and have demonstrated the ability to detect metal ions such as Mercury (II), Iron (III), Copper (II), Arsenic (III), Chromium (VI), Lead (II), Cobalt (II), *etc.* in range of micro to nanomolar concentrations, frequently accompanied by dopants such as nitrogen and sulfur, or as composites to facilitate easy recovery and reuse

[50, 51]. For instance, Zhang *et al.* reported the use of CDs generated from hydrothermal treatment of citric acid (CA) and N-aminoethylpiperazine (AEP) to detect hazardous Hg^{2+} in real water samples with a limit of detection (LOD) of 0.09 μM [52]. Yoon and group produced carbon dots from mushrooms (*Pleurotus* spp.) and MCDs and utilized them for label-free sensing of Hg^{2+} ions (limit of detection: 4.13 nM). Further, they attached dihydrolipoicacid acid (DHLA) to the surface of MCDs, which led to a two-fold increase in Hg^{2+} ion sensing, with a limit of detection as low as 17.4 pM. This probe was successfully applied to sense Hg^{2+} in real environmental samples (river and tap water) with a limit of detection of 274 pM and 192 pM, respectively [53]. Using the fluorescent quenching effect, Yan *et al.* developed N-CDs from crown daisy leaves that could selectively and sensitively sense Cu^{2+} with a 1.0 nM detection limit due to the complexation interface among Cu^{2+} and carboxyl and amino groups of the N-CDs [54]. Bandi and group made CDs made from Lantana camara berries and used them to identify Pb^{2+} with LODs in the nanomolar range in actual water (tap, pond, and wastewater) and human samples (serum, urine) [55]. Renuka's team used CDs made from table sugar in turbidimetric analysis to measure the presence of Pb^{2+} ions in water samples. Table sugar dots assemble when lead ions are present, allowing turbidity measurements to determine the amount of Pb^{2+} present [56]. Kolekar's group showed that adding Fe^{3+} to CDs made of sucrose not only produces fluorescence quenching but also boosts the absorption signal, allowing for measurement using straightforward UV-VIS spectroscopy [57]. Senol's group prepared novel fluorescent CDs from hydrothermal treatment of cranberry and utilized them for the quantitative determination of Fe^{3+} ion (LOD 0.75 nM) and hypochlorite (ClO^-) [REMOVED HYPERLINK FIELD] ion (LOD 1.6 nM) in tap water samples and real natural spring [58]. Table **1** includes further examples of CDs synthesized from various sources, as well as the associated features and metals detected.

Table 1. CDs synthesized from various sources, associated features and metals detected.

Carbon Dot Source	Method of Synthesis	Quantum Yield (%)	Metal Detected	Detection Limit	Refs.
N-aminoethyl-piperazine (AEP) and citric acid (CA)	Hydrothermal	56	Hg^{2+}	0.09 μM	[52]
common edible mushrooms (*Pleurotus spp.*)	Thermal treatment	25	Hg^{2+}	4.13 nM	[53]
Sugarcana baggage	Hydrothermal	21	Hg^{2+}	2 nM	[59]
p-phenylenediamine	Hydrothermal	21.4	Hg^{2+}	0.14 nM	[60]
Sucrose	Acidic Carbonization	5.77	Fe^{3+}	0.56μM	[57]

Carbon Dot Source	Method of Synthesis	Quantum Yield (%)	Metal Detected	Detection Limit	Refs.
Cranberry	Hydrothermal		Fe^{3+}	0.75	[58]
Dwarf banana peel + aq. ammonia	Hydrothermal	23	Fe^{3+}	0.66μM	[61]
Ripe banana Peels + ethylene diamine + L-cysteine	Hydrothermal	27	Fe^{3+}	121 pM	[62]
Lantanacamara berries	Hydrothermal	33.15	Pb^{2+}	9.64 nM	[55]
Table sugar	Microwave assisted	2.5	Pb^{2+}	14 ppb	[56]
Thioctic acid and Triethylene tetramine	Hydrothermal	35.2	Pb^{2+}	10.9 nM	[63]
Ocimum sanctum	Hydrothermal	9.3	Pb^{2+}	0.59 nM	[64]
Crown Daisy Leaves	Hydrothermal	-	Cu^{2+}	1.0 nM	[54]
Aminophenylboronic acid	Combustion	1.6	Cu^{2+}	0.3 μM	[65]
Adenosine	Hydrothermal	-	Cu^{2+}	23 nM	[66]
Citric acid and Glutamic acid	Pyrolysis	54	Cr^{6+}	5 nM	[67, 68]
Tulsi leaves	Hydrothermal	3.06	Cr^{6+}	229 nM	[69]
Citric acid & Cysteamine	Microwave	54	As^{3+}	17.2 μM	[70]
Prickly pear cactus	Hydrothermal	12.7	As^{3+}	2.3 nM	[71]

CDs for Detection of Emerging Contaminants

Besides heavy metal ions, other harmful environmental wastes from a diverse range of products, including surfactants, pharmaceuticals, personal care products, cosmetics, disinfectants and odor-enhancing agents pose serious hazards to human health and the environment [72]. They are often termed as 'emerging contaminants'. According to a review, there are around 166 chemicals classified as 'emerging contaminants' in Indian waters [73]. CD-based sensors have been successfully explored for recognition and quantification of these emerging contaminants, namely, herbicide and pesticide residues in farming, unabsorbed antibiotics and other poisonous compounds such as hydrogen peroxide, hydrazines and pyridines [44, 74 - 78]. Chang *et al.* used CDs produced from sucrose by acid carbonization to detect paraoxon-ethyl pesticide, which reduced the PL intensity of the sucrose dots in a concentration-dependent mode up to 5.80 mM, with 0.22 M detection limit [74]. Carbon dots synthesized from Brassica oleracea by hydrothermal synthesis displayed an impressive quantum yield of 43% and could sense amicarbazone, diazinon, and glyphosate with LODs of 0.5, 0.25 and 2 ng/mL, respectively [79]. Carbon dots from maize cob with a quantum yield of 53.46% were capable of detecting DNA and paracetamol (linear ranges:

1-10 ng/mL and 0.1-0.3 mg/mL, respectively) and could be used to create white light-emitting thin sheets [80]. Guo *et al.* used carbon dots produced from solid pyrolysis of DL- glycine and malic acid (QY = 20.5%) and developed a fluorescence "turn-on" assay to perceive enrofloxacin [44]. The fluorescence of CDs was initially shut off *via* electron transfer using Cu^{2+} ions as the quencher and when enrofloxacin was added, the fluorescence was activated due to the dislocation of Cu^{2+} from the CDs. When evaluated on antibiotic-spiked actual water samples, the optimized sensing probe demonstrated a linear detection range from 1.0-15.0 g mL^{-1}. Tan and coworkers devised a versatile approach for creating CDs silver nanoparticle (AgNP) hybrids as colorimetric and fluorescent-based dual optical sensors for the detection of H_2O_2 and H_2O_2-producing species, for example, cholesterol and glucose in complicated mixtures. Fluorescent Ser-His dots derived from serine and histamine precursors were first synthesized and applied for sensing of Ag^+ ions, which were then irradiated with UV light to form AgNPs on the carbon dots, resulting in fluorescence quenching *via* the mechanism of inner filter effects [77]. Similarly, CDs synthesized from dextrose, sewage sludge, water hyacinth, and human hair have been successfully used to detect explosives such as 2, 4, 6-trinitrophenol and TNT, as well as chloroform and herbicides in water samples [34, 81 - 83].

Adsorption of Contaminants

Adsorption-based techniques have long been used for water treatment and environmental pollutant removal [84]. Traditionally, activated carbon was the most commonly utilized adsorbent. However, the high expenditure of producing activated carbon and the difficulty in reproducibility limited its widespread utilization [85]. CDs have been applied as a promising alternative to typical adsorbent materials. Because of their compact size and huge specific surface area, CDs can provide a huge number of adsorption sites on the surface. Facile synthesis of CDs from abundant and inexpensive basic ingredients and the occurrence of a huge number of hydrophilic functional groups endow them with significant benefits for absorbing pollutants [10, 16]. Sabet and colleagues demonstrated the use of nitrogen-doped CDs for the adsorption of Cd^{2+} and Pb^{2+} ions from wastewater. Large specific surface area promotes heavy metal ion adsorption from water [86]. Wang and colleagues reported carbon dots modified mesoporous organosilica as an adsorbent for the exclusion of inorganic metal ions like Cu^{2+}, Hg^{2+} and Pb^{2+} [87]. Carbon dots modified mesoporous organosilica showed better adsorption capacity for different metal ions compared with pristine mesoporous silica due to a huge number of functional groups and a variety of polar moieties on the surface of CDs, for example, amine and carboxyl groups.

Photocatalytic Degradation of Dyes and Organic Pollutants

Photocatalytic degradation of dyes and organic pollutants by CDs has gained significant gained interest in the current years. CDs are appropriate candidates for photocatalytic applications due to the abundance of surface functional groups, uniform dispersion, intense photoluminescence and optical absorption in UV-visible regions [88 - 90]. They can serve as an independent photocatlytic material, an electron donor or acceptor in photocatalyst systems or a photosensitizer. Omer *et al.* synthesized nitrogen-doped carbon dots, loaded them onto ZnO nanoparticles and applied them for the photocatalytic degradation of methylene blue (MB) dye. In the UV region (365 nm), the nanocomposites of ZnO-CDs displayed 95% (for 100 minutes) enhanced photodegradation efficiency as compared to pure ZnO [91]. Further, many CD-based composites have been created to break down chemicals and pesticides found in wastewater. Rong *et al.* synthesized $CD@Fe_3O_4$ particles by incorporating Fe_3O_4 into hydrothermally prepared CDs from ethylenediamine and citric acid. It was found that these CD particles shortened the half-life of the fungicide hexaconazole by 96% in daylight [92]. Endocrine-disrupting compounds such as 2, 4-dichlorophenol were destroyed after 90 minutes of light coverage *via* a metal-free composite of CDs derived from Arabian dates and graphitic carbon nitride (g-C_3N_4) [93]. CD/GCN gel-like composites have also been shown to be effective at degrading a variety of chemicals and dyes, including phenol [94]. Surfactants were also remedied using a combination of three materials: CDs produced from hyperbranched polyurethane polymers, Colocasia esculenta leaf extracts and ZnO [95]. Free radical cascades induced by CD exposure to sunshine were capable of degrading dodecyl-benzene sulfonate (96% in 110 minutes) and detergent (94% in 150 minutes) [95].

Water Treatment by Membrane-based Separation

In recent years, a lot of research has been conducted to assess the applicability of CDs in filtration membranes, for example, ultrafiltration (UF), reverse osmosis (RO), nanofiltration (NF), pressure retarded osmosis (PRO), forward osmosis (FO), organic solvent nanofiltration (OSN) and membrane distillation (MD), as a result of their good biocompatibility, ultra-small sizes, non-toxic, steady photoluminescence, tunable hydrophilicity, abundant surface functional groups, antifouling characteristics, exceptional light capture capability, and other excellent physicochemical properties [96]. For membrane separation, CDs are incorporated into either thin-film nanocomposite (TFN) membranes or polymer composite membranes. In general, the performance of membranes is considerably improved when they are integrated with CDs (CDs@membranes). Song's group reported a chlorine-resistant TFN membrane adapted with CDs [97]. The component of the -COOH groups on CDs initially reacted with the -NH$_2$ groups of m-phenylenediamine (MPD), and then the remaining -COOH groups established covalent connections with the terminal -COOH groups of trimesoyl chloride (TMC) in condensation reactions during interfacial polymerization. At 16 bar pressure, these TFN membranes demonstrated exceptionally stable performance with a 98.8% NaCl rejection rate. More significantly, the TFN membrane treated with CDs demonstrated long-term stability over 120 hours of reverse osmosis (RO) testing and a 51.8% increase in permeate flux compared to the unmodified [97]. Similarly, interfacial polymerization of m-phenylenediamine (MPD) and trimethyl chloride (TMC) monomers produced the CDs/polyamide thin-film nanocomposite (TFN) RO membrane on the microporous polysulfone substrate [98]. With an initial NaCl concentration of 2000 mg/L, the reports showed a filtration efficiency of up to 99.0%. Furthermore, in the presence of CDs, the water flux and stability of RO membranes dramatically boosted. Yang and colleagues described innovative nanofiltration (NF) membranes using hydrophilic CDs as the interlayer [99]. The -COOH groups in CDs react with the polyethyleneimine (PEI), resulting in a decrease in the degree of cross-linking of the NF membrane's interface polymerized layer. Furthermore, hydrophilic CDs were capable of forming rapid ion and water transport channels in NF membranes. The water permeability of membranes increased dramatically from 3.6 to 9.7 LMH bar^{-1} [99].

Antimicrobial Agents

Microbes have been reported to be killed or inhibited by CDs. The basic mechanism is that reactive oxygen species are produced under irradiation-excited circumstances. They then cling to and degrade cell walls and cell membranes, causing damage to DNA and altering the structure of genetically organized cells [100 - 102]. Tan and colleagues created the first biomimetic nanodots from PEI and serine precursors using a simple one-step hydrothermal method. The bioinspired antimicrobial nanodots (BAM-dots) were found to be ultrasmall (3.8 ± 0.7 nm) and featured unique zwitterionic (0.52± 0.09 mV) and amphiphilic properties similar to natural antimicrobial peptides. The BAM dots were extremely effective adjacent to a broad spectrum of bacterial strains (including Gram+tive *Staphylococcus aureus*, Gram-tive E. coli, and multidrug resistance bacteria *Pseudomonas aeruginosa*) due to the effective bacteria membrane permeabilization mechanism, resulting in a quick bactericidal effect (eradication > 97%) without drug resistance development in superbugs [103]. In certain cases, antibacterial chemicals derived from herb-based precursors were maintained on CDs, boosting their potential to kill fungi and bacteria. A. Gedanken *et al.* used hydrothermal treatment to synthesize carbon dots from turmeric leaves (*Curcuma longa*) [104]. After 8 hours of exposure, the inclusive bactericidal effectiveness of

CDs was 0.25 mg/mL for S. aureus and E.coli, while for S. epidermidis and K. pneumoniae, the CDs at 0.5 mg/mL had a good antibacterial impact within 8 hours and absolute eradication after 24 hours. The release of reactive oxygen species led to the demise of bacterial cells [104].

CONCLUSION AND FUTURE RECOMMENDATIONS

In conclusion, this chapter discusses the classification, synthesis, properties and potential of CDs for different domains of applications, with special emphasis on environmental applications. Recent studies on CDs' role in sensing heavy metals in water, detection and adsorption of emerging contaminants, and photocatalytic degradation of organic pollutants and dyes have been discussed. Further, water treatments by membrane-based separation using CDs and the application of CDs as antimicrobial agents have been highlighted. Environmental applications of CDs are still in their infancy, and there are still many problems and challenges that need to be addressed, such as large-scale synthesis of CDs and better integration with technology to enhance their environmental impact. Thus, CDs are estimated to bring even more intriguing potential for environmental applications. Their future impact on environmental science and engineering will grow as they develop and integrate with other domains of nanotechnology.

REFERENCES

[1]   Cvitanovic C, Hobday AJ. Building optimism at the environmental science-policy-practice interface through the study of bright spots. Nat Commun 2018; 9(1): 3466.
       [http://dx.doi.org/10.1038/s41467-018-05977-w] [PMID: 30154434]

[2]   Zhang Q, Uchaker E, Candelaria SL, Cao G. Nanomaterials for energy conversion and storage. Chem Soc Rev 2013; 42(7): 3127-71.
       [http://dx.doi.org/10.1039/c3cs00009e] [PMID: 23455759]

[3]   Sharma S, Dutta V, Singh P, *et al.* Carbon quantum dot supported semiconductor photocatalysts for efficient degradation of organic pollutants in water: A review. J Clean Prod 2019; 228: 755-69.
       [http://dx.doi.org/10.1016/j.jclepro.2019.04.292]

[4]   Yang W, Ratinac KR, Ringer SP, Thordarson P, Gooding JJ, Braet F. Carbon nanomaterials in biosensors: should you use nanotubes or graphene? Angew Chem Int Ed 2010; 49(12): 2114-38.
       [http://dx.doi.org/10.1002/anie.200903463] [PMID: 20187048]

[5]   Li F, Jiang X, Zhao J, Zhang S. Graphene oxide: A promising nanomaterial for energy and environmental applications. Nano Energy 2015; 16: 488-515.
       [http://dx.doi.org/10.1016/j.nanoen.2015.07.014]

[6]   Song B, Xu P, Zeng G, *et al.* Carbon nanotube-based environmental technologies: the adopted properties, primary mechanisms, and challenges. Rev Environ Sci Biotechnol 2018; 17(3): 571-90.
       [http://dx.doi.org/10.1007/s11157-018-9468-z]

[7]   Zhao DL, Chung TS. Applications of carbon quantum dots (CQDs) in membrane technologies: A review. Water Res 2018; 147: 43-9.
       [http://dx.doi.org/10.1016/j.watres.2018.09.040] [PMID: 30296608]

[8]   Long C, Jiang Z, Shangguan J, Qing T, Zhang P, Feng B. Applications of carbon dots in environmental pollution control: A review. Chem Eng J 2021; 406: 126848.

[http://dx.doi.org/10.1016/j.cej.2020.126848]

[9]   Hebbar A, Selvaraj R, Vinayagam R, *et al.* A critical review on the environmental applications of carbon dots. Chemosphere 2023; 313: 137308.
[http://dx.doi.org/10.1016/j.chemosphere.2022.137308] [PMID: 36410502]

[10]   Lim SY, Shen W, Gao Z. Carbon quantum dots and their applications. Chem Soc Rev 2015; 44(1): 362-81.
[http://dx.doi.org/10.1039/C4CS00269E] [PMID: 25316556]

[11]   Cayuela A, Soriano ML, Carrillo-Carrión C, Valcárcel M. Semiconductor and carbon-based fluorescent nanodots: the need for consistency. Chem Commun 2016; 52(7): 1311-26.
[http://dx.doi.org/10.1039/C5CC07754K] [PMID: 26671042]

[12]   Haque E, Kim J, Malgras V, *et al.* Recent advances in graphene quantum dots: synthesis, properties, and applications. Small Methods 2018; 2(10): 1800050.
[http://dx.doi.org/10.1002/smtd.201800050]

[13]   Yan Y, Gong J, Chen J, *et al.* Recent advances on graphene quantum dots: from chemistry and physics to applications. Adv Mater 2019; 31(21): 1808283.
[http://dx.doi.org/10.1002/adma.201808283] [PMID: 30828898]

[14]   Kumar R, Kumar VB, Gedanken A. Sonochemical synthesis of carbon dots, mechanism, effect of parameters, and catalytic, energy, biomedical and tissue engineering applications. Ultrason Sonochem 2020; 64: 105009.
[http://dx.doi.org/10.1016/j.ultsonch.2020.105009] [PMID: 32106066]

[15]   Li H, Yan X, Kong D, *et al.* Recent advances in carbon dots for bioimaging applications. Nanoscale Horiz 2020; 5(2): 218-34.
[http://dx.doi.org/10.1039/C9NH00476A]

[16]   Mehta A, Mishra A, Basu S, *et al.* Band gap tuning and surface modification of carbon dots for sustainable environmental remediation and photocatalytic hydrogen production – A review. J Environ Manage 2019; 250: 109486.
[http://dx.doi.org/10.1016/j.jenvman.2019.109486] [PMID: 31518793]

[17]   Rani UA, Ng LY, Ng CY, Mahmoudi E. A review of carbon quantum dots and their applications in wastewater treatment. Adv Colloid Interface Sci 2020; 278: 102124.
[http://dx.doi.org/10.1016/j.cis.2020.102124] [PMID: 32142942]

[18]   Xu X, Ray R, Gu Y, *et al.* Electrophoretic analysis and purification of fluorescent single-walled carbon nanotube fragments. J Am Chem Soc 2004; 126(40): 12736-7.
[http://dx.doi.org/10.1021/ja040082h] [PMID: 15469243]

[19]   Sun YP, Zhou B, Lin Y, *et al.* Quantum-sized carbon dots for bright and colorful photoluminescence. J Am Chem Soc 2006; 128(24): 7756-7.
[http://dx.doi.org/10.1021/ja062677d] [PMID: 16771487]

[20]   Hutton GAM, Martindale BCM, Reisner E. Carbon dots as photosensitisers for solar-driven catalysis. Chem Soc Rev 2017; 46(20): 6111-23.
[http://dx.doi.org/10.1039/C7CS00235A] [PMID: 28664961]

[21]   Liu J, Li R, Yang B. Carbon dots: A new type of carbon-based nanomaterial with wide applications. ACS Cent Sci 2020; 6(12): 2179-95.
[http://dx.doi.org/10.1021/acscentsci.0c01306] [PMID: 33376780]

[22]   Xia C, Zhu S, Feng T, Yang M, Yang B. Evolution and synthesis of carbon dots: from carbon dots to carbonized polymer dots. Adv Sci 2019; 6(23): 1901316.
[http://dx.doi.org/10.1002/advs.201901316] [PMID: 31832313]

[23]   Wei X, Yang J, Hu L, *et al.* Recent advances in room temperature phosphorescent carbon dots: preparation, mechanism, and applications. J Mater Chem C Mater Opt Electron Devices 2021; 9(13): 4425-43.

[http://dx.doi.org/10.1039/D0TC06031C]

[24] Song Y, Zhu S, Shao J, Yang B. Polymer carbon dots—a highlight reviewing their unique structure, bright emission and probable photoluminescence mechanism. J Polym Sci A Polym Chem 2017; 55(4): 610-5.
[http://dx.doi.org/10.1002/pola.28416]

[25] Mandal S, Das P. Are carbon dots worth the tremendous attention it is getting: Challenges and opportunities. Appl Mater Today 2022; 26: 101331.
[http://dx.doi.org/10.1016/j.apmt.2021.101331]

[26] Nguyen V, Yan L, Si J, Hou X. Femtosecond laser-induced size reduction of carbon nanodots in solution: Effect of laser fluence, spot size, and irradiation time. J Appl Phys 2015; 117(8): 084304.
[http://dx.doi.org/10.1063/1.4909506]

[27] Zhou J, Booker C, Li R, *et al.* An electrochemical avenue to blue luminescent nanocrystals from multiwalled carbon nanotubes (MWCNTs). J Am Chem Soc 2007; 129(4): 744-5.
[http://dx.doi.org/10.1021/ja0669070] [PMID: 17243794]

[28] Ming H, Ma Z, Liu Y, *et al.* Large scale electrochemical synthesis of high quality carbon nanodots and their photocatalytic property. Dalton Trans 2012; 41(31): 9526-31.
[http://dx.doi.org/10.1039/c2dt30985h] [PMID: 22751568]

[29] Li X, Ge F, Li X, *et al.* Rapid and large-scale production of carbon dots by salt-assisted electrochemical exfoliation of graphite rods. J Electroanal Chem 2019; 851: 113390.
[http://dx.doi.org/10.1016/j.jelechem.2019.113390]

[30] Huston M, DeBella M, DiBella M, Gupta A. Green synthesis of nanomaterials. Nanomaterials 2021; 11(8): 2130.
[http://dx.doi.org/10.3390/nano11082130] [PMID: 34443960]

[31] Kurian M, Paul A. Recent trends in the use of green sources for carbon dot synthesis–A short review. Carbon Trends 2021; 3: 100032.
[http://dx.doi.org/10.1016/j.cartre.2021.100032]

[32] Zhang L, Yang X, Yin Z, Sun L. A review on carbon quantum dots: Synthesis, photoluminescence mechanisms and applications. Luminescence 2022; 37(10): 1612-38.
[http://dx.doi.org/10.1002/bio.4351] [PMID: 35906748]

[33] de Medeiros TV, Manioudakis J, Noun F, Macairan JR, Victoria F, Naccache R. Microwave-assisted synthesis of carbon dots and their applications. J Mater Chem C Mater Opt Electron Dev 2019; 7(24): 7175-95.
[http://dx.doi.org/10.1039/C9TC01640F]

[34] Singh A, Eftekhari E, Scott J, *et al.* Carbon dots derived from human hair for ppb level chloroform sensing in water. Sustainable Materials and Technologies 2020; 25: e00159.
[http://dx.doi.org/10.1016/j.susmat.2020.e00159]

[35] Zhu S, Song Y, Zhao X, Shao J, Zhang J, Yang B. The photoluminescence mechanism in carbon dots (graphene quantum dots, carbon nanodots, and polymer dots): current state and future perspective. Nano Res 2015; 8(2): 355-81.
[http://dx.doi.org/10.1007/s12274-014-0644-3]

[36] Bao L, Liu C, Zhang ZL, Pang DW. Photoluminescence-tunable carbon nanodots: surface-state energy-gap tuning. Adv Mater 2015; 27(10): 1663-7.
[http://dx.doi.org/10.1002/adma.201405070] [PMID: 25589141]

[37] Jiang K, Sun S, Zhang L, *et al.* Red, green, and blue luminescence by carbon dots: full-color emission tuning and multicolor cellular imaging. Angew Chem Int Ed 2015; 54(18): 5360-3.
[http://dx.doi.org/10.1002/anie.201501193] [PMID: 25832292]

[38] Zhu S, Meng Q, Wang L, *et al.* Highly photoluminescent carbon dots for multicolor patterning, sensors, and bioimaging. Angew Chem Int Ed 2013; 52(14): 3953-7.

[http://dx.doi.org/10.1002/anie.201300519] [PMID: 23450679]

[39]   Ding H, Yu SB, Wei JS, Xiong HM. Full-color light-emitting carbon dots with a surface-stat--controlled luminescence mechanism. ACS Nano 2016; 10(1): 484-91.
[http://dx.doi.org/10.1021/acsnano.5b05406] [PMID: 26646584]

[40]   Pan L, Sun S, Zhang A, *et al.* Truly fluorescent excitation-dependent carbon dots and their applications in multicolor cellular imaging and multidimensional sensing. Adv Mater 2015; 27(47): 7782-7.
[http://dx.doi.org/10.1002/adma.201503821] [PMID: 26487302]

[41]   Liu Y, Zhou Q, Yuan Y, Wu Y. Hydrothermal synthesis of fluorescent carbon dots from sodium citrate and polyacrylamide and their highly selective detection of lead and pyrophosphate. Carbon 2017; 115: 550-60.
[http://dx.doi.org/10.1016/j.carbon.2017.01.035]

[42]   Rangel M, Saluja S, Barba V, Pérez-Huerta JS, Agarwal V. Dual-emissive waste oil based S-doped carbon dots for acetone detection and Cr(VI) detection/reduction/removal. J Environ Chem Eng 2023; 11(2): 109438.
[http://dx.doi.org/10.1016/j.jece.2023.109438]

[43]   Baragau IA, Lu Z, Power NP, *et al.* Continuous hydrothermal flow synthesis of S-functionalised carbon quantum dots for enhanced oil recovery. Chem Eng J 2021; 405: 126631.
[http://dx.doi.org/10.1016/j.cej.2020.126631]

[44]   Guo X, Zhang L, Wang Z, *et al.* Fluorescent carbon dots based sensing system for detection of enrofloxacin in water solutions. Spectrochim Acta A Mol Biomol Spectrosc 2019; 219: 15-22.
[http://dx.doi.org/10.1016/j.saa.2019.02.017] [PMID: 31030043]

[45]   Ahmed, HM, Ghali, M, Zahra, WK, Ayad, M. Optical sensing of pyridine based on green synthesis of passivated carbon dots. Mater. Today Proc. 2020; 33: 1845–1848.
[http://dx.doi.org/10.1016/j.matpr.2020.05.185]

[46]   Jing HH, Bardakci F, Akgöl S, Kusat K, Adnan M, Alam MJ, Gupta R, Sahreen S, Chen Y, Gopinath SCB, *et al.* Green carbon dots: synthesis, characterization, properties and biomedical applications. J. Funct. Biomater. 2023; 14(1):27.
[http://dx.doi.org/10.3390/jfb14010027]

[47]   Rossini, EL, Milani, MI, Pezza, HR. Green synthesis of fluorescent carbon dots for determination of glucose in biofluids using a paper platform. Talanta 2019; 201: 503–510.
[http://dx.doi.org/10.1016/j.talanta.2019.04.045]

[48]   Reddy KR, Hassan M, Gomes VG. Hybrid nanostructures based on titanium dioxide for enhanced photocatalysis. Appl Catal A Gen 2015; 489: 1-16.
[http://dx.doi.org/10.1016/j.apcata.2014.10.001]

[49]   Jaishankar M, Tseten T, Anbalagan N, Mathew BB, Beeregowda KN. Toxicity, mechanism and health effects of some heavy metals. Interdiscip Toxicol 2014; 7(2): 60-72.
[http://dx.doi.org/10.2478/intox-2014-0009] [PMID: 26109881]

[50]   Torres Landa SD, Reddy Bogireddy NK, Kaur I, Batra V, Agarwal V. Heavy metal ion detection using green precursor derived carbon dots. iScience 2022; 25(2): 103816.
[http://dx.doi.org/10.1016/j.isci.2022.103816] [PMID: 35198881]

[51]   Cheng Y, Chen Z, Wang Y, Xu J. Continuous synthesis of N, S co-coped carbon dots for selective detection of CD (II) ions. J Photochem Photobiol Chem 2022; 429: 113910.
[http://dx.doi.org/10.1016/j.jphotochem.2022.113910]

[52]   Zhang H, You J, Wang J, Dong X, Guan R, Cao D. Highly luminescent carbon dots as temperature sensors and "off-on" sensing of Hg^{2+} and biothiols. Dyes Pigments 2020; 173: 107950.
[http://dx.doi.org/10.1016/j.dyepig.2019.107950]

[53]   Venkateswarlu S, Viswanath B, Reddy AS, Yoon M. Fungus-derived photoluminescent carbon

nanodots for ultrasensitive detection of Hg^{2+} ions and photoinduced bactericidal activity. Sens Actuators B Chem 2018; 258: 172-83.
[http://dx.doi.org/10.1016/j.snb.2017.11.044]

[54]   Xiao-Yan W, Xue-Yan H, Tian-Qi W, Xu-Cheng F. Crown daisy leaf waste–derived carbon dots: A simple and green fluorescent probe for copper ion. Surf Interface Anal 2020; 52(4): 148-55.
[http://dx.doi.org/10.1002/sia.6733]

[55]   Bandi R, Dadigala R, Gangapuram BR, Guttena V. Green synthesis of highly fluorescent nitrogen – Doped carbon dots from *Lantana camara* berries for effective detection of lead(II) and bioimaging. J Photochem Photobiol B 2018; 178: 330-8.
[http://dx.doi.org/10.1016/j.jphotobiol.2017.11.010] [PMID: 29178994]

[56]   Ansi VA, Renuka NK. Table sugar derived Carbon dot – a naked eye sensor for toxic Pb^{2+} ions. Sens Actuators B Chem 2018; 264: 67-75.
[http://dx.doi.org/10.1016/j.snb.2018.02.167]

[57]   Naik VM, Gunjal DB, Gore AH, *et al.* Quick and low cost synthesis of sulphur doped carbon dots by simple acidic carbonization of sucrose for the detection of Fe^{3+} ions in highly acidic environment. Diamond Related Materials 2018; 88: 262-8.
[http://dx.doi.org/10.1016/j.diamond.2018.07.018]

[58]   Şenol AM, Onganer Y. A novel "turn-off" fluorescent sensor based on cranberry derived carbon dots to detect iron (III) and hypochlorite ions. J Photochem Photobiol Chem 2022; 424: 113655.
[http://dx.doi.org/10.1016/j.jphotochem.2021.113655]

[59]   John BK, Korah BK, Mathew S, Thara C, Chacko AR, Mathew B. Nitrogen and sulphur co-doped carbon quantum dots as a dual-mode sensor for mercuric ions and as efficient antimicrobial agents. Biomass Convers Biorefin 2023.
[http://dx.doi.org/10.1007/s13399-023-04232-7]

[60]   Wang Y, Yang L, Liu B, Yu S, Jiang C. A colorimetric paper sensor for visual detection of mercury ions constructed with dual-emission carbon dots. New J Chem 2018; 42(19): 15671-7.
[http://dx.doi.org/10.1039/C8NJ03683G]

[61]   Atchudan R, Edison TNJI, Perumal S, Muthuchamy N, Lee YR. Hydrophilic nitrogen-doped carbon dots from biowaste using dwarf banana peel for environmental and biological applications. Fuel 2020; 275: 117821.
[http://dx.doi.org/10.1016/j.fuel.2020.117821]

[62]   Das M, Thakkar H, Patel D, Thakore S. Repurposing the domestic organic waste into green emissive carbon dots and carbonized adsorbent: A sustainable zero waste process for metal sensing and dye sequestration. J Environ Chem Eng 2021; 9(5): 106312.
[http://dx.doi.org/10.1016/j.jece.2021.106312]

[63]   Tang M, Zhu B, Wang Y, *et al.* Nitrogen- and sulfur-doped carbon dots as peroxidase mimetics: colorimetric determination of hydrogen peroxide and glutathione, and fluorimetric determination of lead(II). Mikrochim Acta 2019; 186(9): 604.
[http://dx.doi.org/10.1007/s00604-019-3710-4] [PMID: 31385065]

[64]   Kumar A, Chowdhuri AR, Laha D, Mahto TK, Karmakar P, Sahu SK. Green synthesis of carbon dots from *Ocimum* sanctum for effective fluorescent sensing of Pb^{2+} ions and live cell imaging. Sens Actuators B Chem 2017; 242: 679-86.
[http://dx.doi.org/10.1016/j.snb.2016.11.109]

[65]   Rong MC, Zhang KX, Wang YR, Chen X. The synthesis of B, N-carbon dots by a combustion method and the application of fluorescence detection for Cu 2+. Chin Chem Lett 2017; 28(5): 1119-24.
[http://dx.doi.org/10.1016/j.cclet.2016.12.009]

[66]   Zhang WJ, Liu SG, Han L, Luo HQ, Li NB. A ratiometric fluorescent and colorimetric dual-signal sensing platform based on N-doped carbon dots for selective and sensitive detection of copper(II) and pyrophosphate ion. Sens Actuators B Chem 2019; 283: 215-21.

[http://dx.doi.org/10.1016/j.snb.2018.12.012]

[67]   Zhang Q, Du H, Xie S, *et al.* Preparation of one emission nitrogen-fluorine-doped carbon quantum dots and their applications in environmental water samples and living cells for ClO⁻ detection and imaging. J Anal Methods Chem 2023; 2023: 1-9.
[http://dx.doi.org/10.1155/2023/7515979]

[68]   Zhang Y, Fang X, Zhao H, Li Z. A highly sensitive and selective detection of Cr(VI) and ascorbic acid based on nitrogen-doped carbon dots. Talanta 2018; 181: 318-25.
[http://dx.doi.org/10.1016/j.talanta.2018.01.027] [PMID: 29426518]

[69]   Bhatt S, Bhatt M, Kumar A, Vyas G, Gajaria T, Paul P. Green route for synthesis of multifunctional fluorescent carbon dots from Tulsi leaves and its application as Cr(VI) sensors, bio-imaging and patterning agents. Colloids Surf B Biointerfaces 2018; 167: 126-33.
[http://dx.doi.org/10.1016/j.colsurfb.2018.04.008] [PMID: 29635135]

[70]   D P, Saini S, Thakur A, Kumar B, Tyagi S, Nayak MK. A "Turn-On" thiol functionalized fluorescent carbon quantum dot based chemosensory system for arsenite detection. J Hazard Mater 2017; 328: 117-26.
[http://dx.doi.org/10.1016/j.jhazmat.2017.01.015] [PMID: 28103487]

[71]   Radhakrishnan K, Panneerselvam P. Green synthesis of surface-passivated carbon dots from the prickly pear cactus as a fluorescent probe for the dual detection of arsenic (III) and hypochlorite ions from drinking water. RSC Advances 2018; 8(53): 30455-67.
[http://dx.doi.org/10.1039/C8RA05861J] [PMID: 35546865]

[72]   Akash S, Sivaprakash B, Rajamohan N, Govarthanan M, Elakiya BT. Remediation of pharmaceutical pollutants using graphene-based materials - A review on operating conditions, mechanism and toxicology. Chemosphere 2022; 306: 135520.
[http://dx.doi.org/10.1016/j.chemosphere.2022.135520] [PMID: 35780979]

[73]   Philip JM, Aravind UK, Aravindakumar CT. Emerging contaminants in Indian environmental matrices – A review. Chemosphere 2018; 190: 307-26.
[http://dx.doi.org/10.1016/j.chemosphere.2017.09.120] [PMID: 28992484]

[74]   Chang MMF, Ginjom IR, Ng SM. Single-shot 'turn-off' optical probe for rapid detection of paraoxon-ethyl pesticide on vegetable utilising fluorescence carbon dots. Sens Actuators B Chem 2017; 242: 1050-6.
[http://dx.doi.org/10.1016/j.snb.2016.09.147]

[75]   Campos BB, Abellán C, Zougagh M, *et al.* Fluorescent chemosensor for pyridine based on N-doped carbon dots. J Colloid Interface Sci 2015; 458: 209-16.
[http://dx.doi.org/10.1016/j.jcis.2015.07.053] [PMID: 26225491]

[76]   Wang L, Bi Y, Hou J, *et al.* Facile, green and clean one-step synthesis of carbon dots from wool: Application as a sensor for glyphosate detection based on the inner filter effect. Talanta 2016; 160: 268-75.
[http://dx.doi.org/10.1016/j.talanta.2016.07.020] [PMID: 27591613]

[77]   Xu HV, Zhao Y, Tan YN. Nanodot-directed formation of plasmonic-fluorescent nanohybrids toward dual Optical detection of glucose and cholesterol *via* hydrogen peroxide sensing. ACS Appl Mater Interfaces 2019; 11(30): 27233-42.
[http://dx.doi.org/10.1021/acsami.9b08708] [PMID: 31282641]

[78]   Sha R, Jones SS, Vishnu N, Soundiraraju B, Badhulika S. A novel biomass derived carbon quantum dots for highly sensitive and selective detection of hydrazine. Electroanalysis 2018; 30(10): 2228-32.
[http://dx.doi.org/10.1002/elan.201800255]

[79]   Ashrafi Tafreshi F, Fatahi Z, Ghasemi SF, Taherian A, Esfandiari N. Ultrasensitive fluorescent detection of pesticides in real sample by using green carbon dots. PLoS One 2020; 15(3): e0230646.
[http://dx.doi.org/10.1371/journal.pone.0230646] [PMID: 32208468]

[80]   Jagannathan M, Dhinasekaran D, Soundharraj P, *et al.* Green synthesis of white light emitting carbon quantum dots: Fabrication of white fluorescent film and optical sensor applications. J Hazard Mater 2021; 416: 125091.
[http://dx.doi.org/10.1016/j.jhazmat.2021.125091] [PMID: 33866289]

[81]   Siddique AB, Pramanick AK, Chatterjee S, Ray M. Amorphous carbon dots and their remarkable ability to detect 2,4,6-trinitrophenol. Sci Rep 2018; 8(1): 9770.
[http://dx.doi.org/10.1038/s41598-018-28021-9] [PMID: 29950660]

[82]   Hu Y, Gao Z. Sewage sludge in microwave oven: A sustainable synthetic approach toward carbon dots for fluorescent sensing of para-Nitrophenol. J Hazard Mater 2020; 382(Jan): 121048.
[http://dx.doi.org/10.1016/j.jhazmat.2019.121048] [PMID: 31476723]

[83]   Deka MJ, Dutta P, Sarma S, Medhi OK, Talukdar NC, Chowdhury D. Carbon dots derived from water hyacinth and their application as a sensor for pretilachlor. Heliyon 2019; 5(6): e01985.
[http://dx.doi.org/10.1016/j.heliyon.2019.e01985] [PMID: 31338457]

[84]   Uddin MK. A review on the adsorption of heavy metals by clay minerals, with special focus on the past decade. Chem Eng J 2017; 308: 438-62.
[http://dx.doi.org/10.1016/j.cej.2016.09.029]

[85]   Jain A, Balasubramanian R, Srinivasan MP. Hydrothermal conversion of biomass waste to activated carbon with high porosity: A review. Chem Eng J 2016; 283: 789-805.
[http://dx.doi.org/10.1016/j.cej.2015.08.014]

[86]   Sabet M, Mahdavi K. Green synthesis of high photoluminescence nitrogen-doped carbon quantum dots from grass *via* a simple hydrothermal method for removing organic and inorganic water pollutions. Appl Surf Sci 2019; 463: 283-91.
[http://dx.doi.org/10.1016/j.apsusc.2018.08.223]

[87]   Wang L, Cheng C, Tapas S, *et al.* Carbon dots modified mesoporous organosilica as an adsorbent for the removal of 2,4-dichlorophenol and heavy metal ions. J Mater Chem A Mater Energy Sustain 2015; 3(25): 13357-64.
[http://dx.doi.org/10.1039/C5TA01652E]

[88]   Ramachandran A, J S AN, Karunakaran Yesodha S. Polyaniline-derived nitrogen-doped graphene quantum dots for the ultratrace level electrochemical detection of trinitrophenol and the effective differentiation of nitroaromatics: structure matters. ACS Sustain Chem& Eng 2019; 7(7): 6732-43.
[http://dx.doi.org/10.1021/acssuschemeng.8b05996]

[89]   Mehta A, D P, Thakur A, Basu S. Enhanced photocatalytic water splitting by gold carbon dot core shell nanocatalyst under visible/sunlight. New J Chem 2017; 41(11): 4573-81.
[http://dx.doi.org/10.1039/C7NJ00933J]

[90]   Qin J, Zeng H. Photocatalysts fabricated by depositing plasmonic Ag nanoparticles on carbon quantum dots/graphitic carbon nitride for broad spectrum photocatalytic hydrogen generation. Appl Catal B 2017; 209: 161-73.
[http://dx.doi.org/10.1016/j.apcatb.2017.03.005]

[91]   Omer KM, Mohammad NN, Baban SO, Hassan AQ. Carbon nanodots as efficient photosensitizers to enhance visible-light driven photocatalytic activity. J Photochem Photobiol Chem 2018; 364: 53-8.
[http://dx.doi.org/10.1016/j.jphotochem.2018.05.041]

[92]   Rong S, Tang X, Liu H, *et al.* Synthesis of carbon dots@Fe_3O_4 and their photocatalytic degradation properties to hexaconazole. NanoImpact 2021; 22: 100304.
[http://dx.doi.org/10.1016/j.impact.2021.100304] [PMID: 35559982]

[93]   Sim LC, Tai JY, Leong KH, *et al.* Metal free and sunlight driven g-C3N4 based photocatalyst using carbon quantum dots from Arabian dates: Green strategy for photodegradation of 2,4-dichlorophenol and selective detection of Fe3+. Diamond Related Materials 2021; 120: 108679.
[http://dx.doi.org/10.1016/j.diamond.2021.108679]

[94] Zhou Y, ElMetwally AE, Chen J, *et al.* Gel-like carbon dots: A high-performance future photocatalyst. J Colloid Interface Sci 2021; 599: 519-32. b
[http://dx.doi.org/10.1016/j.jcis.2021.04.121] [PMID: 33964697]

[95] Wei W, Huang J, Gao W, Lu X, Shi X. Carbon dots fluorescence-based colorimetric sensor for sensitive detection of aluminum ions with a smart phone. Chemosensors 2021; 9(2): 25.
[http://dx.doi.org/10.3390/chemosensors9020025]

[96] Tran NA, Hien NT, Hoang NM, Dang HL, Van Quy T, Hanh NT, Vu NH, Dao VD. Carbon dots in environmental treatment and protection applications. Desalination. 2023 Feb 15;548:116285.
[http://dx.doi.org/10.1016/j.desal.2022.116285]

[97] Song X, Zhou Q, Zhang T, Xu H, Wang Z. Pressure-assisted preparation of graphene oxide quantum dot-incorporated reverse osmosis membranes: antifouling and chlorine resistance potentials. J Mater Chem A Mater Energy Sustain 2016; 4(43): 16896-905.
[http://dx.doi.org/10.1039/C6TA06636D]

[98] Park SJ, Kwon SJ, Kwon HE, *et al.* Aromatic solvent-assisted interfacial polymerization to prepare high performance thin film composite reverse osmosis membranes based on hydrophilic supports. Polymer 2018; 144: 159-67.
[http://dx.doi.org/10.1016/j.polymer.2018.04.060]

[99] Yang WJ, Shao DD, Zhou Z, *et al.* Carbon quantum dots (CQDs) nanofiltration membranes towards efficient biogas slurry valorization. Chem Eng J 2020; 385: 123993.
[http://dx.doi.org/10.1016/j.cej.2019.123993]

[100] Saravanan A, Maruthapandi M, Das P, *et al.* Applications of N-doped carbon dots as antimicrobial agents, antibiotic carriers, and selective fluorescent probes for nitro explosives. ACS Appl Bio Mater 2020; 3(11): 8023-31.
[http://dx.doi.org/10.1021/acsabm.0c01104] [PMID: 35019541]

[101] Xin Q, Shah H, Nawaz A, *et al.* Antibacterial carbon-based nanomaterials. Adv Mater 2019; 31(45): 1804838.
[http://dx.doi.org/10.1002/adma.201804838] [PMID: 30379355]

[102] Li H, Huang J, Song Y, *et al.* Degradable carbon dots with broad-spectrum antibacterial activity. ACS Appl Mater Interfaces 2018; 10(32): 26936-46.
[http://dx.doi.org/10.1021/acsami.8b08832] [PMID: 30039700]

[103] Xu HV, Zheng XT, Wang C, Zhao Y, Tan YN. Bioinspired antimicrobial nanodots with amphiphilic and zwitterionic-like characteristics for combating multidrug-resistant bacteria and biofilm removal. ACS Appl Nano Mater 2018; 1(5): 2062-8.
[http://dx.doi.org/10.1021/acsanm.8b00465]

[104] Saravanan A, Maruthapandi M, Das P, Luong JHT, Gedanken A. Green synthesis of multifunctional carbon dots with antibacterial activities. Nanomaterials 2021; 11(2): 369.
[http://dx.doi.org/10.3390/nano11020369] [PMID: 33540607]

CHAPTER 6

Green Synthesized Nanoparticles and Different Domains of their Applications

Nakul Kumar[1], Pankaj Kumar[2,*], Snigdha Singh[2], Virendra Kumar Yadav[3], Deepankshi Shah[2], Mohd. Tariq[4], Ramesh Kumar[5] and **Sunil Soni[6]**

[1] *Gandhinagar Institute of Science, Gandhinagar University, Gandhinagar, Gujarat, India*

[2] *Department of Environmental Science, Parul Institute of Applied Sciences, Parul University, Vadodara, Gujarat, India*

[3] *Department of Life Sciences, Hemchandracharya North Gujarat University, Matarvadi Part, Gujarat, India*

[4] *Department of Life Sciences, Parul Institute of Applied Sciences, Parul University, Vadodara, Gujarat, India*

[5] *Department of Environmental Science, School of Earth Sciences, Central University of Rajasthan, Ajmer, Rajasthan, India*

[6] *School of Environment and Sustainable Development, Central University of Gujarat, Gandhinagar, India*

Abstract: Science has undergone a revolution with the development of nanotechnology. The vast applications of nanoparticles (NPs) have greatly helped every area of technology. Nanomaterials (NMs) can be created *via* a range of physical and chemical practices along with the use of ultrasound and microwave heating processes, but green synthesis has drawn great attention, especially when it involves the use of microbes or plant extracts. Green synthesis is a recent and advanced method to make NPs because it is simpler, cheaper, more reproducible and environmentally friendly than other approaches. When compared to other classical methods of NP synthesis, plants produce NPs that are more stable and simpler to scale up and have a variety of applications. This chapter has reviewed and discussed various applications of green synthesized NPs along with the latest developments in the eco-friendly synthesis of gold, silver, copper, palladium, iron, and iron oxide NPs. Due to the widespread use of nanoscale metals in different industries, including engineering, medicine, and the environment, the topic of nanoscale metal synthesis is currently relevant. The bulk of nanoscale metals is currently produced chemically, which has unintended effects like environmental contamination and serious health issues. To overcome these challenges, green synthesis can be used as a commercial chemical method to decrease metal ions from the environment. Green synthesis is more advantageous than classical chemical synthesis because it improves environmental quality and is also safe for human health.

* **Corresponding author Pankaj Kumar:** Department of Environmental Science, Parul Institute of Applied Sciences, Parul University, Vadodara, Gujarat, India; Tel: +918460571814
E-mail: pankaj.kumar25135@paruluniversity.ac.in

Neha Agarwal, Vijendra Singh Solanki and Sreekantha B. Jonnalagadda (Eds.)

Keywords: Biological method, Environmental sustainability, Green synthesis, Human health, Nanomaterials, Plant extracts.

INTRODUCTION

With the advent of nanotechnology, it has become possible to create and evaluate objects on a molecular scale, between 1-100 nanometers, which are known as NMs. NMs have vital applications in many fields, including optics, electronics, mechanics, medicine, biotechnology, microbiology, environmental cleanup, multiple engineering disciplines, and material science. Their most significant attribute is their enormous surface area to their respective volume ratios. Different production procedures have been developed for the synthesis of metallic NPs, such as bottom-up and top-down approaches [1]. In a nutshell, the top-down strategy includes the size decrease of bulk material *via* lithographic techniques and mechanical methods like grinding and milling, whereas the bottom-up approach involves the gathering of small building blocks into bigger structures, such as chemical synthesis. However, the bottom-up strategy, in which an NP is "grown" from simpler molecules known as reaction precursors, is an acceptable and efficient technique for creating NPs [2]. By varying precursor amounts and reaction conditions, it may be feasible to regulate NP size and form depending on its intended use (temperature, pH, *etc.*). Fig. (**1**) is an illustration of the generic manufacturing of green NPs and their different uses.

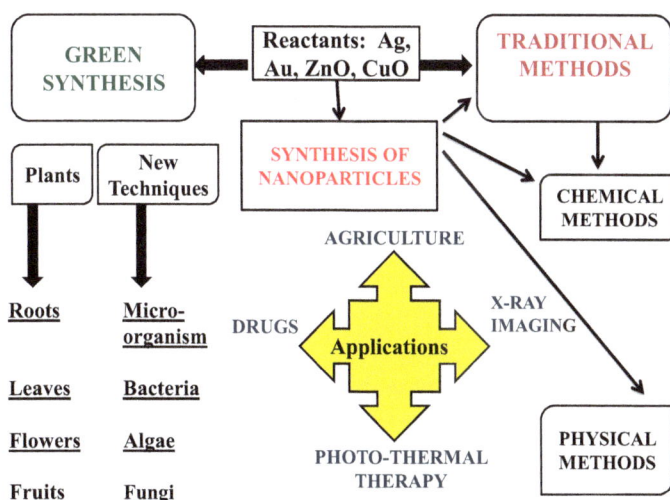

Fig. (1). Illustration of the generic manufacturing of green NPs and their use [3].

Both physical and chemical methods utilize very poisonous and reactive reductants like sodium borohydride and hydrazine hydrate that have unfavorable effects on living beings and the environment. Researchers are still working to create simple, reliable, and efficient green chemistry procedures for making NMs. Numerous species serve as safe, environmentally acceptable, and long-lasting precursors to create stable, well-functionalized NPs from bacteria, actinomycetes, fungi, yeast, and viruses. Therefore, it is crucial to look for a more dependable and long-lasting method of producing NMs. NM production raises issues on their commercial feasibility, environmental sustainability, social adaptability, and accessibility to local resources (Fig. **1**). To restrict the overall costs of finished products of nanotechnology-based materials and to increase affordability, industries must strive hard to maintain a balance between eco-friendly operations and sustainability. Nano-based ecologically friendly production techniques are solutions that work without the usage of dangerous chemicals.

Apart from environmental remediation, NPs also have multiple biological applications. For instance, new viruses are constantly evolving, creating a problem that is becoming more widespread. Therefore, in addition to developing novel antiviral medications, inventive procedures must be improved to maximize the efficacy of already available medications and other methods for restricting viral spread [4]. New approaches are frequently being tried to boost the effectiveness of vaccines. Recently, intriguing antibacterial and antiviral techniques based on nanotechnology have come under investigation [5]. These might include antiviral facemasks, clothing, and other coatings that, when in touch with a surface, could potentially destroy the virus. A huge number of studies have confirmed that NPs are also considered agents that enhance cellular immunity and extend the impact of antigens. Due to their multiple targeted mechanisms of action, silver NPs (AgNPs), in particular, synthesized utilizing plant extracts, are a promising contender for novel antiviral medicines [6]. These NPs have been demonstrated to be antiviral against the influenza virus, hepatitis B virus, herpes simplex type 1, chikungunya virus, and HIV [7]. A similar method was used to create gold NPs (AuNPs-As), which work as a reducing agent and have antiviral action against the measles virus. Additionally, CuO NPs were created using *Syzygium alternifolium* fruit extract and function as an antiviral agent against the Newcastle Disease Virus (NDV) [8]. NPs preserve antigens from deprivation and are concentrated on delivery in antigen-introducing cells and sub-cellular regions of interest because of their significant bio-physicochemical properties.

BIOTIC COMPONENTS FOR "GREEN SYNTHESIS"

In medicinal and nanoscience, many physico-chemical synthesis techniques need higher radiation, extremely lethal reductants, and alleviating agents, all of which can be harmful to living organisms. In contrast, the environmentally friendly bioreduction technique used in the biosynthesis of metallic NPs only requires a tiny amount of energy to accomplish the reaction process. The advantage of this reduction method is its low-cost procedure [9]. Different biotic components for green synthesis are described below.

Bacteria

Nowadays, different bacteria are widely used in a variety of commercial biotechnological and environmental applications, such as phytoremediation, bioremediation, genetic engineering, and bioleaching [10]. The ability of bacteria to reduce metal ions makes them vital applicants during the synthesis of NPs. In the creation of new metallic and non-metallic NPs, several bacterial species are employed [11]. Prokaryotic bacteria and actinomycetes are being utilized in the manufacture of NPs. The synthesis of bacterial NPs has now become popular due to the easy handling process. The manufacture of AuNPs has also frequently utilized different species of bacteria such as *Bacillus megaterium, Desulfovibrio desulfuricans*, E. coli, *Bacillus subtilis, Shewanella alga, Rhodopseudomonas capsulate*, and *Plectonema boryanum* [12].

Fungi

Metal/metal oxide biosynthesis of NPs by fungi is a highly effective way of generating monodisperse NPs with clearly defined morphologies. They serve as more effective biotic agents for the production of metallic NPs because they have various intracellular enzymes [13]. Competent fungi can manufacture more NPs than bacteria because fungi have enzymes, proteins, and decreasing agents on their cell surfaces.

Yeast

Yeasts are single-celled microbes in eukaryotic cells. There are more than 1500 different types of yeast in existence. Numerous research teams have reported successful yeast-based NP/NM production. *Saccharomyces cerevisiae* and a silver-tolerant yeast strain have both been shown to biosynthesize Ag and AuNPs [14].

Plants

In varied concentrations, heavy metals can accumulate in various parts of plants. Plant extract-based biosynthetic approaches have gained greater attention. As a result, they are a wonderful alternative to conventional NP creation techniques and are simple, efficient, inexpensive, and feasible. Several plants may be employed in a "one-pot" manufacturing procedure to decrease and stabilize metallic NPs. Numerous investigators have employed green synthesis techniques to produce the particles utilizing plant leaf extracts to better examine the numerous advantages of metal/metal oxide NPs in green chemistry.

Proteins, coenzymes, and carbohydrates are examples of biomolecules found in plants that have a remarkable capability to transform metal ions into NPs. Initial research on plant extract-assisted synthesis and other biosynthesis processes used Au and AgNPs. *Aloe barbadensis Miller* (Aloe vera), *Avena sativa* (oats), *Medicago sativa* (alfalfa), *Ocimum sanctum* (tulsi), *Citrus limon* (lemon), *Azadirachta indica* (neem), *Coriandrum sativum* (coriander) and other plants have all been used in the creation of AgNPs and AuNPs [15]. Coenzymes, proteins, and carbohydrates are examples of biomolecules present in plants that have a noteworthy capability to change metal salts into NPs. Initial research on plant extract-supported synthesis and other processes used Au and AgNPs.

NATURAL EXTRACTS USED FOR SYNTHESIS OF NPS

Algae have emerged as a favorable multipurpose microbial arrangement for the biosynthesis of NPs with the welfare of wastewater bioremediation, bioenergy generation, and large-scale manufacture of valued marketable commodities like dyes and pharmaceuticals [16]. Algae-based NPs (AgNPs, AuNPs, *etc.*) have many characteristics, including antifouling, antibacterial, and photocatalytic activities, which make them suitable for nanophotonic, diagnostics, optical biosensing, and bioremediation, among others [17]. AgNPs can interact aggressively with microbial surfaces and perhaps act as antibacterial agents due to their larger area and smaller size. AuNPs, which are employed in many different applications, are the least toxic NPs. More studies are essential to fully explore and harness the ability of algae-based NPs for greener and eco-friendly applications. Fig. (**2**) is the detailed scheme of the reduction of Au and AgNPs.

There are many pros and cons of using proteins as eco-friendly agents for metal ions. They have been successful in regulating NP size and form [19]. The toxic effects of chemical solvents are lessened when water is used as a solvent. The extraordinary sensitivity of proteins, their shifting structure, and lower performance are all disadvantages of employing these features. It exhibits resilience to high temperatures because proteins are also temperature-sensitive [20].

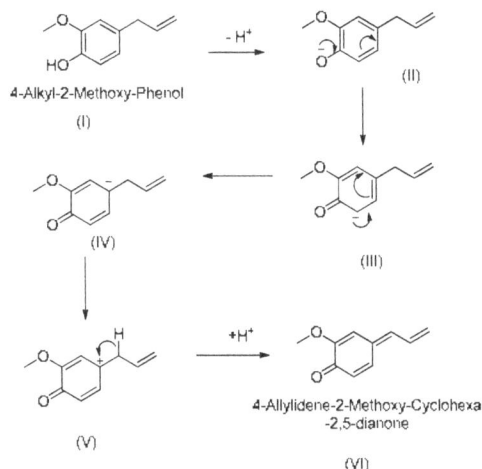

4-Alkyl-2-Methoxy-Phenol
(I)

(II)

(III)

(IV)

(V)

4-Allylidene-2-Methoxy-Cyclohexa
-2,5-dianone

(VI)

Fig. (2). Schematic diagram of silver and gold ions reduction [18].

Plant extracts are utilized in a variety of ways to create NPs. These methods avoid cumbersome procedures and are economical and environmentally friendly [21]. Modern studies have been dedicated heavily to metal NPs made from plant extract and recognized as biocompatible and harmless. These NPs have the potential to be antiviral, antibacterial, and anticancer and also offer a great deal of potential for target drug delivery [22]. The list of plants and other species employed for the unique generation of NPs is shown in Table **1**.

Table 1. Plant extracts used in the production of NPs.

Extracts from Plants	NPs	References
Acacia gum, Tree gum	Scandium	[23]
Cinnamomum cassia (Lauraceae)	Silver	[24]
Alternaria, F. oxysporum, Curvularia, C. Indicum, and Phoma	Silver	[25]
Aspergillus (Trichocomaceae)	Silver	[26]
Tragacanth gum	Ni–Mg ferrite, Ni–Cu–Mg ferrite, Ni–Cu–Zn ferrite, ZnO	[27]

Methodologies for green synthesis that use biological precursors are reliant on numerous physicochemical reactions (acidic, basic, or neutral). Due to the accessibility of powerful phytochemicals such as ketones, aldehydes, flavones, amides, terpenoids, carboxylic acids, phenols, and ascorbic acids in numerous plant extracts, mostly in leaves, plant biodiversity has drawn a lot of attention for the creation of NPs.

GREEN SYNTHESIS

Green synthesis offers a very promising future for generating nanostructured materials compared to conventional methods. We recognize that when the phrase "green synthesis" is employed, it indirectly alludes to a low-cost, easy-to-follow technique without the usage of pricey or difficult machinery. These qualities have set green synthesis apart from other methods for creating and advancing metallic nanostructures. Soon, it is projected that more and more natural substances will be used to collect, stabilize, and produce metallic NPs, including fresh extracts, fruits, microbes, and other green materials [28]. These include environmentally friendly substances that have never been used for these purposes, especially those that may be used as drugs, bactericides, fungicides, and other substances, as well as environmentally friendly substances like sugar, ascorbic acid, starch, and other substances. Similar to this, we would increasingly anticipate results that emphasize synthesis while still achieving acceptable control of particle size by adjusting macroscale variables during the experimental process. Table **2** summarizes some important green synthesized NPs from plants and their applications.

Table 2. Green synthesis of NPs using plants and their uses [29].

Plant	Source	Metal Oxide	Size(nm)	Use	Refs.
Ficus carica	Leaf	Iron (II, III) oxide	43–57	Antioxidant	[30]
Azadirachta indica	Leaf	Copper(II) oxide	NA	Anticancer	[31]
Peltophorum pterocarpum	Leaf	Iron (II, III) oxide	85	Rhodamine deprivation	[32]
Terminalia chebula	Seed	Iron (II, III) oxide	NA	Methylene blue degradation	[33]
Punica granatum	Peel	Zinc oxide	118.6	Antibacterial	[34]
Lactuca serriols	Seed	Nickel oxide	NA	Dye Deprivation	[35]
Vitis rotundifolia	Fruit	Cobalt oxide	NA	Blue dye Deprivation	[36]

FACTORS AFFECTING SYNTHESIS OF NPS

The formation of stabilized NPs and their nucleation is governed by several factors. These characteristics (shape and magnitude) depend mostly on the plant extract's processing constraints, as well as the response of the metal salts, pH, reaction duration, temperature, and the amount of plant extract to metal salts [37]. Each of these elements is briefly covered in the following sections during the organism's growth phase. The synthesis of NPs has experimentally gained momentum; for instance, a group of workers investigated how the biomass growth phase affected the production of Ag NPs [38]. It has been investigated that

Bacillus sp. formed excessively higher NPs during the inactive phase when compared to the biomass derived from other phases. Conferring to the previous studies, the fungus' capacity to tolerate metals is increased by the release of enzymes and other materials metabolites during the inactive phase, consequently reducing the metal stress. The capacity to tolerate metal is also based on the microbes' type and the metal in question. *Aspergillus sp.*, for instance, has been shown to undergo a prolonged mid-log phase when nickel is included in the growth medium [39]. However, it has been noted that the same organism's stationary phase can be prolonged by the addition of chromium in the medium. However, the majority of research in the literature contends that microorganisms in their stationary phase should produce NPs more frequently.

Additionally, it has been noted that crucial variables that affect the NP size in chemical synthesis processes include the molar ratios of the reactants. It is well known that during chemical synthesis, the concentration of the reactants can have a direct impact on the products. In this regard, the reactant concentration can systematically regulate the form of biosynthesized silver nanocrystals by $AgNO_3$ and extraction of the citrus leaf. The authors claim that spherical NPs were produced with silver nitrate and citric acid in the ratio of 1:4 by volume. The creation of bio-organics from plant extract, however, was also said to have enhanced the particle size of silver nanoparticles. Even while a direct correlation between the precursor concentrations and the structure of the nanocrystal could not be established, it was revealed that precursors with larger molar ratios significantly influenced the NPs' shapes [40]. Additionally, it was claimed that the pH had a significant impact on the metallic ion reduction reaction. Pandian examined the impact of various pH levels on *Brevibacterium* species' production of CdS nanocrystallites. Using 1 M HCl and 1 M NaOH solutions, the pH of the incubation mixtures was altered. It was discovered that NOs' size varied with pH. Overall, the potential for available functional groups in the reaction was increased by an alkaline pH, which favored nucleation and NP formation. In the past, it was found that an alkaline environment encouraged the production of a variety of NPs alongside protein molecules [41]. It was noted that particle agglomeration occurred when the nanocrystallites were acidified from pH 7 to pH 6. In a different investigation, AgNPs created from *Cinnamon zeylanicum* bark extract improved in number with accumulative extraction of bark and its concentrations and at higher pH levels (more than pH 5) [42]. Additionally, nanocrystals precipitated from the solution at pH levels lower than 6. Biomanufacturing techniques for metallic NPs with methodical nanostructures and programmable functions are crucial in both basic research and real-world applications because of their low toxicity, reduced creation of hazardous elements, and energy conservation.

NP'S STABILITY AND TOXICITY

The dispersion and mobility of released NPs in the atmosphere are governed by their eco-friendly nature to form metastable aqueous suspensions or aerosols in surrounding fluids. One can determine how stable NPs are in an environment by determining how likely it is for them to mix or interact with the environment [44]. The ability of the suspension in terms of stability is generally affected by the particle size and their empathy for other ecological constituents. Later, when applied to the aquatic environment, the "green" synthesis of AgNPs made from tea leaf extract was revealed to be steady [45]. The obtained material also served to demonstrate the strength of AgNPs (in an aqueous media) produced utilizing plant extracts and metabolites. It has been suggested that surface complexation influences the inherent steadiness of NPs by controlling their colloidal strength.

MULTIPLE APPLICATIONS OF GREEN SYNTHESIZED NPS

Some applications of green synthesized NPs are discussed below (Fig. **3**).

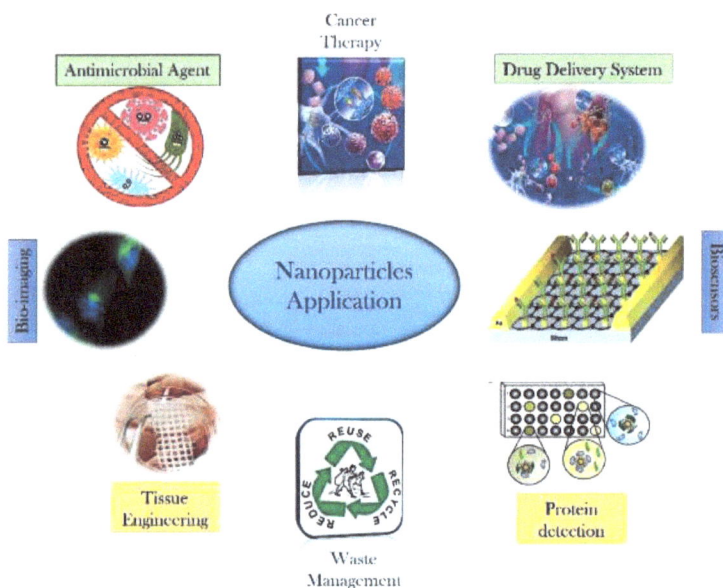

Fig. (3). Applications of green synthesized NPs in environmental and biomedical fields [43].

Antimicrobial Activity

Numerous studies were conducted to improve antimicrobial activity as a result of the growth in bacterial resistance to conventional antibiotics and antiseptics. *In vitro,* antimicrobial experiments show that the metallic NPs successfully inhibit many bacterial species [46]. Two important aspects that affect the potential of metallic NPs are the material utilized to make them and their particle size. Bacterial resistance to antibiotics has grown over time, endangering the public's health in a significant way. For example, methicillin, sulfonamides, penicillin, and vancomycin resistance are characteristics of antimicrobial drug-resistant microorganisms.

Catalytic Activity

As common organic contaminants of wastewater, 4-nitrophenol, and its derivatives, which are generally utilized to make pesticides and synthetic colors, can seriously harm the environment. The environmental impact of 4-nitrophenol is significant due to its poisonous and inhibitive properties. Therefore, it is essential to reduce these contaminants [47]. 4-nitrophenol has been used in a variety of sectors, such as paracetamol intermediate, rubber antioxidants, corrosion inhibitors, and sulfur dyes. The introduction of $NaBH_4$ as a metal catalyst (such as gold, silver, copper, or palladium NPs) is the quickest and the most efficient technique to reduce 4-nitrophenol [48].

Removal of Pollutant Dyes

One major category of organic contaminants utilized in a variety of applications is cationic and anionic dyes. Due to their enormous need in the paper, textile, leather, plastic, culinary, printing, and pharmaceutical industries, organic dyes play a very important role in causing pollution [49]. In the textile industry, the pigmentation process used to create various fabrics consumes roughly 60% of the colors used. Due to their recalcitrant character, roughly 15% of dyes that are wasted during the fabric-making processes are released into the hydrosphere and constitute a substantial pollution source [50]. The most significant sources of environmental contamination are the pollutants from these manufacturing facilities.

Heavy Metal Ion Sensing

The ability of heavy metals to contaminate the air, soil, and water is widely acknowledged. These metals include Zn, Co, Cd, Pb, Ni, Cu, Mn, Fe, Cr, Cr, and Hg. Heavy metal contamination is primarily caused by paper, plastic, coal, and dye sectors as well as mining waste and automotive emissions. Several metals, including Pb, Co, Cd, and Hg ions, have tremendous toxicity even at extremely low ppm concentrations [51]. To carry out efficient remediation methods, lethal metals in the aquatic ecosystems must, therefore, be detected. Traditional approaches often provide more sensitivity in multi-element examination [52]. The experimental sets required for this analysis, however, are exceedingly costly, slow, and labor-intensive. The benefits of using metal NPs as colorimetric sensors for the said metal ions in the atmosphere comprise their easiness, affordability, and high sensitivity at sub-ppm concentrations. Synthesized AgNPs were produced using a variety of plant extracts as colorimetric sensors for metal ions such as Cd^{2+}, Cr^{3+}, Hg^{2+}, Ca^{2+}, and Zn^{2+} in water, and synthesized AgNPs were made utilizing a variety of plant extracts [53]. They demonstrated colorimetric detection of Zn^{2+} and Hg^{2+} in their synthesized AgNPs. Similar to fresh AgNPs (MF-AgNPs) and sun-dried AgNPs (MD-AgNPs), fresh and dried mango leaves both demonstrated selective sensitivity for Hg^{2+} and Pb^{2+}.

CONCLUSION AND FUTURE RECOMMENDATIONS

The "green" manufacturing of nanomaterials has gained a lot of attention over the last decade. Natural extracts come in numerous forms and have been utilized successfully for the synthesis and/or manufacturing of many different kinds of materials. The plant extracts have been recognized to be very efficient alleviating and reducing agents for the production of regulated chemicals. The present chapter discusses the latest investigation on the environmentally friendly development of metals and metal oxide nanomaterials and their applications. The role of solvents in synthesis has been comprehensively examined, and comprehensive synthesis methods have been researched to resolve the current glitches in green synthesis. Moving laboratory-based investigation to an industrial scale while addressing the interpretation of historical and recent challenges, particularly the environment and human health implications, should be the primary focus of future studies and in the synthesis of potential green nanomaterials. However, it is projected that 'green' synthesis based on materials/NPs derived from biocomponents will be extensively used in eco-remediation and other critical fields like the pharmaceutical, cosmetic, and food industries. The green synthesis of NPs using marine plants and algae is an area that is mostly studied. Therefore, plenty of chances are still there to conduct research on novel biogenic synthesis-based green preparation techniques.

ACKNOWLEDGEMENTS

The authors are grateful to the Department of Environmental Science, Parul Institute of Applied Sciences, Parul University, Vadodara, and Gandhinagar Institute of Science, Gandhinagar University, Gandhinagar, for providing facilities and financial support.

REFERENCES

[1]   Saif S, Tahir A, Chen Y. Green synthesis of iron nanoparticles and their environmental applications and implications. Nanomaterials 2016; 6(11): 209.

[http://dx.doi.org/10.3390/nano6110209]

[2]   Jebali A, Ramezani F, Kazemi B. Biosynthesis of Silver Nanoparticles by *Geotricum* sp. J Cluster Sci 2011; 22(2): 225-32.
[http://dx.doi.org/10.1007/s10876-011-0375-5]

[3]   Naikoo GA, Mustaqeem M, Hassan IU, *et al.* Bioinspired and green synthesis of nanoparticles from plant extracts with antiviral and antimicrobial properties: A critical review. J Saudi Chem Soc 2021; 25(9): 101304.
[http://dx.doi.org/10.1016/j.jscs.2021.101304]

[4]   Dickey SW, Cheung GYC, Otto M. Different drugs for bad bugs: antivirulence strategies in the age of antibiotic resistance. Nat Rev Drug Discov 2017; 16(7): 457-71.
[http://dx.doi.org/10.1038/nrd.2017.23] [PMID: 28337021]

[5]   Haque A, Akçeşme FB, Pant AB. A review of Zika virus: hurdles toward vaccine development and the way forward. Antivir Ther 2018; 23(4): 285-93.
[http://dx.doi.org/10.3851/IMP3215] [PMID: 29300166]

[6]   Singh L, Kruger HG, Maguire GEM, Govender T, Parboosing R. The role of nanotechnology in the treatment of viral infections. Ther Adv Infect Dis 2017; 4(4): 105-31.
[http://dx.doi.org/10.1177/2049936117713593] [PMID: 28748089]

[7]   Baram-Pinto D, Shukla S, Perkas N, Gedanken A, Sarid R. Inhibition of herpes simplex virus type 1 infection by silver nanoparticles capped with mercaptoethane sulfonate. Bioconjug Chem 2009; 20(8): 1497-502.
[http://dx.doi.org/10.1021/bc900215b] [PMID: 21141805]

[8]   Yugandhar P, Vasavi T, Jayavardhana Rao Y, Uma Maheswari Devi P, Narasimha G, Savithramma N. Cost effective, green synthesis of copper oxide nanoparticles using fruit extract of *Syzygium alternifolium* (Wt.) walp., characterization and evaluation of antiviral activity. J Cluster Sci 2018; 29(4): 743-55.
[http://dx.doi.org/10.1007/s10876-018-1395-1]

[9]   Dahoumane SA, Yéprémian C, Djédiat C, *et al.* Improvement of kinetics, yield, and colloidal stability of biogenic gold nanoparticles using living cells of *Euglena gracilis* microalga. J Nanopart Res 2016; 18(3): 79.
[http://dx.doi.org/10.1007/s11051-016-3378-1]

[10]   Gericke M, Pinches A. Microbial production of gold nanoparticles. Gold Bull 2006; 39(1): 22-8.
[http://dx.doi.org/10.1007/BF03215529]

[11]   Iravani S. Bacteria in nanoparticle synthesis: Current status and future prospects. Int Sch Res Not. Labrenz M 2014; 2014: p. 359316.

[12]   Thakkar KN, Mhatre SS, Parikh RY. Biological synthesis of metallic nanoparticles. Nanomedicine Nanotechnology. Biol Med 2010; 6(2): 257-62.

[13]   Mohanpuria P, Rana NK, Yadav SK. Biosynthesis of nanoparticles: technological concepts and future applications. J Nanopart Res 2008; 10(3): 507-17.
[http://dx.doi.org/10.1007/s11051-007-9275-x]

[14]   Yurkov AM, Kemler M, Begerow D. Species accumulation curves and incidence-based species richness estimators to appraise the diversity of cultivable yeasts from beech forest soils. PLoS One 2011; 6(8): e23671.
[http://dx.doi.org/10.1371/journal.pone.0023671] [PMID: 21858201]

[15]   Gunalan S, Sivaraj R, Rajendran V. Green synthesized ZnO nanoparticles against bacterial and fungal pathogens. Prog Nat Sci 2012; 22(6): 693-700.
[http://dx.doi.org/10.1016/j.pnsc.2012.11.015]

[16]   Chaudhary R, Nawaz K, Khan AK, Hano C, Abbasi BH, Anjum S. An overview of the algae-mediated biosynthesis of nanoparticles and their biomedical applications. Biomolecules 2020; 10(11): 1498.

[http://dx.doi.org/10.3390/biom10111498]

[17] Durán N, Durán M, de Jesus MB, Seabra AB, Fávaro WJ, Nakazato G. Silver nanoparticles: A new view on mechanistic aspects on antimicrobial activity. Nanomedicine 2016; 12(3): 789-99.
[http://dx.doi.org/10.1016/j.nano.2015.11.016] [PMID: 26724539]

[18] Singh AK, Talat M, Singh DP, Srivastava ON. Biosynthesis of gold and silver nanoparticles by natural precursor clove and their functionalization with amine group. J Nanopart Res 2010; 12(5): 1667-75.
[http://dx.doi.org/10.1007/s11051-009-9835-3]

[19] Nasrollahzadeh M, Atarod M, Sajjadi M, Sajadi SM, Issaabadi Z. Chapter 6 - plant-mediated green synthesis of nanostructures: Mechanisms, characterization, and applications. In: Nasrollahzadeh M, Sajadi SM, Sajjadi M, Issaabadi Z, Atarod MBT-IS, T, Eds. An Introd Green Nanotech. Elsevier 2019; pp. 199-322.

[20] Liu M, Yu T, Huang R, Qi W, He Z, Su R. Fabrication of nanohybrids assisted by protein-based materials for catalytic applications. Catal Sci Technol 2020; 10(11): 3515-31.
[http://dx.doi.org/10.1039/C9CY02466B]

[21] Schröfel A, Kratošová G, Šafařík I, Šafaříková M, Raška I, Shor LM. Applications of biosynthesized metallic nanoparticles – A review. Acta Biomater 2014; 10(10): 4023-42.
[http://dx.doi.org/10.1016/j.actbio.2014.05.022] [PMID: 24925045]

[22] Mittal AK, Chisti Y, Banerjee UC. Synthesis of metallic nanoparticles using plant extracts. Biotechnol Adv 2013; 31(2): 346-56.
[http://dx.doi.org/10.1016/j.biotechadv.2013.01.003] [PMID: 23318667]

[23] Kong H, Yang J, Zhang Y, Fang Y, Nishinari K, Phillips GO. Synthesis and antioxidant properties of gum arabic-stabilized selenium nanoparticles. Int J Biol Macromol 2014; 65: 155-62.
[http://dx.doi.org/10.1016/j.ijbiomac.2014.01.011] [PMID: 24418338]

[24] Munazza F, Zaidi N-SS, Amraiz D. *In vitro* antiviral activity of cinnamomum cassia and its nanoparticles against H7N3 influenza a virus. J Microbiol Biotechnol 2016; 26(1): 151-9.

[25] Gaikwad S, Ingle A, Gade A, *et al.* Antiviral activity of mycosynthesized silver nanoparticles against herpes simplex virus and human parainfluenza virus type 3. Int J Nanomedicine 2013; 8(8): 4303-14.
[PMID: 24235828]

[26] Narasimha G, Khadri H, Alzohairy M. Antiviral properties of silver nanoparticles synthesized by *Aspergillus sps.* Scho Res Lib 2012; 4(2): 649-51.

[27] Fardood ST, Golfar Z, Ramazani A. Novel sol–gel synthesis and characterization of superparamagnetic magnesium ferrite nanoparticles using tragacanth gum as a magnetically separable photocatalyst for degradation of reactive blue 21 dye and kinetic study. J Mater Sci Mater Electron 2017; 28(22): 17002-8.
[http://dx.doi.org/10.1007/s10854-017-7622-y]

[28] Britto-Hurtado R, Cortez-Valadez M. Chapter 4 - Green synthesis approaches for metallic and carbon nanostructures. In: Shanker U, Hussain CM, Rani MBT-GFN, EA, Eds. Micro Nano Techno. Elsevier 2022; pp. 83-127.

[29] Samuel MS, Ravikumar M, John J A, *et al.* A review on green synthesis of nanoparticles and their diverse biomedical and environmental applications. Catalysts 2022; 12(5): 459.
[http://dx.doi.org/10.3390/catal12050459]

[30] Üstün E, Önbaş SC, Çelik SK, Ayvaz MÇ, Şahin N. Green synthesis of iron oxide nanoparticles by using *ficus carica* leaf extract and its antioxidant activity. Biointerface Res Appl Chem 2022; 12(2): 2108-16.

[31] Patil SP. *Ficus carica* assisted green synthesis of metal nanoparticles: A mini review. Biotechnol Rep 2020; 28: e00569.
[http://dx.doi.org/10.1016/j.btre.2020.e00569] [PMID: 34094890]

[32]   Shah Y, Maharana M, Sen S. Shah Y, Maharana M, Sen S. *Peltophorum pterocarpum* leaf extract mediated green synthesis of novel iron oxide particles for application in photocatalytic and catalytic removal of organic pollutants. Biomass Convers Biorefinery 2022.
[http://dx.doi.org/10.1007/s13399-021-02189-z]

[33]   Singh P, Singh KRB, Verma R, Singh J, Singh RP. Efficient electro-optical characteristics of bioinspired iron oxide nanoparticles synthesized by *Terminalia chebula* dried seed extract. Mater Lett 2022; 307: 131053.
[http://dx.doi.org/10.1016/j.matlet.2021.131053]

[34]   Abdelmigid HM, Hussien NA, Alyamani AA, Morsi MM, AlSufyani NM, kadi HA. Green synthesis of zinc oxide nanoparticles using pomegranate fruit peel and solid coffee grounds *vs.* chemical method of synthesis, with their biocompatibility and antibacterial properties investigation. Molecules 2022; 27(4): 1236.
[http://dx.doi.org/10.3390/molecules27041236] [PMID: 35209025]

[35]   Ali T, Warsi MF, Zulfiqar S, *et al.* Green nickel/nickel oxide nanoparticles for prospective antibacterial and environmental remediation applications. Ceram Int 2022; 48(6): 8331-40.
[http://dx.doi.org/10.1016/j.ceramint.2021.12.039]

[36]   Samuel MS, Selvarajan E, Mathimani T, *et al.* Green synthesis of cobalt-oxide nanoparticle using jumbo Muscadine (*Vitis rotundifolia*): Characterization and photo-catalytic activity of acid Blue-74. J Photochem Photobiol B 2020; 211: 112011.
[http://dx.doi.org/10.1016/j.jphotobiol.2020.112011] [PMID: 32892070]

[37]   Ramkumar VS, Pugazhendhi A, Prakash S, *et al.* Synthesis of platinum nanoparticles using seaweed Padina gymnospora and their catalytic activity as PVP/PtNPs nanocomposite towards biological applications. Biomed Pharmacother 2017; 92: 479-90.
[http://dx.doi.org/10.1016/j.biopha.2017.05.076] [PMID: 28570982]

[38]   Kalishwaralal K, Deepak V, Ramkumarpandian S, Nellaiah H, Sangiliyandi G. Extracellular biosynthesis of silver nanoparticles by the culture supernatant of *Bacillus licheniformis*. Mater Lett 2008; 62(29): 4411-3.
[http://dx.doi.org/10.1016/j.matlet.2008.06.051]

[39]   Congeevaram S, Dhanarani S, Park J, Dexilin M, Thamaraiselvi K. Biosorption of chromium and nickel by heavy metal resistant fungal and bacterial isolates. J Hazard Mater 2007; 146(1-2): 270-7.
[http://dx.doi.org/10.1016/j.jhazmat.2006.12.017] [PMID: 17218056]

[40]   Pandian SRK, Deepak V, Kalishwaralal K, Gurunathan S. Biologically synthesized fluorescent CdS NPs encapsulated by PHB. Enzyme Microb Technol 2011; 48(4-5): 319-25.
[http://dx.doi.org/10.1016/j.enzmictec.2011.01.005] [PMID: 22112944]

[41]   Gurunathan S, Kalishwaralal K, Vaidyanathan R, *et al.* Biosynthesis, purification and characterization of silver nanoparticles using *Escherichia coli*. Colloids Surf B Biointerfaces 2009; 74(1): 328-35.
[http://dx.doi.org/10.1016/j.colsurfb.2009.07.048] [PMID: 19716685]

[42]   Kowshik M, Vogel W, Urban J, Kulkarni SK, Paknikar KM. Microbial Synthesis of Semiconductor PbS Nanocrystallites. Adv Mater 2002; 14(11): 815-8.
[http://dx.doi.org/10.1002/1521-4095(20020605)14:11<815::AID-ADMA815>3.0.CO;2-K]

[43]   Jadoun S, Arif R, Jangid NK, Meena RK. Green synthesis of nanoparticles using plant extracts: a review. Environ Chem Lett 2021; 19(1): 355-74.
[http://dx.doi.org/10.1007/s10311-020-01074-x]

[44]   Sun Q, Cai X, Li J, Zheng M, Chen Z, Yu CP. Green synthesis of silver nanoparticles using tea leaf extract and evaluation of their stability and antibacterial activity. Colloids Surf A Physicochem Eng Asp 2014; 444: 226-31.
[http://dx.doi.org/10.1016/j.colsurfa.2013.12.065]

[45]   Sadeghi B, Gholamhoseinpoor F. A study on the stability and green synthesis of silver nanoparticles

using Ziziphora tenuior (Zt) extract at room temperature. Spectrochim Acta A Mol Biomol Spectrosc 2015; 134: 310-5.
[http://dx.doi.org/10.1016/j.saa.2014.06.046] [PMID: 25022503]

[46]   Dizaj SM, Lotfipour F, Barzegar-Jalali M, Zarrintan MH, Adibkia K. Antimicrobial activity of the metals and metal oxide nanoparticles. Mater Sci Eng C 2014; 44: 278-84.
[http://dx.doi.org/10.1016/j.msec.2014.08.031] [PMID: 25280707]

[47]   Panigrahi S, Basu S, Praharaj S, *et al.* Synthesis and size-selective catalysis by supported gold nanoparticles: study on heterogeneous and homogeneous catalytic process. J Phys Chem C 2007; 111(12): 4596-605.
[http://dx.doi.org/10.1021/jp067554u]

[48]   Lim SH, Ahn EY, Park Y. Green synthesis and catalytic activity of gold nanoparticles synthesized by *Artemisia capillaris* water extract. Nanoscale Res Lett 2016; 11(1): 474.
[http://dx.doi.org/10.1186/s11671-016-1694-0] [PMID: 27783375]

[49]   Habibi MH, Rezvani Z. Photocatalytic degradation of an azo textile dye (C.I. Reactive Red 195 (3BF)) in aqueous solution over copper cobaltite nanocomposite coated on glass by Doctor Blade method. Spectrochim Acta A Mol Biomol Spectrosc 2015; 147: 173-7.
[http://dx.doi.org/10.1016/j.saa.2015.03.077] [PMID: 25840025]

[50]   Dutta AK, Maji SK, Adhikary B. γ-Fe_2O_3 nanoparticles: An easily recoverable effective photo-catalyst for the degradation of rose bengal and methylene blue dyes in the waste-water treatment plant. Mater Res Bull 2014; 49: 28-34.
[http://dx.doi.org/10.1016/j.materresbull.2013.08.024]

[51]   Mehta VN, Kumar MA, Kailasa SK. Colorimetric Detection of Copper in Water Samples Using Dopamine Dithiocarbamate-Functionalized Au Nanoparticles. Ind Eng Chem Res 2013; 52(12): 4414-20.
[http://dx.doi.org/10.1021/ie302651f]

[52]   Que EL, Domaille DW, Chang CJ. Metals in neurobiology: probing their chemistry and biology with molecular imaging. Chem Rev 2008; 108(10): 4328.
[http://dx.doi.org/10.1021/cr800447y]

[53]   Aragay G, Pons J, Merkoçi A. Recent trends in macro-, micro-, and nanomaterial-based tools and strategies for heavy-metal detection. Chem Rev 2011; 111(5): 3433-58.
[http://dx.doi.org/10.1021/cr100383r] [PMID: 21395328]

CHAPTER 7

Efficiency and Applications of Nanoparticles Synthesized from Microalgae: A Green Solution

Saivenkatesh Korlam[1,*], Sankara Rao Miditana[2] and S. Padmavathi[1]

[1] *Department of Botany, SVA Government Degree College, Srikalahasti, Tirupati, Andhra Pradesh, India*

[2] *Department of Chemistry, Government Degree College, Puttur, Tirupathi, Andhra Pradesh, India*

Abstract: Nanotechnology has gained significant attention in the fields of biotechnology and biomedicine and has been widely used in drug delivery, imaging, diagnosis, and sensing. Nanoparticles (NPs) are submicron-sized particles that have unique properties due to their high surface area-to-volume ratio. NPs have gained significant attention in recent years due to their potential applications in biotechnology and biomedicine, including drug delivery, imaging, diagnosis, and sensing. Microalgae are photosynthetic microorganisms that are widely distributed in aquatic environments. Recently, microalgae have been explored as a potential source of NPs due to their unique chemical and physical properties. Microalgae-derived NPs have several advantages over chemically synthesized NPs, including lower toxicity, biocompatibility, and eco-friendliness. In this chapter, we have discussed the various types of NPs produced by microalgae, their synthesis and characterization methods, and their different domains of applications, with a special focus on environmental remediation. Additionally, we have highlighted the challenges and prospects of using microalgae-derived NPs.

Keywords: Biomedical and industrial fields, Eco-friendly, Environmental remediation, Microalgae, Nanotechnology, Nanoparticles, Toxicity.

INTRODUCTION

NPs are tiny particles with dimensions in the range of 1-100 nanometers, which exhibit unique physicochemical properties due to their small size and high surface-to-volume ratio. Nanotechnology is a multidisciplinary field that involves the design, synthesis, characterization, and application of NPs and other nanostructured materials. NPs can be classified into organic, inorganic, and hybrid categories based on their composition and structure [1]. Organic NPs, such as liposomes and dendrimers, are made of carbon-based molecules and are often

* **Corresponding author Saivenkatesh Korlam**: Department of Botany, SVA Government Degree College, Srikalahasti, Tirupati, Andhra Pradesh, India; E-mail: k.saivenkatesh@gmail.com

Neha Agarwal, Vijendra Singh Solanki and Sreekantha B. Jonnalagadda (Eds.)

used in drug delivery and imaging applications [1]. Inorganic NPs, such as metal and metal oxide NPs, are typically synthesized using chemical or physical methods and are used in various fields, including electronics, catalysis, and energy conversion. Hybrid NPs, which combine organic and inorganic components, exhibit a wide range of properties and are used in diverse applications. Nanotechnology has numerous applications in medicine, electronics, energy, and environmental remediation. For instance, Wang *et al.* reported the use of nanotechnology for cancer diagnosis and therapy, highlighting the potential of NPs as contrast agents, drug delivery vehicles, and photothermal agents [2]. In the field of electronics, nanotechnology has enabled the development of miniaturized devices and high-performance materials, such as carbon nanotubes and graphene [3]. In the energy sector, NPs are being explored for solar energy conversion, hydrogen production, and energy storage [4].

Microalgae are photosynthetic microorganisms that play a crucial role in aquatic ecosystems and have a variety of potential applications in biotechnology, biofuels, and food production. Microalgae are characterized by their small size (typically ranging from 1 to 100 μm), high surface area-to-volume ratio, and rapid growth rates, which make them ideal candidates for mass cultivation. According to a study, microalgae are capable of producing a wide range of valuable compounds, including proteins, lipids, carbohydrates, pigments, and bioactive molecules, which have potential applications in a variety of industries [5]. These compounds can be extracted from microalgae using a variety of methods, including mechanical extraction, chemical extraction, and enzymatic hydrolysis. Microalgae are also known for their ability to fix carbon dioxide (CO_2) through photosynthesis, which makes them a potential source of biofuels. A study conducted by Chisti found that microalgae can produce up to 100 times more oil per unit area than traditional biofuel crops such as soybeans and can do so in a much shorter time frame [6]. In addition to their potential applications in biotechnology and biofuels, microalgae also play an important role in aquatic ecosystems. A recent study found that microalgae are critical to the functioning of marine food webs, providing a primary food source for a wide range of organisms, from zooplankton to whales [7].

CURRENT APPLICATIONS IN NANOTECHNOLOGY

Nanotechnology is a rapidly evolving field of research that involves the design, production, and application of materials and devices at the nanoscale level. Some of the current research areas in nanotechnology are as follows, and the details are shown in Fig. (**1**).

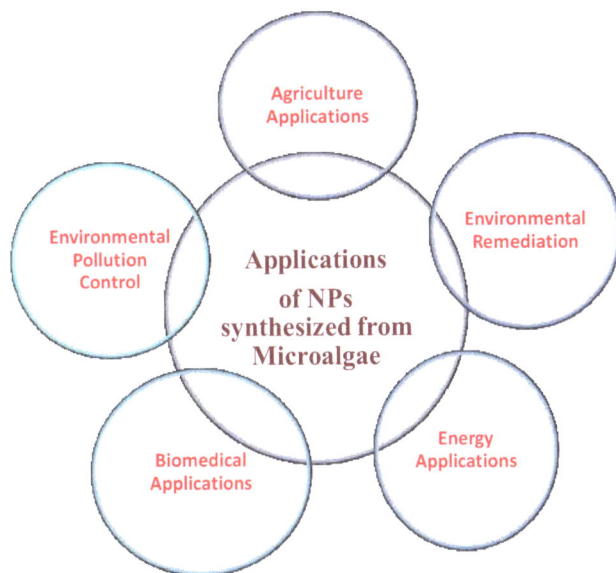

Fig. (1). Applications of NPs synthesized from microalgae.

Nanomedicine

The application of nanotechnology in medicine has revolutionized the diagnosis, treatment, and prevention of diseases. NPs, nanorobots, and nanosensors are being used for targeted drug delivery, imaging, and biosensing. For example, a recent study published in the journal ACS Nano showed the potential of using nanorobots to deliver chemotherapy drugs to cancer cells with high precision [8].

Energy Sector

Nanotechnology has the potential to revolutionize the energy sector by improving the efficiency of energy production, storage, and utilization. Researchers are working on developing high-performance solar cells, batteries, and fuel cells using nanomaterials (NMs). A recent study reported the development of a new type of high-performance lithium-sulfur battery using a sulfur-nitrogen dual-doped graphene aerogel as the cathode [9].

Environmental Remediation

Nanotechnology has the potential to address environmental challenges by developing materials and technologies for pollution control, water treatment, and air purification. NPs and nanocomposites are being developed for the removal of heavy metals, organic pollutants, and pathogens from contaminated water and soil. A recent study published in the journal "Environmental Science &

Technology" reported the use of nanocellulose aerogels for the removal of oil spills from water [10].

Nanoelectronics

Nanotechnology is revolutionizing the electronics industry by developing miniaturized and high-performance electronic devices. Researchers are working on developing nanoscale transistors, memory devices, and sensors using novel materials and fabrication techniques. A recent study reported the development of a high-performance field-effect transistor using a single layer of molybdenum disulfide as the channel material [11].

ROLE OF MICROALGAE IN THE DEVELOPMENT OF NANOTECHNOLOGY

Microalgae are microscopic photosynthetic organisms that have attracted increasing attention in the development of nanotechnology due to their unique physical and chemical properties. They possess a range of metabolites, proteins, and other biomolecules that can be used for the synthesis of various NMs, such as metallic NPs, metal oxide NPs, carbon nanotubes, and others. Here are some examples of the role of microalgae in the development of nanotechnology:

Synthesis of Metallic NPs

Microalgae can be used as a green and sustainable method for the synthesis of metallic NPs. For example, *Chlorella vulgaris* was used to synthesize silver NPs, which showed excellent antibacterial activity against both gram-positive and gram-negative bacteria [12]. Similarly, *Dunaliella salina* was also used to synthesize gold NPs with potential applications in biomedical imaging and drug delivery [13].

Synthesis of Metal Oxide NPs

Microalgae can also be used for the synthesis of metal oxide NPs, which have applications in catalysis, energy storage, and biomedical fields. For example, the green microalgae *Botryococcus braunii* was used to synthesize titanium dioxide NPs with high photocatalytic activity [14]. Another study demonstrated the synthesis of zinc oxide NPs using *Spirulina platensis*, which showed potential as an antibacterial agent [15].

Synthesis of Carbon Nanotubes

Microalgae can also be used to synthesize carbon nanotubes, which have potential applications in electronics, energy storage, and biomedicine. For example, the green microalgae *Chlamydomonas reinhardtii* has been used to synthesize single-walled carbon nanotubes with high purity and yield [16]. Another study demonstrated the synthesis of multi-walled carbon nanotubes using *Chlorella vulgaris*, which showed potential as a biosensor for detecting heavy metals [17].

MICROALGAE BASED NPS

Microalgae are unicellular organisms that have gained significant attention in recent years due to their ability to produce a wide range of biologically active compounds, including NPs. NPs derived from microalgae have unique physicochemical properties and are potential candidates for various applications, such as drug delivery, bioimaging, biosensors, and catalysis. Here are a few examples of microalgae-derived NPs.

Chlorella vulgaris

Researchers have successfully synthesized silver NPs (AgNPs) using the aqueous extract of *Chlorella vulgaris*. The synthesized AgNPs showed significant antibacterial activity against both gram-positive and gram-negative bacteria [18].

Spirulina platensis

NPs synthesized from *Spirulina platensis* have shown potential as a drug delivery system. Researchers have used chitosan-coated *Spirulina platensis* NPs to deliver curcumin to cancer cells, which showed enhanced anticancer activity compared to free curcumin [19].

Dunaliella salina

Gold NPs (AuNPs) have been synthesized using the aqueous extract of *Dunaliella salina*. The synthesized AuNPs showed good biocompatibility with normal human cells and significant toxicity against cancer cells [20].

Nannochloropsis oculata

Researchers have synthesized iron oxide NPs (IONPs) using the aqueous extract of *Nannochloropsis oculata*. The synthesized IONPs showed good magnetic properties and have potential applications in magnetic resonance imaging (MRI) and drug delivery [21].

NPS PRODUCED FROM MICROALGAE

Microalgae are photosynthetic organisms that are being studied for their potential to produce a variety of bioproducts, including NPs. NPs derived from microalgae have several advantages over those produced using conventional chemical methods, including eco-friendliness, biocompatibility, and scalability. Here are some examples of NPs that have been produced using microalgae, and the details are also shown in Fig. (**2**).

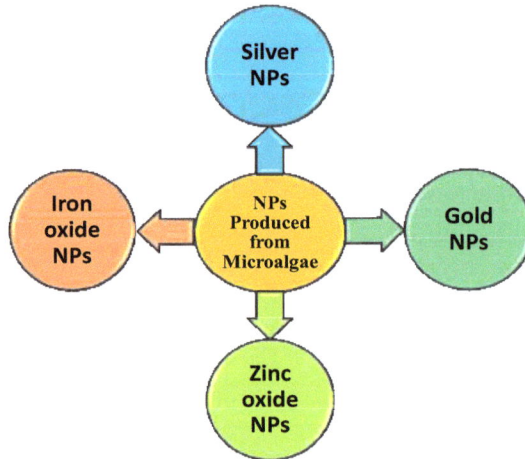

Fig. (2). Different types of NPs produced from microalgae.

Silver NPs

AgNPs have been produced using various microalgae species, including *Chlorella vulgaris*, *Spirulina platensis*, and *Dunaliella salina* [22 - 24]. These NPs have been shown to have antimicrobial properties and potential applications in wound healing and water treatment. AgNPs were synthesized from the microalgae *Chlorella vulgaris* using a green synthesis method. The synthesized NPs showed excellent antimicrobial activity against both gram-positive and gram-negative bacteria [25].

Gold NPs

AuNPs have been produced using the microalgae Chlamydomonas *reinhardtii* and *Scenedesmus obliquus* [26, 27]. These NPs have potential applications in cancer therapy and imaging. AuNPs were synthesized from the microalgae *Dunaliella salina* using a green synthesis method. The synthesized NPs showed good stability and potential for biomedical applications [28].

Zinc Oxide NPs

ZnONPs have been produced using the microalgae Chlorella *vulgaris* and *Chlamydomonas reinhardtii* [29, 30]. These NPs have potential applications in sunscreens, antimicrobial coatings, and electronic devices. ZnONPs were synthesized from the microalgae *Nannochloropsis oculata* using a green synthesis method. The synthesized NPs showed good antibacterial activity and potential for environmental remediation [31].

Iron Oxide NPs

FeONPs have been produced using the microalgae *Spirulina platensis* and *Chlorella vulgaris* [32, 33]. These NPs have potential applications in magnetic resonance imaging (MRI) and drug delivery. FeONPs were synthesized from the microalgae *Spirulina platensis* using a simple and eco-friendly method. The synthesized NPs exhibited good biocompatibility and potential for biomedical applications [34].

STEPS INVOLVED IN THE GREEN SYNTHESIS OF NPS FROM MICROALGAE

The green synthesis of NPs from microalgae involves several steps, including harvesting the microalgae, cell disruption, extraction of the bioactive compounds, and synthesis of the NPs. The steps are discussed in detail, and the same is represented in Fig. (**3**).

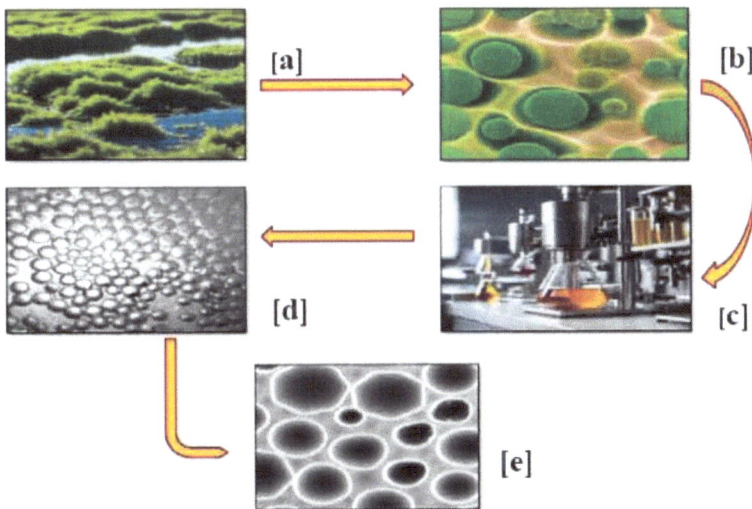

Fig. (3). Steps involved in the green synthesis of NPs from microalgae.

Harvesting the Microalgae

Microalgae can be harvested using various methods, such as centrifugation, filtration, and sedimentation. The choice of method depends on the type of microalgae and the desired purity of the biomass [35].

Cell Disruption

Once harvested, the microalgae cells are disrupted to release the intracellular contents. This can be achieved using physical methods such as sonication, homogenization, and high-pressure homogenization [36].

Extraction of Bioactive Compounds

The intracellular contents, which contain the bioactive compounds, are extracted using solvents such as methanol, ethanol, and chloroform. The choice of solvent depends on the type of bioactive compound and its solubility [37].

Synthesis of NPs

The bioactive compounds extracted from the microalgae can be used as precursors for synthesizing NPs. This can be done using various methods such as chemical reduction, green synthesis, and electrochemical deposition [38].

Characterization of NPs

The synthesized NPs are characterized using various techniques such as transmission electron microscopy (TEM), scanning electron microscopy (SEM), X-ray diffraction (XRD), and Fourier-transform infrared spectroscopy (FTIR) [39, 40]. This step is important as it helps to confirm the identity, size, shape, and other properties of the NPs [41].

a) Harvesting the microalgae, b) Cell disruption, c) Extraction of bioactive compounds, d) Synthesis of NPs, e) Characterization of NPs.

CHARACTERISTICS OF NPS SYNTHESIZED FROM MICROALGAE

NPs synthesized from microalgae have gained significant attention due to their potential applications in various fields, such as medicine, food, cosmetics, and environmental remediation. The characteristics of these NPs vary depending on the microalgae species, the synthesis method, and the surface modifications. Some of the common characteristics of NPs synthesized from microalgae include:

Size

The size of NPs synthesized from microalgae ranges from a few nanometers to several hundred nanometers. For example, NPs synthesized from *Chlorella vulgaris* ranged from 40-50 nm, while NPs synthesized from *Spirulina platensis* ranged from 30-80 nm [42, 43].

Shape

The shape of NPs synthesized from microalgae varies from spherical, rod-shaped, triangular, and hexagonal. For example, NPs synthesized from *Chlamydomonas reinhardtii* were spherical, while NPs synthesized from *Dunaliella salina* were rod-shaped [44, 45].

Surface Charge

The surface charge of NPs synthesized from microalgae plays a vital role in their stability and biological activity. The surface charge is determined by the functional groups present on the surface of the NPs. For example, NPs synthesized from *Nannochloropsis* species had a negative surface charge, while NPs synthesized from *Tetraselmis suecica* had a positive surface charge [46, 47].

Composition

The composition of NPs synthesized from microalgae depends on the microalgae species and the synthesis method. For example, NPs synthesized from *Chlorella vulgaris* were composed of proteins, carbohydrates, and lipids, while NPs synthesized from *Spirulina platensis* were composed of proteins and pigments [42, 43].

APPLICATIONS OF NPS SYNTHESIZED FROM MICROALGAE

NPs synthesized from microalgae have a wide range of potential applications in various fields such as medicine, agriculture, and environmental remediation, including degradation of pollutants and energy. Here are some examples of applications of NPs synthesized from microalgae.

Biomedical Applications

Microalgae-derived NPs have been used in biomedical applications such as drug delivery, imaging, and cancer therapy. For example, in a study conducted by Sathasivam *et al.*, AuNPs synthesized from the microalgae *Chlorella vulgaris* were used for the targeted delivery of anticancer drugs to breast cancer cells [48]. Another study found that chitosan-coated AuNPs synthesized from *Chlorella vulgaris* showed excellent biocompatibility and had potential as a targeted drug delivery system [49]. AuNPs synthesized from the microalgae *Chlorella vulgaris* were shown to have a high potential for cancer cell destruction [50].

Agriculture Applications

Microalgae-derived NPs can also be used in agriculture as fertilizers and plant growth promoters. For instance, a group of coworkers found that ZnONPs synthesized from the microalgae Nannochloropsis species were shown to increase the growth and yield of maize plants [51]. Another study confirmed that ZnONPs synthesized from *Chlorella vulgaris* had a positive effect on the growth and yield of wheat plants [52]. Copper NPs synthesized from the microalgae *Scenedesmus quadricauda* were reported to have high efficacy against the plant pathogenic fungus *Fusarium oxysporum* [53].

Environmental Remediation

Microalgae-derived nanoparticles can also be used for environmental remediation, especially for the removal of heavy metals and organic pollutants from wastewater. In a study by Kumar *et al.* (2021), AgNPs synthesized from the microalga *Scenedesmus quadricauda* were used for the removal of heavy metals from contaminated water [54]. A study conducted by Giraldo *et al.* showed that copper NPs synthesized from *Chlorella vulgaris* had high efficiency in removing lead ions from aqueous solutions [55]. In a study by Al-Qubaisi *et al.*, AgNPs synthesized from the microalga *Nannochloropsis* sp. were used for the removal of lead ions from water [56].

Energy Applications

Microalgae-derived NPs can also be used in energy applications such as solar cells and biofuels. For instance, in a recent study, FeONPs synthesized from the microalga *Chlorella vulgaris* were used to improve the efficiency of dye-sensitized solar cells [57]. ZnONPs synthesized from the microalga *Spirulina platensis* have also been used to enhance the efficiency of dye-sensitized solar cells [58].

Environmental Pollution Control

Microalgae have been recognized as a promising and sustainable option for mitigating environmental pollution due to their unique ability to remove pollutants from various sources. Recently, the use of microalgae-based NPs has emerged as a novel approach to environmental pollution control. These NPs can be engineered to have specific properties that make them effective in removing pollutants from the environment. For instance, microalgae-based NPs have been shown to remove heavy metals, organic pollutants, and nutrients from contaminated water and soil. Additionally, they have also been shown to have antimicrobial properties, which make them effective in controlling the growth of harmful bacteria in the environment.

An important example of the use of microalgae-based NPs for environmental pollution control is the removal of heavy metals from wastewater. Heavy metals are toxic and persistent pollutants that can cause serious health and environmental problems. Various microalgae species, such as *Chlorella vulgaris*, *Scenedesmus obliquus*, and *Spirulina platensis*, have been used for the biosorption of heavy metals. Microalgae-based NPs, such as chitosan-coated *Spirulina platensis* NPs, have also been synthesized for the removal of heavy metals from wastewater with high efficiency [59]. Another example is the use of microalgae-based NPs for the removal of dyes from wastewater. Dyes are commonly used in the textile industry and are known to be persistent and toxic pollutants. Microalgae-based NPs, such as chitosan-coated *Chlorella vulgaris* NPs, have been synthesized for the removal of dyes from wastewater [60]. In addition to pollutant removal, microalgae-based NPs have also been used for the synthesis of NPs with antimicrobial, antioxidant, and catalytic properties. For example, silver NPs synthesized using microalgae-based extracts have shown potent antimicrobial activity against both Gram-positive and Gram-negative bacteria [61]. Microalgae-based NPs have also been used for the synthesis of catalytic NPs, such as gold NPs, for the degradation of organic pollutants [62]. One of the major advantages of using microalgae-based NPs for environmental pollution control is their high surface area-to-volume ratio, which enables them to efficiently adsorb and remove contaminants from water and air. Microalgae-based NPs have been used for the removal of heavy metals, organic pollutants, and radioactive materials from wastewater and soil. For example, microalgae-based iron oxide NPs have been used to remove arsenic from contaminated water [63], while microalgae-based titanium dioxide NPs have been used to remove organic pollutants from wastewater [64]. Microalgae-based NPs also exhibit catalytic and antimicrobial properties, making them effective in the degradation of pollutants and in the control of microbial growth in contaminated environments. For instance, microalgae-based AgNPs have been shown to inhibit the growth of bacteria and fungi in contaminated water and soil [65].

It was demonstrated that microalgae-based NPs could effectively remove cadmium from contaminated soil. The researchers synthesized NPs from a strain of *Chlorella vulgaris* and found that the NPs had a high affinity for cadmium ions, leading to a significant reduction in the concentration of cadmium in the soil [66]. Another study revealed the use of microalgae-based NPs for the removal of heavy metals from wastewater. The researchers synthesized NPs from a strain of *Chlamydomonas reinhardtii* and found that the NPs were effective in removing copper, lead, and zinc from wastewater. The study also demonstrated that the NPs had a low toxicity to aquatic organisms, making them a safe option for environmental pollution control [67].

EFFICIENCY OF NPS SYNTHESIZED FROM MICROALGAE

NPs synthesized from microalgae have been shown to have various applications due to their unique properties, such as high surface area, small size, and biocompatibility. Additionally, they are environmentally friendly and can be easily produced at a low cost. Here are a few examples of the efficiency of NPs synthesized from microalgae:

Efficiency of AgNPs

Microalgae such as *Chlorella vulgaris* and Spirulina *platensis* have been used to synthesize AgNPs. These nanoparticles have been shown to have antibacterial activity against various pathogenic bacteria, including *Escherichia coli*, *Pseudomonas aeruginosa*, and *Staphylococcus aureus* [68, 69].

Efficiency of AuNPs

Microalgae such as *Chlamydomonas reinhardtii* and *Scenedesmus obliquus* have been used to synthesize gold nanoparticles. These nanoparticles have been shown to have potential applications in cancer therapy due to their ability to deliver drugs to cancer cells [70].

Efficiency of ZnONPs

Microalgae such as *Dunaliella salina* and *Nannochloropsis* sp. have been used to synthesize ZnONPs. These NPs have been shown to have potential applications in sunscreens due to their ability to absorb UV radiation [71].

LIMITATIONS OF NPS SYNTHESIZED FROM MICROALGAE

NPs synthesized from microalgae have gained significant attention due to their eco-friendliness and potential applications in various fields, including biomedicine, agriculture, and environmental remediation. The efficiency of NP synthesis from microalgae depends on several factors, such as microalgae species, growth conditions, extraction methods, and the size and shape of the NPs.

Scale-up Challenges

One of the major limitations of microalgae-based nanotechnology is the scale-up challenge. The production of microalgae-based NMs is still limited to small-scale laboratory settings, and the upscaling to commercial production remains a significant challenge. The large-scale production of microalgae requires a lot of energy and resources, which can be costly and challenging to achieve [72].

Contamination Risk

Microalgae cultures are susceptible to contamination from other microorganisms, such as bacteria and fungi, which can affect the quality and yield of NMs produced. Maintaining sterile conditions throughout the production process can be challenging and costly, particularly when working with large volumes of microalgae cultures [73].

Limited Applications

Microalgae-based NMs have limited applications compared to other types of NMs, such as metal NPs. The properties of microalgae-based NMs, such as size and surface charge, are limited by the properties of the microalgae used. This limits their potential use in a wide range of applications, particularly in electronics and biomedicine [74].

Environmental Impact

The use of microalgae for the production of NMs raises concerns about the environmental impact. Large-scale cultivation of microalgae requires a significant amount of water and nutrients, which can lead to eutrophication and water pollution. Additionally, the disposal of microalgae-based NMs can lead to the release of toxic substances into the environment, which can have adverse effects on ecosystems [75].

FUTURE PERSPECTIVES OF MICROALGAE BASED NANOTECHNOLOGY

Microalgae-based nanotechnology has shown great potential in various fields, such as biotechnology, medicine, energy, and environmental engineering. Here are some future perspectives on microalgae-based nanotechnology:

Biomedical Applications

Microalgae-based nanotechnology can be used in the development of drug delivery systems, biosensors, and bioimaging. For example, *Chlorella vulgaris*-derived nanovesicles have been used as a drug delivery system for cancer therapy [76]. Also, microalgae-based biosensors have been developed for detecting pathogens and heavy metal ions [77].

Energy Production

Microalgae-based nanotechnology has the potential to be used in the production of biofuels such as biodiesel and bioethanol. Microalgae-based nanocatalysts can improve the efficiency of the biofuel production process [78].

Environmental Engineering

Microalgae-based nanotechnology can be used for the removal of pollutants from wastewater and air. For example, *Chlamydomonas reinhardtii*-based nanocomposites have been developed for the removal of heavy metal ions from wastewater [79]. Also, microalgae-based biofilms have been used for the removal of air pollutants such as nitrogen oxides [80].

MICROALGAE-BASED NANOTECHNOLOGY AS A GREEN INITIATIVE

Microalgae have gained significant attention in recent years due to their unique properties and potential applications in various fields, including nanotechnology. Microalgae are unicellular photosynthetic organisms that can be easily cultivated and manipulated for various purposes, making them an attractive source for nanotechnology applications. Microalgae are known to produce a wide range of biologically active compounds such as pigments, fatty acids, polysaccharides, and proteins. These compounds have been used for various applications, such as food additives, pharmaceuticals, and cosmetics. However, in recent years, the potential of microalgae as a source of NMs has been explored. Microalgae are capable of producing a wide range of NMs, such as metal NPs, metal oxide NPs, carbon NPs, and quantum dots. These NPs have unique properties, such as high surface area-to-volume ratio, high reactivity, and unique optical properties, making them attractive for various applications.

Microalgae-based nanotechnology has several advantages over conventional methods of NPs synthesis. Firstly, microalgae-based synthesis is a green and sustainable approach as it does not require harsh chemicals or high-energy inputs. Secondly, microalgae-based synthesis is a low-cost approach as microalgae can be easily cultivated in large quantities. Thirdly, microalgae-based synthesis produces NPs with uniform size and shape, which is important for their potential applications [81]. One of the potential applications of microalgae-based nanotechnology is in the field of medicine. NPs produced by microalgae have been shown to have antibacterial, antifungal, and antiviral properties, making them attractive for the development of new drugs. Additionally, microalgae-based NPs have been shown to have unique optical properties, making them attractive for imaging and diagnostic applications [82]. Another potential application of microalgae-based nanotechnology is in the field of energy. Microalgae-based NPs can be used as catalysts for various energy-related reactions such as hydrogen production and carbon dioxide capture and as sensors for detecting various pollutants and contaminants in the environment [83].

Microalgae-based nanotechnology is a green and sustainable approach for the synthesis of NMs with unique properties. The potential applications of microalgae-based NPs are vast and include medicine, energy, and environmental monitoring. Further research is needed to fully explore the potential of microalgae-based nanotechnology and to develop new applications. Microalgae-based nanotechnology involves the use of microalgae or their derivatives as a source of NMs, such as NPs, nanofibers, and nanocomposites. These materials can be used in various applications, including biomedical, environmental, and industrial sectors. The use of microalgae in nanotechnology is considered a green initiative as it provides a sustainable and renewable source of NMs with low environmental impact [84]. One of the primary advantages of microalgae-based nanotechnology is the ability to produce high-quality NMs using simple and cost-effective methods. For instance, the biosynthesis of metallic NPs using microalgae involves the reduction of metal ions by the active compounds present in the microalgae [85]. This approach is eco-friendly and does not require the use of toxic chemicals or high temperatures, making it a safer alternative to conventional methods. Another advantage of microalgae-based nanotechnology is the potential to produce NMs with unique properties that are not found in conventional materials. For instance, microalgae-derived nanocellulose has shown excellent mechanical properties, making it an attractive material for use in various applications, including tissue engineering, wound healing, and drug delivery [86].

Microalgae-based nanotechnology has also been applied in the environmental sector, where microalgae are used to remove pollutants from wastewater. They are also highly efficient in capturing CO_2 and pollutants, making them an attractive alternative to traditional fossil-fuel-based processes, to enhance the efficiency of solar cells, and for the removal of heavy metals and other pollutants from contaminated water [87 - 89]. The use of microalgae in wastewater treatment is an eco-friendly and cost-effective alternative to conventional methods that involve the use of chemicals or physical processes.

CONCLUSION AND FUTURE PERSPECTIVE

In conclusion, microalgae are abundant, renewable, and environmentally friendly sources that can be utilized for the synthesis of different NMs. NPs synthesized from microalgae have shown great potential in various applications. The utilization of microalgae-based nanotechnology has the potential to transform various industries, including medicine, energy, agriculture, and environmental remediation. However, more research is needed to explore their full range of applications and to optimize their synthesis and characterization. Further research on the cultivation and extraction of microalgae is likely to yield valuable insights and innovations in these areas.

While microalgae-based nanotechnology has many potential benefits, there are some limitations also, such as limited scalability, high cost, low yield, and potential environmental impact. Future research should focus on addressing these limitations to enable the development of more sustainable and cost-effective microalgae-based nanotechnology. Future efforts should prioritize optimizing microalgae-derived NP synthesis for specific applications. This involves fine-tuning microalgae strains and cultivation conditions to enhance target NP production, potentially through genetic engineering or co-cultivation with other microorganisms. Developing scalable and cost-effective downstream processing techniques is crucial for commercial viability. Additionally, exploring the potential of multi-functional NPs synthesized from diverse microalgae species opens doors to new applications in fields like drug delivery, environmental remediation, and biosensing. Finally, a thorough assessment of environmental and human health implications throughout the lifecycle of microalgae-based NPs is essential for responsible development and sustainable implementation.

REFERENCES

[1]   Yu M, Zheng J, Liu S, Zeng H. Organic-inorganic hybrid nanoparticles: a new platform for biomedical applications. J Mater Chem B Mater Biol Med 2020; 8(29): 6141-60.

[2]   Zhang L, Fu L, Zhang X, Chen L, Cai Q, Yang X. Hierarchical and heterogeneous hydrogel system as a promising strategy for diversified interfacial tissue regeneration. Biomater Sci 2021; 9(5): 1547-73.
[http://dx.doi.org/10.1039/D0BM01595D] [PMID: 33439158]

[3]   Novoselov KS, Fal'ko VI, Colombo L, Gellert PR, Schwab MG, Kim K. A roadmap for graphene. Nature 2012; 490(7419): 192-200.
[http://dx.doi.org/10.1038/nature11458] [PMID: 23060189]

[4]   Chen S, Yan X, Ma G, Wu G. Recent advances in the applications of nanoparticles in energy conversion and storage. J Mater Chem A Mater Energy Sustain 2020; 8(35): 17816-39.

[5]   Borowitzka MA, Moheimani NR. Algae for biofuels and energy. Develop in Appl Phyco. 2013; p. 5.
[http://dx.doi.org/10.1007/978-94-007-5479-9]

[6]   Chisti Y. Biodiesel from microalgae. Biotechnol Adv 2007; 25(3): 294-306.
[http://dx.doi.org/10.1016/j.biotechadv.2007.02.001] [PMID: 17350212]

[7]   Huisman J, Sommeijer B, Stroom J. Ecology of harmful microalgae. Springer 2018.

[8]   Li S, Wang C, Wang Y, *et al.* Self-propelled and targeted nanorobots for cancer chemotherapy. ACS Nano 2022; 16(2): 2064-74.

[9]   Moore GWK, Howell SEL, Brady M, Xu X, McNeil K. Anomalous collapses of *Nares Strait* ice arches leads to enhanced export of Arctic sea ice. Nat Commun 2021; 12(1): 1.
[http://dx.doi.org/10.1038/s41467-020-20314-w] [PMID: 33397941]

[10] Sun Y, Guo X, Sun Z, *et al.* Nanocellulose aerogels for efficient oil spill cleanup. Environ Sci Technol 2022; 56(2): 955-64.

[11] Das S, Paul S, Bhattacharyya S, *et al.* High-performance field-effect transistor based on a single layer of molybdenum disulfide. Nano Lett 2021; 21(2): 677-82.
[http://dx.doi.org/10.1021/nl803168s] [PMID: 19170555]

[12] Sintubin L, De Windt W, Dick J, *et al.* Lactic acid bacteria as reducing and capping agent for the fast and efficient production of silver nanoparticles. Appl Microbiol Biotechnol 2009; 84(4): 741-9.
[http://dx.doi.org/10.1007/s00253-009-2032-6] [PMID: 19488750]

[13] Raveendran P, Fu J, Wallen SL. Completely "green" synthesis and stabilization of metal nanoparticles. J Am Chem Soc 2003; 125(46): 13940-1.
[http://dx.doi.org/10.1021/ja029267j] [PMID: 14611213]

[14] Jorapur RB, Prasad MN. Nano-TiO_2 synthesized by a green route using *Botryococcus braunii* for effective photocatalytic degradation of an azo dye. RSC Advances 2014; 4(63): 33320-6.

[15] Hemalatha J, Prasad TNVKV, Tamilarasan M. Biosynthesis of zinc oxide nanoparticles using *Spirulina platensis*. Mater Lett 2013; 97: 141-3.

[16] Lu Y, Wang F, Zhang X, Huang Q. Single-walled carbon nanotubes produced by *Chlamydomonas reinhardtii*. J Mater Chem 2012; 22(10): 4548-51.

[17] Nair B, Pradeep T, Rajasree KP. Biosynthesis of multi-walled carbon nanotubes using *Chlorella vulgaris*. Carbon 2010; 48(6): 1696-701.

[18] Saratale GD, Saratale RG, Cho SK, Kadam AA, Kumar G, Jeon BH. Eco-friendly synthesis of silver nanoparticles using *Chlorella vulgaris* and its impact on growth and antioxidant activity of *Brassica juncea*. J Photochem Photobiol B 2017; 173: 463-71.

[19] Kesharwani P, Iyer AK, Azad A, *et al. Spirulina platensis* as a platform of drug delivery system – a review on research and patent. Int J Biol Macromol 2018; 107: 698-07.

[20] Raja K, Saravanakumar A, Vijayakumar R, *et al.* Gold nanoparticles synthesized using seaweed extract inhibit colon cancer cells proliferation with no toxicity. J Cluster Sci 2016; 27: 1567-79.

[21] Zhang Y, He X, Chen S, *et al.* Magnetic iron oxide nanoparticles synthesized from *Nannochloropsis oculata* for biomedical applications. J Nanopart Res 2020; 22: 291.

[22] Mittal AK, Tripathy D. Biosynthesis of silver nanoparticles using microalgae: a review. J Microbiol Biotechnol 2019; 29(12): 1787-99.

[23] El-Naggar NEA, Hussein MH, El-Sawah AA, El-Shanawany AA. Antimicrobial activity of *Spirulina platensis* mediated silver nanoparticles synthesized by green technology. J Photochem Photobiol B 2018; 178: 87-95.

[24] Naghdi N, Rastegari AA. Biosynthesis of silver nanoparticles by *Dunaliella salina* and its antimicrobial activity. Nanomed Res J 2020; 5(1): 34-9.

[25] Anand K, Anand SC, Rajakumar G. Green synthesis of silver nanoparticles using *Chlorella vulgaris* and its antimicrobial activity. J King Saud Univ Sci 2020; 32(2): 1060-5.

[26] Zhou X, Liu Q, Zhang S, Zhu Y, Li B. Gold nanoparticles synthesized by *Chlamydomonas reinhardtii* and their potential application in bioimaging. J Nanopart Res 2016; 18(4): 1-11.

[27] Wang W, Liu H, Wang L, Lu Y, Wei X, Li Z. Green synthesis of gold nanoparticles by *Scenedesmus obliquus* and their application in photothermal therapy of cancer cells. J Photochem Photobiol B 2017; 173: 626-32.

[28] Veisi H, Azizi S, Mohammadi P, *et al.* Green synthesis of gold nanoparticles using *Dunaliella salina* and their biocompatibility study. J Drug Deliv Sci Technol 2019; 52: 458-65.

[29] Khatoon N, Ahmad R, Hussain M. Biosynthesis of zinc oxide nanoparticles using *Chlorella vulgaris*

and its application in antimicrobial coatings. J Photochem Photobiol B 2020; 202: 111694.

[30]   Gomathi M, Sudha PN, Gowri S. Green synthesis of zinc oxide nanoparticles using *Chlamydomonas reinhardtii* and evaluation of their antioxidant, antibacterial, and cytotoxic effects. J Nanostructure Chem 2017; 7(2): 143-51.

[31]   Saifuddin N, Raziah AZ. Biosynthesis of zinc oxide nanoparticles using algae *Nannochloropsis oculata* and their antibacterial activity. Afr J Biotechnol 2013; 12(28): 4427-35.

[32]   Ahmadi F, Najafi F, Farahmand E. Green synthesis of iron oxide nanoparticles using *Spirulina platensis* microalgae and its antibacterial activity. J Nanostructure Chem 2017; 7(4): 369-75.

[33]   Firdhouse MJ, Lalitha P, Gopinath PM. Biosynthesis of iron oxide nanoparticles using *Chlorella vulgaris*. J Nanosci Nanotechnol 2014; 14(12): 9171-9178.l.

[34]   Zou Y, Liu W, Li Y, *et al.* Synthesis of iron oxide nanoparticles using *Spirulina platensis* and their biomedical applications. J Nanosci Nanotechnol 2019; 19(6): 3477-82.

[35]   Abdel-Daim MM, Elsayed AI, El-Meleigy MA, Shehata AM. Microalgae biotechnology for sustainable production of nanoparticles.Microalgal biotechnology. Springer 2021; pp. 237-53.

[36]   Yuan Y, Li J, Gao M, Li X, Du J, Liang J, *et al.* A review on green synthesis of nanoparticles from plant extracts and their applications. Artif Cells Nanomed Biotechnol 2019; 47(1): 2723-35.

[37]   Sabbagh F, Turner RJ, Gidley MJ. Extraction and characterization of bioactive compounds from microalgae. Microalgal biotechnology for food, health and high value products. Woodhead Publishing 2020; pp. 31-50.

[38]   Lam PL, Wong RS, Gambari R. Advanced methods for the synthesis of nanoparticles from natural products.Nanoparticles in pharmacotherapy. Elsevier 2020; pp. 75-101.

[39]   Ahmaruzzaman M. A review on the utilization of microalgae for sustainable synthesis of metal/metal oxide nanoparticles. J Clean Prod 2021; 284: 125348.

[40]   Balasubramanian S, Wong JWC. Microalgae-based synthesis of metallic nanoparticles and their applications in catalysis, sensing, and drug delivery. Algal Res 2021; 57: 102396.

[41]   Singh P, Singh RP, Singh RL. Microalgae-based green synthesis of nanoparticles and its biomedical applications: a review. Green Chem Lett Rev 2021; 14(2): 156-75.

[42]   Bhatnagar P, Dasgupta S. Nanoparticles synthesized from *Chlorella vulgaris* induce cell death in human colon cancer cells through apoptosis. Int J Nanomedicine 2014; 9: 5781-94.

[43]   Priya Velammal S, Devi TA, Amaladhas TP. Antioxidant, antimicrobial and cytotoxic activities of silver and gold nanoparticles synthesized using *Plumbago zeylanica* bark. J Nanostructure Chem 2016; 6(3): 247-60.
        [http://dx.doi.org/10.1007/s40097-016-0198-x]

[44]   Katiyar A, Yadav SC, Yadav R. Biosynthesis of silver nanoparticles from *Desmodium adscendens* (Sw.) DC: an investigation of the anticancer and antibacterial activity. J Cluster Sci 2015; 26(6): 1919-34.

[45]   Avnir D, Levy D, Reisfeld R. Surface modification of nanoparticles produced by *Dunaliella salina*. J Colloid Interface Sci 1984; 97(2): 520-6.

[46]   Subramanian G, Krishnamoorthy P, Selvam K. Microalgal biomass as a Nano factory for synthesizing silver nanoparticles with antimicrobial activity. Biotechnol Rep 2018; 17: 1-5.

[47]   Kim T, Kang S, Sung JH, Kang YK, Kim YH, Jang JK. Characterization of polyester cloth as an alternative separator to nafion membrane in microbial fuel cells for bioelectricity generation using swine wastewater. J Microbiol Biotechnol 2016; 26(12): 2171-8.
        [http://dx.doi.org/10.4014/jmb.1608.08040] [PMID: 27666990]

[48]   Sathasivam R, Baskar G, Kadirvelu K. Biosynthesis of gold nanoparticles using marine microalga, *Chlorella vulgaris* and its application in anticancer drug delivery. Int J Nanomedicine 2015; 10: 7197-

207.

[49] Yuan Q, Wang J, Liu H, Liu H, Liu J. Chitosan-coated gold nanoparticles synthesized from microalgae for targeted drug delivery. J Microbiol Biotechnol 2020; 30(4): 523-32.

[50] El-Sheekh MM, Alaraidh IA, Almuqrin AH, Alharbi SA, Al-Qahtani KM, El-Sharouny HM. Green synthesis of gold nanoparticles using *Chlorella vulgaris* and evaluation of their potential anticancer activity. Saudi J Biol Sci 2020; 27(1): 267-73.

[51] Shukla S, *et al.* Green synthesis of zinc oxide nanoparticles from microalgae: potential application as fertilizer for maize plants. Environ Sci Pollut Res Int 2020; 27(6): 5892-902.

[52] Navarro E, Baunthiyal M, Ramírez JP. Microalgae: a potential source of bioactive compounds for agriculture and related industries.Microbial biotechnology in agriculture and aquaculture. Elsevier 2018; pp. 305-22.

[53] Hajipour H, Barzegar M, Najafi F, Zabihi E, Khosravi-Darani K. Copper nanoparticles synthesized by microalgae as an effective fungicide against *Fusarium oxysporum*. Biol Trace Elem Res 2019; 188(1): 110-9.

[54] Kumar S, Dhawan G, Mohan D, *et al.* Scenedesmus quadricauda derived silver nanoparticles for removal of heavy metals from contaminated water. J Environ Manage 2021; 282: 111957.

[55] Giraldo LF, Lobo-Rodero A, Guerrero-Beltrán CE, Durán JDG. Microalgae-derived copper nanoparticles: synthesis and application in lead adsorption. Environ Sci Pollut Res Int 2019; 26(8): 7886-94.
[PMID: 31889272]

[56] Al-Qubaisi MS, Rasedee A, Flaifel MH, Ahmad SH, *et al.* Biosynthesis and characterization of silver nanoparticles using marine microalgae, Nannochloropsis sp. and its inhibitory effects on human cancer cell lines. J Photochem Photobiol B 2020; 212: 112034.

[57] Nicolai L, Gačević Ž, Calleja E, Trampert A. Electron Tomography of Pencil-Shaped GaN/(In,Ga)N Core-Shell Nanowires. Nanoscale Res Lett 2019; 14(1): 232.
[http://dx.doi.org/10.1186/s11671-019-3072-1] [PMID: 31300916]

[58] Ahmadian-Fard-Fini S, Noorazar M, Afshar Taromi F. Green synthesis of ZnO nanoparticles using *Spirulina platensis* and their application in dye-sensitized solar cells. Mater Res Express 2021; 8(3): 0350a8.

[59] Thangaraju S, Rajendran K, Madhan B, Anbalagan G. Chitosan-coated *Spirulina platensis* nanoparticles for the efficient removal of heavy metals from wastewater. J Env Mng 2020; 261: 110221.

[60] Li X, Xu Z, Chen H, *et al.* Chitosan-modified microalgae-based biosorbents for the removal of reactive dyes from wastewater. J Appl Phys 2017; 29(5): 2395-05.

[61] Selvam K, Sudhakar C, Govarthanan M, *et al.* Eco-friendly biosynthesis and characterization of silver nanoparticles using *Tinospora cordifolia* (Thunb.) Miers and evaluate its antibacterial, antioxidant potential. J Rad Res App Sci 2017; 10(1): 6-12.
[http://dx.doi.org/10.1016/j.jrras.2016.02.005]

[62] Bhuvaneshwari M, Nachiyar CV, Subramanian P, *et al.* Green synthesis of gold nanoparticles using *Chlorella vulgaris* and their catalytic degradation of 4-nitrophenol. Environ Sci Pollut Res Int 2018; 25(4): 3814-21.
[PMID: 30539399]

[63] Lee H, Shim E, Yun HS, *et al.* Biosorption of Cu(II) by immobilized microalgae using silica: kinetic, equilibrium, and thermodynamic study. Environ Sci Pollut Res Int 2016; 23(2): 1025-34.
[http://dx.doi.org/10.1007/s11356-015-4609-1] [PMID: 25953610]

[64] Kumar S, Nehra M, Dilbaghi N. Synthesis of titanium dioxide nanoparticles using microalgae for the removal of organic pollutants from wastewater. J Environ Chem Eng 2017; 5(6): 5886-95.

[65] Ibrahim E, Abdeen AO, Salama A, Mahmoud MA, Abd-El-Kader MA. Microalgae-based silver nanoparticles: a sustainable approach towards controlling microbial growth in water and soil environments. Environ Sci Pollut Res Int 2020; 27(5): 4875-86.

[66] Shi Y, Li J, Zhang X, Li D, Zhang H. Synthesis of *Chlorella vulgaris*-based nanoparticles for effective cadmium ion removal from soil. Environ Sci Pollut Res Int 2020; 27(11): 12832-43.
[PMID: 32002836]

[67] Yang Y, Liu X, Fan Y, He H, Guo X, Chen G, *et al. Chlamydomonas reinhardtii*-based nanoparticles for heavy metal removal from wastewater: synthesis, characterization and toxicity evaluation. J Hazard Mater 2021; 406: 124664.

[68] Sathishkumar M, Sneha K, Won SW, Cho CW, Kim S, Yun YS. *Cinnamon zeylanicum* bark extract and powder mediated green synthesis of nano-crystalline silver particles and its bactericidal activity. Colloids Surf B Biointerfaces 2009; 73(2): 332-8.
[http://dx.doi.org/10.1016/j.colsurfb.2009.06.005] [PMID: 19576733]

[69] Yehia RS, Dhabi NA. Biogenic synthesis of silver nanoparticles using microalgae *Chlorella vulgaris* and their antibacterial activity against some human pathogens Nanomaterials 2019; 9(2): 253.

[70] Kumar SA, Abyaneh MK, Gosavi SW, Kulkarni SK, Pasricha R, Ahmad A. Nitrate reductase-mediated synthesis of silver nanoparticles from $AgNO_3$. Biotechnol Appl Biochem 2007; 49(1): 63-8.

[71] Raman N, Mohan K. Phycoremediation of industrial effluent using microalgae and nanoparticle synthesis from the algal biomass. J Ind Eng Chem 2015; 21: 1187-92.

[72] Guedes AC, Amaro HM, Malcata FX. Microalgae as sources of high added-value compounds—a brief review of recent work. Biotechnol Prog 2011; 27(3): 597-613.
[http://dx.doi.org/10.1002/btpr.575] [PMID: 21452192]

[73] Sun Y, Wang X, Zhou Z, Zhu X, Wang S. Microalgae as a source of valuable micromolecules-their development as a future feedstock for the chemical industry. Bioresour Technol 2017; 245: 1714-26.

[74] Narasimhan B, Sakthivel N, Gurunathan S. Microalgae as nanomaterial factories for biotechnological applications. Trends Biotechnol 2018; 36(3): 310-21.
[PMID: 30301571]

[75] Jorquera O, Kiperstok A, Sales EA, Embiruçu M, Ghirardi ML. Comparative energy life-cycle analyses of microalgal biomass production in open ponds and photobioreactors. Bioresour Technol 2010; 101(4): 1406-13.
[http://dx.doi.org/10.1016/j.biortech.2009.09.038] [PMID: 19800784]

[76] Gao D, Jin Y, Ju Q. *Chlorella vulgaris*-derived nanovesicles as a novel drug delivery system for cancer therapy. Nanotechnology 2020; 31(40): 405102.

[77] Wu L, Wang Y, Zhu G, Chen X, Liu H. Microalgae-based biosensors for environmental monitoring: a review. Biosens Bioelectron 2021; 173: 112794.

[78] Jiang L, Liang Y, Zhang R, Chen H. Microalgae-based nanocatalysts for biofuel production: a review. Bioresour Technol 2021; 331: 125086.

[79] Anantharaman N, Karthikeyan R, Arumugam M. *Chlamydomonas reinhardtii*-mediated biosynthesis of iron oxide nanoparticles and its application in the removal of heavy metal ions from wastewater. Environ Sci Pollut Res Int 2019; 26(13): 13335-44.

[80] Luo Y, Ye B, Ye J, *et al.* Ca^{2+} and SO_4^{2-} accelerate the reduction of Cr(VI) by Penicillium oxalicum SL2. J Hazard Mater 2020; 382: 121072.
[http://dx.doi.org/10.1016/j.jhazmat.2019.121072] [PMID: 31470304]

[81] Patil YN, Park JB, Kim DH. Microalgae-derived nanomaterials for biomedical and energy applications: a review. J Ind Eng Chem 2019; 69: 1-14.

[82] Pancha I, Chokshi K, George B. Microalgal nanotechnology: A road map towards sustainable

development. Renew Sustain Energy Rev 2019; 101: 143-62.

[83]   Hu Q, Sommerfeld M, Jarvis E, *et al.* Microalgal triacylglycerols as feedstocks for biofuel production: perspectives and advances. Plant J 2008; 54(4): 621-39.
[http://dx.doi.org/10.1111/j.1365-313X.2008.03492.x] [PMID: 18476868]

[84]   Sharma N, Singh V, Zafar M, *et al.* Microalgae-based nanotechnology: a review of current research and future prospects. Renew Sustain Energy Rev 2021; 137: 110621.

[85]   Vittorio O, Curcio M, Cojoc M, *et al.* Microalgae-based nanotechnology: A new avenue for future green innovations. Trends Biotechnol 2018; 36(2): 174-7.

[86]   Dussud C, Hudec C, George M, *et al.* Microalgae-based processes for the synthesis of metallic nanoparticles and their biomedical applications. Nanomat 2018; 8(10): 807.
[PMID: 30304791]

[87]   Sharma S, Jaiswal S, Gupta S, *et al.* Microalgae-based nanotechnology: a green approach towards diverse applications. RSC Advances 2019; 9(13): 7535-50.

[88]   Kim JS, Cho HR, Lee SH, *et al.* Octopus-like skin has light touch. Nano Today 2016; 11(2): 125-6.
[http://dx.doi.org/10.1016/j.nantod.2016.03.001]

[89]   Gao Y, Jiao Y, Zhang H, *et al.* One-step synthesis of a dual-emitting carbon dot-based ratiometric fluorescent probe for the visual assay of Pb^{2+} and PPi and development of a paper sensor. J Mater Chem B Mater Biol Med 2019; 7(36): 5502-9.
[http://dx.doi.org/10.1039/C9TB01203F] [PMID: 31424064]

CHAPTER 8

Application of Green Synthesized Nanomaterials for Environmental Waste Remediation: A Nano-Bioremediation Strategy

M. Nanda[1], **S. Agrawal**[1] and **S.K. Shahi**[1,*]

[1] *Bioresource Product Research Laboratory, Department of Botany, School of Life Science, Guru Ghasidas Vishwavidyalaya (A Central University), Bilaspur, Chhattisgarh, India*

Abstract: In the current scenario, dangerous refractory organic and inorganic pollutants are continuously released into the environment as a result of industrialization that poses a significant threat on a global scale. One of the greatest challenges that needs to be solved is the effective management of various pollutants and waste. Despite their effectiveness, traditional treatment methods have several drawbacks, such as the fact that they are time-consuming and target-specific. As a result, it directs the search for a suitable replacement. Researchers are paying close attention to the novel technique of nano-bioremediation to remove pollutants from various contaminated locations. This approach combines the benefits of bioremediation and nanotechnology to develop a remediation process that is quicker, more productive, and less harmful to the environment than either approach individually. This chapter summarizes the green synthesis methods of various nanomaterials (NMs) along with an explanation of the remediation methodology, its mechanism, and prospective applications in environmental remediation. Additionally, the removal and valorzsation of waste materials using green nanotechnology supported by microbes and enzymes are highlighted. This chapter also discusses the multiple constraints of nano-bioremediation as well as the factors responsible for the efficiency of NMs.

Keywords: Biofabrication, Green synthesis, Immobilization, Microemulsion, Nano-bioremediation, Nanocomposite, Nanotechnology.

INTRODUCTION

In the present scenario, the rate of urbanization and industrialization is accelerating and releasing unsustainable pollutants into the atmosphere. These xenobiotic pollutants cannot be easily removed from the environment and cause harmful effects on ecological safety and human health [1].

---

* **Corresponding author S.K. Shahi:** Bioresource Product Research Laboratory, Department of Botany, School of Life Science, Guru Ghasidas Vishwavidyalaya (A Central University), Bilaspur, Chhattisgarh, India; E-mail: sushilkshahi@gmail.com

Neha Agarwal, Vijendra Singh Solanki and Sreekantha B. Jonnalagadda (Eds.)

"Environmental waste remediation" refers to the process of addressing and mitigating environmental pollution or contamination by removing, treating, or neutralizing harmful substances in the air, water, soil, or other environmental media. The goal is to restore or improve the quality of the environment by eliminating or reducing the negative impacts of pollutants or hazardous materials. The removal of dangerous pollutants from the environment is performed through biological, chemical, and physical methods. Nevertheless, the widespread use of the conventional approach is constrained by time and energy requirements, high operational costs, and maintenance [2]. Among several existing technologies, the most promising and cost-effective strategy for removing contaminants is bioremediation in combination with nanotechnology. Bio-based technique or bioremediation is a hygienic, sustainable, and an "environmentally suitable" green method for the removal of pollutants. In recent times, the removal of pollutants from contaminated sites in a sustainable manner has become crucial due to the lesser risk involved and for improving the quality of contaminated sites by restoration method [3 - 5].

Nanotechnology has gained significant interest in recent years across a variety of industries, including textiles, electronics, medicine, and pharmaceuticals. The scope of nanotechnology has provided greater opportunity to manage the major environmental challenges such as remediation of environmental contaminants [6, 7]. Nanoparticles (NPs) possess unique shapes and sizes. They have increased adsorption and catalyst properties and high reactivity; therefore, new technologies are gaining more attention in the preparation of NPs by researchers. Nanoparticles are fabricated by chemical or physical methods through the top-down or the bottom-up method. The top-down method includes arc discharge, diffusion, lithographic techniques, high-energy ball milling, and irradiation, where large molecules break down and form nano-sized particles. Meanwhile, the bottom-up method includes chemical and biological approaches; in these methods, atoms combine and form clusters, and then cluster aggregates in the form of nanoparticles. To produce nanoparticles, biological approaches are being used due to the numerous drawbacks of physiochemical methods, including their high cost and potentially harmful by-products [8]. The utilization of fungal, bacterial, and algal cultures, as well as their metabolites and biomolecules, for the long-term synthesis of NMs, is gaining popularity and possesses a potential application in the field of bioremediation [9, 10]. To achieve an effective, economical, and long-lasting solution for a clean environment, nanobiotechnology and bioremediation have been combined [11 - 13].

Another application for nanocomposites or NPs is the removal of dangerous substances from wastewater. For instance, an absorbent made of a reduced amount of graphene and iron oxide was produced to get rid of phenazopyridine

[14]. The ligand-dependent functional materials are appropriate for removing heavy metals and other pollutants from wastewater because they have additional advantages. Nickel had been removed from petroleum-polluted water using an embeddable composite absorbent. To extract caesium from wastewater, a synthetic zeolite-based absorbent was also developed. Carbon and its derivatives are the most often utilized nanoabsorbents for removing heavy metal ions and organic dyes from aqueous solutions. The degradation of organic pollutants in the treatment of organic wastewater has attracted the attention of numerous investigations. These materials are based on carbon nanotubes (CNTs). Some of the NMs that have recently been utilized to remove the dye from wastewater include multiwalled CNTs, chitosan nanoabsorbents, mesoporous silica, and nanocopper oxide made from e-waste [15]. Adsorptive or reactive procedures that are used for on-site (*in-situ*) or off-site (*ex-situ*) treatment of pollutants are at the heart of efforts to achieve "environmental improvement". While the latter method dissolves the organic impurities, leaving no toxic byproducts like CO_2 and H_2O, the former technique involves the removal of heavy metal contaminants through sequestration. The present chapter compiles and summarizes the brief knowledge of nanotechnology in synthesis methodology related to the application in the degradation of pollutants and limitations of NPs in the field of remediation.

HISTORICAL ASPECTS OF NPS

Professor Richard Feynman initially outlined the concept of nanotechnology in his lecture "There's Plenty of Room at the Bottom", and Professor Norio Taniguchi coined the term [16, 17]. The technique is frequently described as the "Next Industrial Revolution" [18]. This technology is characterized by using tiny NPs (<100 nm), and the United States National Nanotechnology Initiative (USNNI) has defined it as "the understanding and control of NMs (dimensions 1-100 nm), where unique phenomena occur, allowing for novel nanotechnology applications". Numerous scientific fields, including agriculture, the removal of pollutants from soil and water, *etc.*, have adopted nanotechnology [19, 20]. Due to its extremely small size, high surface area to volume ratio, ease of usage and flexibility for both *In vivo* and *In vitro* applications, nanotechnology plays a crucial role in the process of bioremediation [20].

CLASSIFICATION OF NPS

According to previous studies, particles (atomic or molecular aggregates) having different shapes like rod, circular, spherical, triangular, polygon and stars with a size range of 1-100 nm are denoted as NPs [21]. The uniformity of the size, shape, and structure of nanoparticles determines their attributes, including magnetic, reactivity, stability, and optical features. The structure of the nanoparticles depends on their dimensions, such as zero dimension (0-D) viz. fullerenes, atomic clusters; one dimension (1D) *viz*, nanofibers and nanowires; two dimensions (D) such as nanodisks, nanolayers [22]. The two groups into which NPs are divided are primarily organic (fullerenes) and inorganic NPs (noble metal NPs such as gold and silver, magnetic nanoparticles, and semiconductor nanoparticles such as titanium dioxide and zinc oxide) [12]. According to a study, there are three different kinds of NPs: natural (such as volcanic dust and mineral composites), accidental (like welding fumes, coal combustion, and diesel exhaust), and manufactured [23].

NPs are manufactured by three methods: chemical, physical, and biological methods. The chemical method includes hydrothermal synthesis, sol-gel method, microemulsions, microwave synthesis, and Langmuir-Blodgett method, whereas the physical method includes high-energy ball milling, laser vaporization, ion beam techniques, laser pyrolysis, *etc.* The biological approach uses DNA, protein templates, plant extracts, and microbes to synthesize NPs. After the synthesis of NPs, a characterization process is performed to examine the size, shape, morphology, surface charge and crystallographic nature. The characterization process is done by transmission electron microscopy (TEM), energy dispersive X-ray spectroscopy (EDS), X-ray photoelectron microscopy (XPS), scanning electron microscope (SEM), X-ray diffraction (XRD), atomic force microscopy (AFM), UV-visible spectroscopy, fourier transformed infrared spectroscopy (FTIR), and, nuclear magnetic resonance (NMR) [20, 24, 25].

BIOFABRICATION OF NMS

In many branches of science and technology, including medicine, electrochemistry, biotechnology, chemical, and bioremediation, biological NPs play an important role [26]. They can be used in a variety of soil, air, and water bioremediation processes. They are effective because they have good adsorptive and catalytic capabilities. Biogenic NPs are those that are fabricated by using living organisms such as algae, bacteria, fungi, plants, and yeast, as shown in Fig. (**1**) [9]. NPs are created by microorganisms using two different processes: an internal process and an external one. In the intracellular process, metal ions that are positively charged diffuse with negatively charged cell walls of organisms. The enzymatic reaction that uses enzymes to change metal ions into NPs happens outside of cells [21]. In contrast, plants have a variety of secondary metabolites such as terpenoids, sugars, polyphenols, alkaloids, proteins, and phenolic acids that are essential for the creation of NPs. During the synthesis process, these main metabolites serve as capping, stabilizing, and reducing agents. The size, shape, chemical composition, surface area, and dispersity of the produced biogenic NPs are examined next.

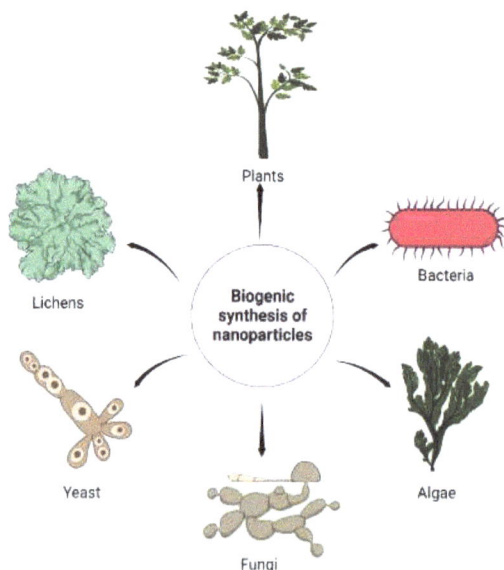

Fig. (1). Biogenic synthesis of NPs [9].

Fabrication of NMs using Bacteria

Various bacterial species have potential bio-factories to produce NPs, such as copper, palladium, gold, zinc, silver, manganese, titanium, cadmium sulfide, *etc.* Bacterial species can immobilize, mobilize, and precipitate metals; they reduce the metal ions and precipitate metals into nanometre scale for the synthesis of NPs. Studies reported that *Pseudomonas stutzeri* AG259 is used in the synthesis of silver NPs (Ag NPs); it reduces Ag ions from silver nitrate solution and accumulates Ag NPs inside the periplasmic space of the cell [28]. *Rhodopseudomonas capsulata* synthesized spherical gold NPs (Au NPs) in an acidic environment [29]. The synthesis of iron oxide NPs by *Bacillus subtilis* has also been examined [30]. A new, rational approach to biosynthesis makes use of bacterial vitamins, enzymes, biodegradable polymers, polysaccharides, and biological systems. The advantage of producing many NPs in the range of 100-200 nm in size in a pure state, free of other cellular protein contaminants, and then purifying those NPs by filtration is provided by the extracellular secretion of bacterial enzymes. Bacterial cells are useful for bioremediation and biotechnological applications due to their ability to bind metals. Table **1** lists bacterial species that were employed in the production of NPs. By optimizing bacterial growth, biological functions, and enzymatic reactions, bacterial NP qualities are enhanced. The large-scale manufacture of bacterial NPs is more advantageous because cheaper and harmless ingredients are needed for the synthesis or stabilization [31].

Table 1. Types of microorganisms used for the synthesis of various NPs.

Types of NPs	Microorganism	References
Bacteria		
Gold [Au]	*Streptomyces* sp. VITDDK3	[48]
	Klebsiella pneumonia	[49]
	Streptomyces viridogens	[50]
	Rhodopseudomonas capsulate	[51]
	Brevibacterium casei	[52]
	RhodoPseudomonas aeruginosa	[53]
Silver [Ag]	*Brevibacterium casei*	[54]
	Actinomycete, Nocardiopsis sp. MBRC-1	[55]
	Streptomyces rochei	[56]
	Streptomyces albidoflavus	[57]
	Streptomyces naganishii	[58]
	Streptomyces sp. ERI-3	[59]
	Bacillus thruringiensis	[60]
	Staphylococcus aureus	[61]
Iron [Fe]	*Bacillus subtilis*	[30]
	Klebsiella oxytoca	[62]
	Shewanella oneidensis	[63]
	Klebsiella oxytoca	[64]
Copper [Cu]	*Shewanella oneidensis*	[65]
	Pseudomonas fluorescens	[66]
	Streptomyces sp.	[67]
	Pseudomonas stutzeri	[68]
Zinc [Zn]	*Lactobacillus plantarum* VITES07	[69]
	Aeromonas hydrophila	[70]
	Lactobacillus	[71]
Bio-palladium	*Desulfovibrio vulgaris*	[72]
Algae		
Gold [Au]	*Euglena gracilis*	[73]
	Sargassum muticum	[74]
	Chlorella vulgaris	[75]
	Turbinaria conoides	[76]
	Padina gymnospora	[77]

(Table 1) cont.....

Types of NPs	Microorganism	References
Silver [Ag]	*Scenedesmus obliquus*	[78]
	Dictyosphaerum, Pectinodesmus	[79]
	Spirulina	[80]
	Caulerpa racemosa	[81]
	Cystophora moniliformis	[82]
	Sargassum longifolium	[83]
	Turbinaria conoides	[76]
Iron [Fe]	*Sargassum muticum*	[84]
	Euglena gracilis	[85]
Palladium	*Spirulina platensis*	[86]
	Chlorella vulgaris	[87]
Copper [Cu]	*Bifurcaria bifurcate*	[88]
Zinc [Zn]	*Sargassum muticum*	[89]
Fungus and Yeast		
Gold [Au]	*Cylindrocladium floridanum*	[90]
	Candida albicans	[91]
	Epicoccum nigrum	[92]
	Neurospora crassa	[93]
	Penicillium brevicompactum	[94]
	Hormoconis resinae	[95]
	Aspergillus clavatus	[96]
	Aspergillus oryzae	[29]
Silver [Ag]	*Aspergillus tubingensis*	[97]
	Neurospora crassa	[98]
	Aspergillus oryzae	[99]
	Aspergillus tamarii	[100]
	Aspergillus flavus	[101]
	Trichoderma reesei	[102]
	Cladosporium cladosporiodes	[103]
Iron [Fe]	*Pleurotus* sp.	[104]
	Fusarium oxysporum	[105]

(Table 1) cont.....

Types of NPs	Microorganism	References
Copper [Cu]	*Sterium hirsutum*	[106]
	Hypocrea lixii	[107]
	Fusarium oxysporum	[108]
	Penicilium citrinum	[109]
Zinc [Zn]	*Candida albicans*	[110]
	Saccharomyces cerevisiae	[111]
Plants		
Gold [Au]	*Lansium domesticum*	[112]
	Lawsonia inermis	[113]
	Magnolia kobus	[114]
	Medicago sativa	[115]
	Mirabilis jalapa	[116]
	Hibiscus cannabinus	[117]
	Galaxaura elongata	[118]
Silver[Ag]	*Memecylon edule*	[119]
	Mentha piperita	[120]
	Momordica cymbalaria	[121]
	Mucuna pruriens	[122]
	Hemidesmus indicus	[123]
Copper oxide [cuo]	*Mukia maderaspatana*	[124]
	Prosopis farcta	[125]
	Tridax procumbens	[126]
	Garcinia mangostana	[127]
Zinc oxide[Zno]	*Artocarpus gomezianus*	[128]
	Aloe barbadensis	[129]
	Berberis aristata	[130]
	Calotropis gigantea	[131]
	Grewia flaviscences	[132]
	Hydrastis canadensis	[133]
Manganes oxide[MgO]	*Prunus yedoensis*	[134]
	Enteromorpha flexuosa	[135]
	Cucurbita maxima	[136]
	Eucalyptus chapmaniana	[137]

(Table 1) cont.....

Types of NPs	Microorganism	References
Ferric Oxide[FeO]	*Gardenia jasminoides*	[133]
	Euphorbia helioscopia	[138]

Fabrication of NMs using Algae

Algae are oxygenic, photoautotrophic, eukaryotic microorganisms that have metal-accumulating properties. Algae show several characteristics for the synthesis of NPs, such as metal accumulation, ease of handling, high tolerance, economic viability, and extracellular enzyme secretion [32]. The algae species used for bionanopreparation are listed in Table **1**. Marine brown algae secrete polysaccharides fucoidans from the cell wall, which is crucial for the synthesis of Au NPs because fucoidans have anti-cancer, anti-viral, anti-inflammatory, and anticoagulant properties [33]. Brown algae also have mucilaginous polysaccharides and carboxyl groups in their cell walls, which are used to synthesize NPs by efficiently absorbing metal [34]. The action of nitrate reductase, electron shuttle quinones, or both has been postulated as a potential mechanism for the fungus-mediated synthesis of NPs. In both bacteria and fungi, the production of NPs has been discovered to be mediated by nitrate reductase and α-NADPH-dependent reductases [35]. Few studies have found that the fungus *Verticillium sp.* was employed for the reduction of aqueous $AuCl_4$ ions, resulting in the formation of Au NPs with a reasonably suitable structure and good monodispersity. Further investigation into this has shown that the charged amino acid residues in the fungal enzymes located in the cell wall of the fungal mycelia electrostatically interacted with the metal ions $AuCl_4$ to trap them on the surface of the fungal biomass [36]. The manufacturing of Au NPs was accomplished using the endophytic fungus *Colletotrichum sp.* isolated from *Geranium* leaf [36]. When *Trichothecium sp.* was cultured with gold ions under static conditions, extracellular NPs were produced and under shaking conditions, intracellular AuNPs were produced [37]. *Verticillium sp.* and *Fusarium oxysporum* produced magnetic NPs at room temperature. Additionally, *F. oxysporum* produced a shuttle quinone and a nitrate-dependent reductase that were both capable of extracellularly generating AgNPs or silver hydrosols [35].

Fabrication of NMs using Fungi and Yeast

Fungi are a suitable candidate for the synthesis of nanoparticles because they are easy to handle, have wide diversity, produce large amounts of extracellular enzymes, are economically viable, have high metal tolerance, and have high bio-accumulative ability. Fungal species used for the biogenic nanoparticle synthesis are shown in Table **1**. Studies reported the intracellular synthesis of gold nanoparticles from fungi and the intracellular formation of Ag NPs in *Verticillium* [21, 38]. In comparison to bacteria, the fungus has larger well dimensions and a large quantity of nanoparticles created because fungi release a huge amount of protein, which translates to a higher yield of nanoparticle synthesis [38]. Additionally, acquired proteins were also employed to create nanoparticles in place of fungal culture. In addition, yeast is employed in the creation of nanoparticles. Au NPs are created using *Saccharomyces cerevisiae* [39].

Fabrication of NMs using Lichen

Lichens are composite creatures that coexist with perennial trees, algae, fungi, or cyanobacteria in obligate beneficial symbioses [40, 41]. Lichen cells are advantageous for industrial, pharmacological, biotechnological, medicinal, and cosmetic uses because they contain several secondary metabolites and other bioactive compounds [42]. Some researchers have highlighted the possibility for various lichen species to create distinctive NPs with varying forms, sizes, and physicochemical and biological activity [43]. It has also demonstrated the ability of various lichen species to synthesize multiple kinds of NPs and their potential to function as promising antimicrobial compounds [43].

Fabrication of NMs using Plants

Few works have been done to understand the real mechanism underlying the synthesis of metal NPs, despite several studies being completed by researchers on the screening and identification of plants for the biosynthesis of metallic NPs [44]. To create metallic NPs, several theories have been put forth. Many alternative possibilities regarding the production of metal nanoparticles using plant extract have been offered in the wake of the initial studies, which were published in the 1960s and the late 1990s [45]. Plants contain different types of secondary metabolites like sugars, terpenoids, alkaloids, and polyphenols that play a crucial role during the biosynthesis of metal nanoparticles [45]. According to an FTIR analysis of *Cinnamon zeylanisum* extract, it contains terpenoids that are readily associated with metal ions [46]. Strong antioxidant action is displayed by the wider family of varied organic polymers known as terpenoids, which include five-carbon isoprene chains. *C. zeylanisum* extract can be considered as an example as it contains eugenol, a terpenoid that is essential to the reduction of Au and Ag NPs. Further investigation using FTIR demonstrated that the presence of the -OH group causes the dissociation of the eugenol proton, which results in the production of resonance structures capable of further oxidation, followed by the formation of NPs [46, 47]. Similar to anthocyanins, flavonoids are a subclass of the polyphenolic group, which also includes isoflavonoids, flavones, chalcones, and flavanones. These compounds actively contribute to metal ion reduction and chelation. Reactive hydrogen species (ROS) are released during the tautomeric transition of flavonoids from the enol group to the keto group, and they act as a reducing agent for metal ions [43]. Table **1** explains the production of several NPs employing bacteria, plants, fungi, and algae.

NANO-BIOREMEDIATION: A NEW STRATEGY FOR RESTORATION OF THE ENVIRONMENT

Nanobioremediation has emerged as a new economically feasible and eco-friendly technique for the remediation of organic and inorganic pollutants from contaminated sites directly or indirectly [139]. Microorganisms are moral biological agents for remediation, which are considered easy to handle, cost-effective, renewable sources, and have high metal tolerance; they behave like oxidants or reductants, adsorbents and catalysts [63]. Several *in-situ* and *ex-situ* applications of biogenic NPs can solve environmental problems, as shown in Fig. (**2**). For example, iron-oxidizing and iron-reducing bacteria are widely used for the synthesis of biogenic NPs. Iron oxide biogenic NPs not only remediate metals like arsenic, cobalt and chromium but also eliminate contamination of aromatic and aliphatic compounds *via* oxidation and reduction procedures [140]. Advantages and applications of nano-bioremediation technology are presented in Fig. (**3**). Further, studies confirmed that biogenic AgNPs possess good catalytic and antibacterial properties. Photocatalytic degradation of organic pollutants by AgNPs is an eco-friendly and cost-effective technique [141].

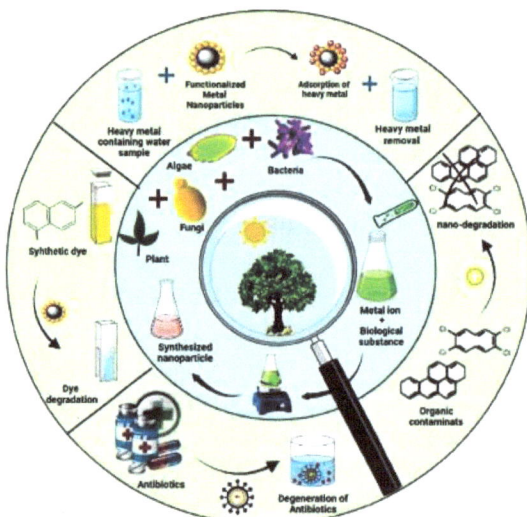

Fig. (2). Overview of biofabrication and application of NMs.

Nanobioremediation

Fig. (3). Advantages and application of nanobioremediation process.

APPLICATION OF BIOGENIC NPS

Applications of biogenic NPs in the remediation of different pollutants are described below and listed in Table **2**.

Table 2. Application of biogenic NPs in pollution control.

Biogenic NPs	Microorganisms	Pollutants Remediation	References
Fe-Mn oxide NPs	*Pseudomonas* sp. QJX-1	Arsenic-contaminated groundwater	[155]
	Sulphate-reducing bacteria	Arsenic-contaminated groundwater	[165]
		Chromium [VI], sulfate and COD removal	[166]
		contamination absorption from mine drainage water	[167]
Biogenic Fe$_3$O$_4$ NPs	*Sphingomonas* sp. strain XLDN2-5 cells	Carbazole	[168]
Biogenic Pd/Fe NPs	*Trametes versicolor*	Triclosan	[169]
Biogenic Pd/Fe NPs	*Sphingomonas wittichii* RW1 [DSM 6014]	2,3,7,8-tetrachlorodibenzo-p-dioxin	[170]
Biogenic Pd/Fe NPs	*Burkholderia xenovorans* LB400	Polychlorinated biphenyl	[171]

(Table 2) cont.....

Biogenic NPs	Microorganisms	Pollutants Remediation	References
Biogenic manganese oxide [BioMnOx]	Manganese-oxidizing bacteria [MOB] C-S1	1,2,4-triazole from chemical industrial effluent	[172]
	Desmodesmus sp.	Bisphenol A	[173]
	Pseudomonas putida MnB1	Heavy metals	[156]
	Pseudomonas putida	Organic micro-pollutants [estrone, 17-α ethinylestradiol, diclofenac, ibuprofen]	[142]
Biogenic palladium [bio-Pd] NPs	*Klebsiella oxytoca* GS-4-08	Azo dyes containing wastewater	[146]
Biogenic palladium NPs	*C. pasteurianum* BC1	Chromium Cr [VI]	[174]
Biogenic palladium [Pd [0]] NPs	*Shewanella oneidensis* MR-1	Polychlorinated biphenyls	[175]
Biogenic palladium [bio-Pd] NPs dropped with gold	*Shewanella oneidensis*	Dechlorination of halogenated micro-pollutants	[176]
Biogenic platinum NPs	*Desulfovibrio vulgaris*	Pharmaceutical products removal	[110]
Biogenic gold NPs	*Saccharomyces cerevisiae, Cylindrocladium floridanum*	Quinclorac degradation Degradation of 4-nitrophenol to 4-aminophenols	[150] [63]
Carbon nanotubes	*Shewanella oneidensis* MR-1	Cr [VI]	[177]
Algae-derived silver NPs	*Chlorella pyrenoidosa*	Methylene blue dye degradation	[141]
Nanogold bioconjugate	*Rhizopus oryzae*	Organophosphorus pesticides	[152]
Biogenic selenium NPs	*Bacillus safensis* JG-B5T	Selenium from wastewater	[178]
	Citrobacter freundii Y9	Mercury from soil	[158]
Biogenic pyrite NPs	Sulfate reducing bacteria	Arsenic from groundwater	[153]

Remediation of Organic Pollutants

The stabilization of transitional metals like chromium and arsenic, as well as the dehalogenation of persistent organic compounds, have been the main benefits of using nanoscale elements like iron, nickel, and palladium on sites contaminated with different kinds of toxic substances [142]. Tetrachloroethylene (TCE), organophosphorus pesticides, dyes, halogenated contaminants, pharmaceutical products, and polyvinyl biphenyls (PCBs) are a few examples of recalcitrant organic compounds that can be remedied using NPs instead of the slow degradation of these compounds by microbial processes [11]. The roles of NPs in the degradation of various organic contaminants have been described in Table **2**.

Remediation of Pharmaceutical Products

The release of pharmaceutical and other organic products into the environment is a serious problem; therefore, the remediation of these pollutants is necessary [143]. Wastewater treatment technology only reduces COD, BOD and nutrients in ppm level; however, it was unable to remove contaminants at ppb values. Biogenic manganese oxide NPs made by *Pseudomonas putida* are utilized to remove a variety of organic micro-pollutants (in ppb levels), including diclofenac, ibuprofen, estrone hormone, and 17-ethinylestradiol, among others through oxidation or reduction technique [142]. Biogenic palladium synthesized by bacteria *Shewanella oneidensis* removes persistent iodinated contrast media (ICM) and dehalogenated compounds by reductive method [144].

Remediation of Dye

Textile effluent is the foulest industrial pollutant, posing a major hazard to soil and water bodies due to its heterogeneity, recalcitrance, and pervasive persistence. Textile wastewater, like all other industrial effluents, contributes significantly to the degradation of the environment. The presence of a large quantity of water-soluble unfixed dyes has produced highly concentrated, colorful, and complicated high-strength wastewater that is resistant to degradation. Excessive use of dyes and other chemicals generates a large quantity of trash, which, when released into bodies of water, creates pollution [145]. Few studies investigated that azo dye can be remediated in fermentative conditions from wastewater by bio-palladium NPs synthesized from *Klebsiella oxytoca* GS-4-08 [146]. Biogenic AgNPs synthesized by *Chlorella pyrenoidosa* algae degraded methylene blue in visible light irradiation through their photocatalytic activity [141]. Nitroaromatic compounds like nitrophenols are commonly present in the effluent released from dye industries, which reach the river water. Nitrophenol possesses carcinogenic and mutagenic properties and causes serious health risks to humans and other animals. According to USEPA, nitrophenols are the priority pollutants, and their maximum permissible limit is 20 ppb. Hence, it is necessary to remediate nitrophenol efficiently. Biogenic AuNPs prepared from fungus *Cylindrocladium floridanum* degrade 4-nitrophenols into 4-aminophenols through their catalytic activity [63]. A study developed bio-Pd (0) (palladium) NPs by precipitating reduced Pd (0) on *C. pasteurianum*. Two azo dyes, Evan's blue (EB) and methyl orange (MO) were effectively reduced by bio-Pd(0) NMs. 10 ppm nano-Pd decomposed MO nearly completely in 7 minutes, but 5 ppm bio-Pd(0) decolorized EB almost completely in 7 min [147].

Remediation of Halogenated Contaminants

Trichloroethylene is a form of halogenated pollutant that enters groundwater through the effluents of the metal and textile industries. Trichloroethylene was remedied using a fixed bed reactor with encapsulated bio-palladium made by Shewanella oneidensis growing on polyacrylamide, silica, or coated on zeolites; 98% remediation of trichloroethylene was then observed [148]. An effective pesticide for weed control in rice and turf fields is called quinclorac. After the paddy crop is harvested, quinclorac finds its way to an aquatic system and has a negative impact on it [149]. For the reductive breakdown of quinclorac, *Saccharomyces cerevisiae* was employed to manufacture AuNPs [150].

Remediation of Organophosphorus Pesticides

Pesticides are used to keep pests and weeds at bay. They are categorized into distinct groups based on their origin and the type of insect they target. Insecticides, herbicides, and fungicides are popular chemical pesticides used in agricultural areas [151]. Furthermore, when used in excess, these compounds have negative environmental consequences, such as diminished insect pollination populations and a threat to endangered species and bird habitats. Chemical pesticides exposed to leaching and their persistence constitute a significant hazard because of their high toxicity, teratogenicity, and ingestion through the food chain. Several strategies were used to remove pesticides from the system, but none of them was completely successful. Nanoremediation focuses on using NPs to alleviate environmental pollution. Titanium silver, dioxide, and zinc oxide (ZnO) NMs are commonly used as photocatalysts for pesticide degradation. *Rhizopus oryzae*-derived nanogold bioconjugate is used as a single-step treatment method to remove organophosphorus pesticides from water [152]. Nanogold bioconjugate also remediated wastewater containing *E. coli*, malathion, parathion, chlorpyrifos and dimethoate pesticides.

Remediation of Heavy Metals

Heavy metals pose substantial health risks to humans and animals due to their carcinogenic and teratogenic qualities. Therefore, it is critical to remove heavy metals from the atmosphere using a sustainable method. Industrial wastewater frequently contains heavy metals like copper and arsenic that come from the metallurgical and mining processing industries. The biogenic iron NPs adsorb these metals, which include arsenic, copper, chromium, and zinc [50]. Biogenic pyrite is synthesized from sulfate-reducing bacteria, which can remediate arsenic from contaminated groundwater in reducing conditions [153]. Biogenic precipitates such as siderite ($FeCO_3$), vivianite ($Fe_3 (PO_4)_2.8H_2O$), and magnetite (Fe_3O_4) have exhibited higher affinity and adsorptive capacity for arsenate. High concentrations of arsenic, Fe (II) and Mn (II) were found in ground and surface water, causing harmful effects on aquatic ecosystems and human health [154]. These metals were oxidized and adsorbed by biogenic Fe-Mn oxides NPs, synthesized from Mn (II) oxidizing *Pseudomonas* sp. QJX-1 [155]. Biogenic manganese oxide NPs, as proven to be excellent adsorbents, possess an amorphous nature, small size, and high surface area, and their adsorption capacity increases when pH and temperature are increased from 3-6 and 15-30 °C, respectively [156]. Due to the combustion of fossil fuels, the concentration of mercury in soil increased [157]. Biogenic selenium NPs prepared from *Citrobacter freundii* Y9 degraded mercury from soil in both aerobic and anaerobic conditions [158].

Remediation of Oil Spills

The economy and ecosystem that depend on these marine habitats are negatively impacted by oil spills and seepage into the ocean. The likelihood of environmental spills rises with each barrel of oil transported across waterways around the world. Although oil spills have been happening for years, cleanup techniques have not changed that much. Post-oil spill response techniques include mechanical ones (booms, sorbents, and skimmers), thermal ones (in situ burning), chemical ones (dispersion), and organic ones (bioremediation) [159]. Microbes have been used historically to clean up hydrocarbon-contaminated wastewater [160]. Following the Santa Barbara, California, oil spill in 1969, George Robinson utilized marine microorganisms to eat oil, which is credited with helping to create the focused use of microbes to break down oil [161]. The ability of local species to digest oil and the rate at which populations can adapt to using oil as their only supply of carbon or energy are the two factors that restrict bioremediation. Each species' abundance is influenced by the composition and concentration of its oil as well as by weather, salinity, temperature, and the availability of nutrients [162]. Considering that materials at the nanoscale can be functionalized to achieve highly specific goals like the creation of Pickering emulsions or magnetically responsive sorbents [163, 164]. Nanoscience could offer a sustainable answer for improving post-oil spill response technologies.

MECHANISM OF REMEDIATION BY NPS

In addition to the several techniques for cleaning up contaminated sites (including chemical and physical remediation), bioremediation is currently a viable and affordable option for eliminating toxins from the environment [9]. Bioremediation incorporates bioaccumulation, biotransformation, biosorption, and biological stabilisation, among others, by utilizing bacteria, plants, and enzymes separately or in combination [179]. It serves as a corrective measure to get rid of emergent, organic, and inorganic pollutants in agricultural soil. Similarly, it promotes plant development and soil health and works to restore degraded land. Metals can be mobilized and immobilized by bacteria, and under some circumstances, the same microbe that can reduce metal ions can precipitate metals at nanoscale sizes. Numerous NPs, including those made of silver, gold, titanium dioxide, magnetite, iron, platinum, titanium, zero-valent iron, palladium, cadmium sulfide, gold nanowire, selenium, and zinc sulfide, are produced by bacteria in a dormant "biofactory" that is being studied [180]. It has also been shown that the interaction of NPs with phytoremediation may eliminate organic pollutants. Here, bioremediation-resistant long-chain hydrocarbons and halocarbons were broken down by an NP-mediated process. Thus, the introduction of biodegradability of living things in pollution control is influenced.

ADVANTAGES OF USING BIOGENIC NPS

A clean and environmentally friendly method of creating NPs is the employment of microbes. Better and more effective alternatives to physical and chemical approaches are provided by the creation of NPs from bacteria, fungi, algae, yeast, and plants. Biogenic NPs are superior and more affordable than those made through physical and chemical means. The biogenic approach uses biomolecules as reducing agents rather than the pricy chemical reductants used in the physical and chemical methods, such as sodium borohydride and hydrazine. Toxic waste from chemically generated NPs is damaging to human health and the environment, whereas biogenic NPs do not generate any toxic material. Biogenic NPs have a huge surface area, which improves the ability of contaminants to bind to them. The lipid bilayer in biological NPs provides stability and solubility. By adjusting the contact time of the reaction, pH, and substrate availability, the size and form of the NPs may be changed [181]. Biogenic NPs are thus a low-cost, highly reactive, low-energy, and environmentally beneficial approach. However, the application areas of nano-bioremediation are constrained, which must be addressed if its full potential is to be realized. These limitations include non-biodegradability, ease of synthesis, potential NP toxicity, and higher recovery costs.

CHALLENGES WITH NANO-BIOREMEDIATION AND FUTURE PROSPECTS

The biosynthesis of NPs is a sluggish process when compared to the physical and chemical methods. Future research is needed to determine how to create biogenic NPs with less retention period. To date, nano-bioremediation has demonstrated an effective technique for the remediation of pollutants from the environment, especially micro pollutants. However, these studies are limited to the lab scale. For industrial purposes, the preparation of engineered NPs with a greater remediation efficiency of pollutants is very necessary. In the field of NPs, the size and shape of particles play a vital role; therefore, to control the NP morphology, proper investigation should be done. In some cases, after the remediation of environmental pollutants, some toxic intermediates are produced that cause harm to the aqueous flora and fauna. Therefore, it is compulsory to check the formation and proper refusal of such types of products. Although nano-bioremediation is a cheap, effective, and feasible technique for the remediation of environmental contaminants, the large-scale production and use of these technologies are associated with health risks and safety problems. To achieve proper results with nano-bioremediation, it is necessary to observe the synthesis method, after degradation toxicity and management of microbial populations. At present, nano bioremediation technology is in the exploratory stage, but existing research suggests that this technique has more potential compared to physicochemical bioremediation techniques. Further molecular-level investigation and the addition of genetic engineering technology are necessary to identify the enzymes or receptors that work with biogenic NPs. Other than this, the effect of environmental factors like pH, ionic strength, and temperature in the process of nanobioremediation should be properly examined along with modification of NM properties to further increase the efficiency of remediation.

CONCLUSION

This chapter highlights the importance of nanotechnology in the bioremediation of contaminants while summarizing current developments in the field of plant extract-based NP production. A novel shift in the field of reducing environmental pollution can be brought about *via* nanobiotechnology. When compared to traditional remediation methods, the use of microbes like bacteria, fungi, algae, and yeast in the preparation of NPs offers excellent removal of recalcitrant pollutants from highly contaminated soil and water. This method is also economically feasible, inexpensive, and environmentally friendly. Biogenic NPs effectively remove polychlorinated biphenyls, dyes, pesticides, heavy metals, pharmaceutical products, *etc.* Nevertheless, there is a need to enhance the investigation of various plant extracts for the green synthesis of metal nanoparticles, particularly in the context of pollutant removal from the environment. Continuous research in this field is crucial to advancing the application of nanotechnology in environmental remediation. In the future, the variation and modification of nanotechnology will improve the efficacy of bioremediation. The innovative, cross-disciplinary, potential research in the field of nano-bioremediation provides a potential recognition in this area.

ACKNOWLEDGEMENT

The authors are grateful for the infrastructural support provided by the host University and acknowledge Dr. Purusottam Banjare for his moral support and guidance.

REFERENCES

[1]   Mohamed EF. Nanotechnology: future of environmental air pollution control. Environmental Management and Sustainable Development 2017; 6(2): 429-54.
        [http://dx.doi.org/10.5296/emsd.v6i2.12047]

[2]   Zelmanov G, Semiat R. Iron(3) oxide-based nanoparticles as catalysts in advanced organic aqueous oxidation. Water Res 2008; 42(1-2): 492-8.
        [http://dx.doi.org/10.1016/j.watres.2007.07.045]

[3]   Bharagava RN, Chowdhary P, Saxena G. Bioremediation: an eco-sustainable green technology: Its applications and limitations. Environ Pollut Biorem Appro. CRC Press 2017; pp. 1-22.

[4]   Bharagava RN, Saxena G, Mulla SI. Introduction to industrial wastes containing organic and inorganic pollutants and bioremediation approaches for environmental management. Bioremediation of industrial waste for environmental safety. Singapore: Springer 2020; pp. 1-18.
        [http://dx.doi.org/10.1007/978-981-13-1891-7_1]

[5]   Saxena G, Purchase D, Mulla SI, Saratale GD, Bharagava RN. Phytoremediation of heavy metal-contaminated sites: eco-environmental concerns, field studies, sustainability issues, and future prospects. Rev Environ Contam Toxicol 2020; 249: 71-131.

[6]   Iverson NM, Barone PW, Shandell M, *et al. In vivo* biosensing *via* tissue-localizable near-infrared fluorescent single-walled carbon nanotubes. Nat Nanotechnol 2013; 8(11): 873-80.
        [http://dx.doi.org/10.1038/nnano.2013.222]

[7]   Singh R, Misra V. Stabilization of zero-valent iron nanoparticles: role of polymers and surfactants. Handbook of Nanoparticles. 2015; pp. 1-19.

[8]   Kumari S, Tyagi M, Jagadevan S. Mechanistic removal of environmental contaminants using biogenic nano-materials. Int J Environ Sci Technol 2019; 16(11): 7591-606.
        [http://dx.doi.org/10.1007/s13762-019-02468-3]

[9]   Yadav KK, Singh JK, Gupta N, Kumar V. A review of nanobioremediation technologies for environmental cleanup: a novel biological approach. J Mater Environ Sci 2017; 8(2): 740-57.

[10]   Sharma D, Kanchi S, Bisetty K. Biogenic synthesis of nanoparticles: A review. Arab J Chem 2019; 12(8): 3576-600.
        [http://dx.doi.org/10.1016/j.arabjc.2015.11.002]

[11]   Cecchin I, Reddy KR, Thomé A, Tessaro EF, Schnaid F. Nanobioremediation: Integration of nanoparticles and bioremediation for sustainable remediation of chlorinated organic contaminants in soils. Int Biodeterior Biodegradation 2017; 119: 419-28.
        [http://dx.doi.org/10.1016/j.ibiod.2016.09.027]

[12]   Tripathi S, Sanjeevi R, Anuradha J, Chauhan DS, Rathoure AK. Nano-bioremediation: Nanotechnology and bioremediation. Biostimulation Remediation Technologies for Groundwater Contaminants. IGI Global 2018; pp. 202-19.

[13]   Mohamed EF, Awad G. Nanotechnology and Nanobiotechnology for Environmental Remediation. Magnetic Nanostructures. Cham: Springer 2019; pp. 77-93.
[http://dx.doi.org/10.1007/978-3-030-16439-3_5]

[14]   Khedri M, Maleki R, Khiavi SG, *et al.* Removal of phenazopyridine as a pharmacological contaminant using nanoporous metal/covalent-organic frameworks (MOF/COF) adsorbent. Appl Mater Today 2021; 25: 101196.
[http://dx.doi.org/10.1016/j.apmt.2021.101196]

[15]   Modi S, Yadav VK, Gacem A, *et al.* Recent and emerging trends in remediation of methylene blue dye from wastewater by using zinc oxide nanoparticles. Water 2022; 14(11): 1749.
[http://dx.doi.org/10.3390/w14111749]

[16]   Feynman R. There's plenty of room at the bottom. California institute of technology. Calt EngineAppl Sci. 1959; 23: pp. 22-36.

[17]   Taniguchi N. On the basic concept of nano-technology. Proc Int Conf Prod Eng. Tokyo: Part II, Japan Society of Precision Engineering 1974.

[18]   Roco MC. International perspective on government nanotechnology funding in 2005. J Nanopart Res 2005; 7(6): 707-12.
[http://dx.doi.org/10.1007/s11051-005-3141-5]

[19]   Sen S, Sen F, Boghossian AA, Zhang J, Strano MS. Effect of reductive dithiothreitol and trolox on nitric oxide quenching of single-walled carbon nanotubes. J Phys Chem C 2013; 117(1): 593-602.
[http://dx.doi.org/10.1021/jp307175f]

[20]   Singh R, Behera M, Kumar S. Nano-bioremediation: An innovative remediation technology for treatment and management of contaminated sites.Bioremediation of Industrial Waste for Environmental Safety. Singapore: Springer 2020; pp. 165-82.
[http://dx.doi.org/10.1007/978-981-13-3426-9_7]

[21]   Menon S, S R, S VK. A review on biogenic synthesis of gold nanoparticles, characterization, and its applications. Resource-Efficient Technologies 2017; 3(4): 516-27.
[http://dx.doi.org/10.1016/j.reffit.2017.08.002]

[22]   Benelmekki M, Vernieres J, Kim JH, Diaz RE, Grammatikopoulos P, Sowwan M. On the formation of ternary metallic-dielectric multicore-shell nanoparticles by inert-gas condensation method. Mater Chem Phys 2015; 151: 275-81.
[http://dx.doi.org/10.1016/j.matchemphys.2014.11.066]

[23]   Monica RC, Cremonini R. Nanoparticles and higher plants. Caryologia 2009; 62(2): 161-5.
[http://dx.doi.org/10.1080/00087114.2004.10589681]

[24]   Sun Y, Giebink NC, Kanno H, Ma B, Thompson ME, Forrest SR. Management of singlet and triplet excitons for efficient white organic light-emitting devices. Nature 2006; 440(7086): 908-12.
[http://dx.doi.org/10.1038/nature04645]

[25]   Ramamurthy AS, Eglal MM. Degradation of TCE by TEOS Coated nZVI in the Presence of Cu(II) for Groundwater Remediation. J Nanomater 2014; 2014: 1-9.
[http://dx.doi.org/10.1155/2014/606534]

[26]   Sharma T, Velmurugan N, Patel P, Chon BH, Sangwai JS. Use of oil-in-water pickering emulsion stabilized by nanoparticles in combination with polymer flood for enhanced oil recovery. Petrol Sci Technol 2015; 33(17-18): 1595-604.
[http://dx.doi.org/10.1080/10916466.2015.1079534]

[27]   Varier KM, Gudeppu M, Chinnasamy A, *et al.* Nanoparticles: antimicrobial applications and its

prospects. Environmental Chemistry for a Sustainable World 2019; 25: 321-55.
[http://dx.doi.org/10.1007/978-3-030-04477-0_12]

[28]   Gudeppu Mounika, Varier Krishnapriya, Chinnasamy Arulvasu, *et al.* Nanobiotechnology approach for the remediation of environmental hazards generated from industrial waste. Emerg Nanost Mater Ener Environ Sci. 2019; pp. 531-61.

[29]   Klaus T, Joerger R, Olsson E, Granqvist CG. Silver-based crystalline nanoparticles, microbially fabricated. Proc Natl Acad Sci USA 1999; 96(24): 13611-4.
[http://dx.doi.org/10.1073/pnas.96.24.13611]

[30]   He S, Guo Z, Zhang Y, Zhang S, Wang J, Gu N. Biosynthesis of gold nanoparticles using the bacteria *Rhodopseudomonas capsulata*. Mater Lett 2007; 61(18): 3984-7.
[http://dx.doi.org/10.1016/j.matlet.2007.01.018]

[31]   Sundaram PA, Augustine R, Kannan M. Extracellular biosynthesis of iron oxide nanoparticles by Bacillus subtilis strains isolated from rhizosphere soil. Biotechnol Bioprocess Eng; BBE 2012; 17(4): 835-40.
[http://dx.doi.org/10.1007/s12257-011-0582-9]

[32]   Iravani S. Bacteria in nanoparticle synthesis: Current status and future prospects. Int Sch Res Notices 2014; 29: 2014-35931.
[http://dx.doi.org/10.1155/2014/359316]

[33]   Thakkar KN, Mhatre SS, Parikh RY. Biological synthesis of metallic nanoparticles. Nanomedicine 2010; 6(2): 257-62.
[http://dx.doi.org/10.1016/j.nano.2009.07.002]

[34]   Lirdprapamongkol K, Warisnoicharoen W, Soisuwan S, Svasti J. Eco-friendly synthesis of fucoidan-stabilized gold nanoparticles. Am J Appl Sci 2010; 7(8): 1038-42.
[http://dx.doi.org/10.3844/ajassp.2010.1038.1042]

[35]   Khandel P, Shahi SK. Microbes mediated synthesis of metal nanoparticles: current status and future prospects. Int J Nanomat Biostruc 2016; 6(1): 1-24.

[36]   Boroumand MA, Namvar F, Moniri M, Tahir P, Azizi S, Mohamad R. Nanoparticles biosynthesized by fungi and yeast: a review of their preparation, properties, and medical applications. Molecules 2015; 20(9): 16540-65.
[http://dx.doi.org/10.3390/molecules200916540]

[37]   Shankar SS, Ahmad A, Pasricha R, Sastry M. Bioreduction of chloroaurate ions by geranium leaves and its endophytic fungus yields gold nanoparticles of different shapes. J Mater Chem 2003; 13(7): 1822-6.
[http://dx.doi.org/10.1039/b303808b]

[38]   Qu Y, Li X, Lian S, *et al.* Biosynthesis of gold nanoparticles using fungus *Trichoderma* sp. WL□Go and their catalysis in degradation of aromatic pollutants. IET Nanobiotechnol 2019; 13(1): 12-7.
[http://dx.doi.org/10.1049/iet-nbt.2018.5177]

[39]   Mukherjee P, Ahmad A, Mandal D, *et al.* Fungus-mediated synthesis of silver nanoparticles and their immobilization in the mycelial matrix: a novel biological approach to nanoparticle synthesis. Nano Lett 2001; 1(10): 515-9.
[http://dx.doi.org/10.1021/nl0155274]

[40]   Sen K, Sinha P, Lahiri S. Time dependent formation of gold nanoparticles in yeast cells: A comparative study. Biochem Eng J 2011; 55(1): 1-6.
[http://dx.doi.org/10.1016/j.bej.2011.02.014]

[41]   Hamida RS, Ali MA, Abdelmeguid NE, Al-Zaban MI, Baz L, Bin-Meferij MM. Lichens—A potential source for nanoparticles fabrication: A review on nanoparticles biosynthesis and their prospective applications. J Fungi (Basel) 2021; 7(4): 291.
[http://dx.doi.org/10.3390/jof7040291]

[42] Yuan X, Xiao S, Taylor TN. Lichen-like symbiosis 600 million years ago. Science 2005; 308(5724): 1017-20.
[http://dx.doi.org/10.1126/science.1111347]

[43] Müller K. Pharmaceutically relevant metabolites from lichens. Appl Microbiol Biotechnol 2001; 56(1-2): 9-16.
[http://dx.doi.org/10.1007/s002530100684]

[44] Rattan R, Shukla S, Sharma B, Bhat M. A mini-review on lichen-based nanoparticles and their applications as antimicrobial agents. Front Microbiol 2021; 12: 633090.
[http://dx.doi.org/10.3389/fmicb.2021.633090]

[45] Makarov VV, Love AJ, Sinitsyna OV, *et al.* "Green" nanotechnologies: synthesis of metal nanoparticles using plants. Acta Nat 2014; 6(1): 35-44.
[http://dx.doi.org/10.32607/20758251-2014-6-1-35-44]

[46] Kuppusamy P, Yusoff M M, Maniam G P, Govindan N. Biosynthesis of metallic nanoparticles using plant derivatives and their new avenues in pharmacological applications - An updated report. Saudi Pharma J : SPJ : the Official Public Saudi Pharma Soc 2016; 24(4): 473-84.

[47] Amrutha LKV, Salini S, Sanjay PM, Sajeet M. Review on terpenoid mediated nanoparticles: significance, mechanism, and biomedical applications. Adv Nat Sci: Nanosci Nanotechnol 2022; 13(3): 033003.

[48] Pirtarighat S, Ghannadnia M, Baghshahi S. Biosynthesis of silver nanoparticles using *Ocimum basilicum* cultured under controlled conditions for bactericidal application. Mater Sci Eng C 2019; 98: 250-5.
[http://dx.doi.org/10.1016/j.msec.2018.12.090]

[49] Vinay Gopal J, Thenmozhi M, Kannabiran K, Rajakumar G, Velayutham K, Rahuman AA. Actinobacteria mediated synthesis of gold nanoparticles using Streptomyces sp. VITDDK3 and its antifungal activity. Mater Lett 2013; 93: 360-2.
[http://dx.doi.org/10.1016/j.matlet.2012.11.125]

[50] Malarkodi C, Rajeshkumar S, Vanaja M, Paulkumar K, Gnanajobitha G, Annadurai G. Eco-friendly synthesis and characterization of gold nanoparticles using *Klebsiella pneumoniae*. J Nanostructure Chem 2013; 3(1): 30.
[http://dx.doi.org/10.1186/2193-8865-3-30]

[51] Balagurunathan R, Radhakrishnan M, Rajendran RB, Velmurugan D. Biosynthesis of gold nanoparticles by actinomycete *Streptomyces viridogens* strain HM10. Indian J Biochem Biophys 2011; 48: 331-5.

[52] Park Y, Hong YN, Weyers A, Kim YS, Linhardt RJ. Polysaccharides and phytochemicals: a natural reservoir for the green synthesis of gold and silver nanoparticles. IET Nanobiotechnol 2011; 5(3): 69-78.
[http://dx.doi.org/10.1049/iet-nbt.2010.0033]

[53] Kalishwaralal K, Deepak V, Ram Kumar Pandian SB, *et al.* Biosynthesis of silver and gold nanoparticles using *Brevibacterium casei*. Colloids Surf B Biointerfaces 2010; 77(2): 257-62.
[http://dx.doi.org/10.1016/j.colsurfb.2010.02.007]

[54] Husseiny MI, El-Aziz MA, Badr Y, Mahmoud MA. Biosynthesis of gold nanoparticles using *Pseudomonas aeruginosa*. Spectrochim Acta A Mol Biomol Spectrosc 2007; 67(3-4): 1003-6.
[http://dx.doi.org/10.1016/j.saa.2006.09.028]

[55] Tripathi V, Fraceto LF, Abhilash PC. Sustainable clean-up technologies for soils contaminated with multiple pollutants: Plant-microbe-pollutant and climate nexus. Ecol Eng 2015; 82: 330-5.
[http://dx.doi.org/10.1016/j.ecoleng.2015.05.027]

[56] Krishnaswamy K, Vali H, Orsat V. Value-adding to grape waste: Green synthesis of gold nanoparticles. J Food Eng 2014; 142: 210-20.

[http://dx.doi.org/10.1016/j.jfoodeng.2014.06.014]

[57]   Selvakumar P, Viveka S, Prakash S, Jasminebeaula S, Uloganathan R. Antimicrobial activity of extracellularly synthesized silver nanoparticles from marine derived Streptomyces rochei. Int J Pharm Biol Sci 2012; 3: 188-97.

[58]   Prakasham RS, , , *et al.* R., Kumar, B. S., Kumar, Y. S., & Shankar, G. G. Characterization of silver nanoparticles synthesized by using marine isolate *Streptomyces albidoflavus*. J Microbiol Biotechnol 2012; 22(5): 614-21.
[http://dx.doi.org/10.4014/jmb.1107.07013]

[59]   Durán N, Marcato PD, Durán M, Yadav A, Gade A, Rai M. Mechanistic aspects in the biogenic synthesis of extracellular metal nanoparticles by peptides, bacteria, fungi, and plants. Appl Microbiol Biotechnol 2011; 90(5): 1609-24.
[http://dx.doi.org/10.1007/s00253-011-3249-8]

[60]   Faghri Zonooz N, Salouti M. Extracellular biosynthesis of silver nanoparticles using cell filtrate of Streptomyces sp. ERI-3. Sci Iran 2011; 18(6): 1631-5.
[http://dx.doi.org/10.1016/j.scient.2011.11.029]

[61]   Jain D, Kachhwaha S, Jain R, Srivastava G, Kothari SL. Novel microbial route to synthesize silver nanoparticles using spore crystal mixture of *Bacillus thuringiensis*. Indian J Exp Biol 2010; 48(11): 1152-6.

[62]   Nanda A, Saravanan M. Biosynthesis of silver nanoparticles from *Staphylococcus aureus* and its antimicrobial activity against MRSA and MRSE. Nanomedicine 2009; 5(4): 452-6.
[http://dx.doi.org/10.1016/j.nano.2009.01.012]

[63]   Anghel L, Balasoiu M, Ishchenko LA, *et al.* Characterization of bio-synthesized nanoparticles produced by *Klebsiella oxytoca*. J Phys Conf Ser 2012; 351(1): 012005.
[http://dx.doi.org/10.1088/1742-6596/351/1/012005]

[64]   Narayanan KB, Sakthivel N. Synthesis and characterization of nano-gold composite using *Cylindrocladium floridanum* and its heterogeneous catalysis in the degradation of 4-nitrophenol. J Hazard Mater 2011; 189(1-2): 519-25.
[http://dx.doi.org/10.1016/j.jhazmat.2011.02.069]

[65]   Binupriya AR, Sathishkumar M, Vijayaraghavan K, Yun SI. Bioreduction of trivalent aurum to nano-crystalline gold particles by active and inactive cells and cell-free extract of *Aspergillus oryzae* var. viridis. J Hazard Mater 2010; 177(1-3): 539-45.
[http://dx.doi.org/10.1016/j.jhazmat.2009.12.066]

[66]   Kim A, Muthuchamy N, Yoon C, Joo S, Park K. MOF-derived Cu@ Cu$_2$O nanocatalyst for oxygen reduction reaction and cycloaddition reaction. Nanomaterials 2018; 8(3): 138.
[http://dx.doi.org/10.3390/nano8030138]

[67]   Shantkriti S, Rani P. Biological synthesis of copper nanoparticles using *Pseudomonas fluorescens*. Int J Curr Microbiol Appl Sci 2014; 3(9): 374-83.

[68]   Umer A, Naveed S, Ramzan N, Rafique MS. Selection of a suitable method for the synthesis of copper nanoparticles. Nano 2012; 7(5): 1230005.
[http://dx.doi.org/10.1142/S1793292012300058]

[69]   Varshney R, Bhadauria S, Gaur M S, Pasricha R. Characterization of copper nanoparticles synthesized by a novel microbiological method. JOM 2010; 62(12): 102-4.
[http://dx.doi.org/10.1007/s11837-010-0171-y]

[70]   Selvarajan E, Mohanasrinivasan V. Biosynthesis and characterization of ZnO nanoparticles using *Lactobacillus plantarum* VITES07. Mater Lett 2013; 112: 180-2.
[http://dx.doi.org/10.1016/j.matlet.2013.09.020]

[71]   Jayaseelan C, Rahuman AA, Kirthi AV, *et al.* Novel microbial route to synthesize ZnO nanoparticles using *Aeromonas hydrophila* and their activity against pathogenic bacteria and fungi. Spectrochim

Acta A Mol Biomol Spectrosc 2012; 90: 78-84.
[http://dx.doi.org/10.1016/j.saa.2012.01.006]

[72] Lee C, Kim JY, Lee WI, Nelson KL, Yoon J, Sedlak DL. Bactericidal effect of zero-valent iron nanoparticles on *Escherichia coli*. Environ Sci Technol 2008; 42(13): 4927-33.
[http://dx.doi.org/10.1021/es800408u]

[73] Martins M, Mourato C, Sanches S, Noronha JP, Crespo MTB, Pereira IAC. Biogenic platinum and palladium nanoparticles as new catalysts for the removal of pharmaceutical compounds. Water Res 2017; 108: 160-8.
[http://dx.doi.org/10.1016/j.watres.2016.10.071]

[74] Dahoumane SA, Yéprémian C, Djédiat C, *et al.* Improvement of kinetics, yield, and colloidal stability of biogenic gold nanoparticles using living cells of *Euglena gracilis* microalga. J Nanopart Res 2016; 18(3): 79.
[http://dx.doi.org/10.1007/s11051-016-3378-1]

[75] Namvar F, Azizi S, Ahmad MB, *et al.* Green synthesis and characterization of gold nanoparticles using the marine macroalgae *Sargassum muticum*. Res Chem Intermed 2015; 41(8): 5723-30.
[http://dx.doi.org/10.1007/s11164-014-1696-4]

[76] Annamalai J, Nallamuthu T. Characterization of biosynthesized gold nanoparticles from aqueous extract of *Chlorella vulgaris* and their anti-pathogenic properties. Appl Nanosci 2015; 5(5): 603-7.
[http://dx.doi.org/10.1007/s13204-014-0353-y]

[77] Rajeshkumar S, Kannan C, Annadurai G. Green synthesis of silver nanoparticles using marine brown algae *Turbinaria conoides* and its antibacterial activity. Int J Pharma Bio Sci 2012; 3(4): 502-10.

[78] Singh M, Kalaivani R, Manikandan S, Sangeetha N, Kumaraguru AK. Facile green synthesis of variable metallic gold nanoparticle using *Padina gymnospora*, a brown marine macroalga. Appl Nanosci 2013; 3(2): 145-51.
[http://dx.doi.org/10.1007/s13204-012-0115-7]

[79] Darwesh O M, Matter I A, Eida M F, Moawad H, Oh Y K. Influence of nitrogen source and growth phase on extracellular biosynthesis of silver nanoparticles using cultural filtrates of *Scenedesmus obliquus*. Appl Sci 2019; 9(70): 1465.

[80] Khalid M, Khalid N, Ahmed I, Hanif R, Ismail M, Janjua HA. Comparative studies of three novel freshwater microalgae strains for synthesis of silver nanoparticles: insights of characterization, antibacterial, cytotoxicity and antiviral activities. J Appl Phycol 2017; 29(4): 1851-63.
[http://dx.doi.org/10.1007/s10811-017-1071-0]

[81] Muthusamy G, Thangasamy S, Raja M, Chinnappan S, Kandasamy S. Biosynthesis of silver nanoparticles from Spirulina microalgae and its antibacterial activity. Environ Sci Pollut Res Int 2017; 24(23): 19459-64.
[http://dx.doi.org/10.1007/s11356-017-9772-0]

[82] Kathiraven T, Sundaramanickam A, Shanmugam N, Balasubramanian T. Green synthesis of silver nanoparticles using marine algae *Caulerpa racemosa* and their antibacterial activity against some human pathogens. Appl Nanosci 2015; 5(4): 499-504.
[http://dx.doi.org/10.1007/s13204-014-0341-2]

[83] Prasad TNVKV, Kambala VSR, Naidu R. Phyconanotechnology: synthesis of silver nanoparticles using brown marine algae *Cystophora moniliformis* and their characterisation. J Appl Phycol 2013; 25(1): 177-82.
[http://dx.doi.org/10.1007/s10811-012-9851-z]

[84] Devi JS, Bhimba BV, Peter DM. Production of biogenic silver nanoparticles using *Sargassum longifolium* and its applications. Indian J Geo-Mar Sci 2013; 42(1): 125-30.

[85] Mahdavi M, Namvar F, Ahmad M, Mohamad R. Green biosynthesis and characterization of magnetic iron oxide [Fe$_3$O$_4$] nanoparticles using seaweed [*Sargassum muticum*] aqueous extract. Molecules

2013; 18(5): 5954-64.
[http://dx.doi.org/10.3390/molecules18055954]

[86] Brayner R, Coradin T, Beaunier P, *et al.* Intracellular biosynthesis of superparamagnetic 2-lines ferri-hydrite nanoparticles using *Euglena gracilis* microalgae. Colloids Surf B Biointerfaces 2012; 93: 20-3.
[http://dx.doi.org/10.1016/j.colsurfb.2011.10.014]

[87] Sayadi MH, Salmani N, Heidari A, Rezaei MR. Bio-synthesis of palladium nanoparticle using *Spirulina platensis* alga extract and its application as adsorbent. Surf Interfaces 2018; 10: 136-43.
[http://dx.doi.org/10.1016/j.surfin.2018.01.002]

[88] Arsiya F, Sayadi M, Sobhani S. Arsenic [III] adsorption using palladium nanoparticles from aqueous solution. J Water and Environ Nanotech 2017; 2(3): 166-73.

[89] Abboud Y, Saffaj T, Chagraoui A, *et al.* Biosynthesis, characterization and antimicrobial activity of copper oxide nanoparticles (CONPs) produced using brown alga extract (*Bifurcaria bifurcata*). Appl Nanosci 2014; 4(5): 571-6. [Bifurcaria bifurcata].
[http://dx.doi.org/10.1007/s13204-013-0233-x]

[90] Azizi S, Ahmad MB, Namvar F, Mohamad R. Green biosynthesis and characterization of zinc oxide nanoparticles using brown marine macroalga *Sargassum muticum* aqueous extract. Mater Lett 2014; 116: 275-7.
[http://dx.doi.org/10.1016/j.matlet.2013.11.038]

[91] Narayanan K B, Park H H, Sakthivel N. Extracellular synthesis of mycogenic silver nanoparticles by *Cylindrocladium floridanum* and its homogeneous catalytic degradation of 4-nitrophenol. Spectro Acta Part A: Molec Biomol Spectro 2013; 116: 485-90.

[92] Chauhan A, Zubair S, Tufail S, *et al.* Fungus-mediated biological synthesis of gold nanoparticles: potential in detection of liver cancer. Int J Nanomedicine 2011; 6: 2305.

[93] Sheikhloo Z, Salouti M, Katiraee F. Biological synthesis of gold nanoparticles by fungus *Epicoccumnigrum*. J Cluster Sci 2011; 22(4): 661-5.
[http://dx.doi.org/10.1007/s10876-011-0412-4]

[94] Castro-Longoria E, Vilchis-Nestor AR, Avalos-Borja M. Biosynthesis of silver, gold and bimetallic nanoparticles using the filamentous fungus *Neurospora crassa*. Colloids Surf B Bioint 2011; 83(1): 42-8.
[http://dx.doi.org/10.1016/j.colsurfb.2010.10.035]

[95] Mishra A, Tripathy SK, Wahab R, *et al.* Microbial synthesis of gold nanoparticles using the fungus *Penicillium brevicompactum* and their cytotoxic effects against mouse mayo blast cancer C2C12 cells. Appl Microbiol Biotechnol 2011; 92(3): 617-30.
[http://dx.doi.org/10.1007/s00253-011-3556-0]

[96] Mishra AN, Bhadauria S, Gaur MS, Pasricha R. Extracellular microbial synthesis of gold nanoparticles using fungus *Hormoconis resinae*. J Miner Met Mater Soc 2010; 62(11): 45-8.
[http://dx.doi.org/10.1007/s11837-010-0168-6]

[97] Saravanan M, Nanda A. Extracellular synthesis of silver bionanoparticles from *Aspergillus clavatus* and its antimicrobial activity against MRSA and MRSE. Colloids Surf B Biointerfaces 2010; 77(2): 214-8.
[http://dx.doi.org/10.1016/j.colsurfb.2010.01.026]

[98] Ottoni CA, Lima Neto MC, Léo P, Ortolan BD, Barbieri E, De Souza AO. Environmental impact of biogenic silver nanoparticles in soil and aquatic organisms. Chemosphere 2020; 239: 124698.
[http://dx.doi.org/10.1016/j.chemosphere.2019.124698]

[99] Calderon B, Fullana A. Heavy metal release due to aging effect during zero valent iron nanoparticles remediation. Water Res 2015; 83: 1-9.
[http://dx.doi.org/10.1016/j.watres.2015.06.004]

[100] Phanjom P, Ahmed G. Biosynthesis of silver nanoparticles by *Aspergillus oryzae* [MTCC No. 1846]

and its characterizations. Nanoscience and Nanotechnology 2015; 5(1): 14-21.

[101] Rajesh Kumar R, Poornima Priyadharsani K, Thamaraiselvi K. Mycogenic synthesis of silver nanoparticles by the Japanese environmental isolate *Aspergillus tamarii*. J Nanopart Res 2012; 14(5): 860.
       [http://dx.doi.org/10.1007/s11051-012-0860-2]

[102] Jain N, Bhargava A, Majumdar S, Tarafdar JC, Panwar J. Extracellular biosynthesis and characterization of silver nanoparticles using Aspergillus flavusNJP08: A mechanism perspective. Nanoscale 2011; 3(2): 635-41.
       [http://dx.doi.org/10.1039/C0NR00656D]

[103] Vahabi K, Mansoori GA, Karimi S. Biosynthesis of silver nanoparticles by fungus *Trichoderma reesei*. Insciences J 2011; 1(1): 65-79. [a route for large-scale production of AgNPs].
       [http://dx.doi.org/10.5640/insc.010165]

[104] Balaji D S, Basavaraja S, Deshpande R, Mahesh D B, Prabhakar B K, Venkataraman A. Extracellular biosynthesis of functionalized silver nanoparticles by strains of *Cladosporium cladosporioides* fungus. Colloids Surf B Bioint 2009; 68(1): 88-92.
       [http://dx.doi.org/10.1016/j.colsurfb.2008.09.022]

[105] Mazumdar H, Haloi N. A study on biosynthesis of iron nanoparticles by *Pleurotus* sp. J Microbiol Biotechnol Res 2011; 1(3): 39-49.

[106] Mirzadeh S, Darezereshki E, Bakhtiari F, Fazaelipoor MH, Hosseini MR. Characterization of zinc sulfide (ZnS) nanoparticles Biosynthesized by *Fusarium oxysporum*. Mater Sci Semicond Process 2013; 16(2): 374-8.
       [http://dx.doi.org/10.1016/j.mssp.2012.09.008]

[107] Cuevas R, Durán N, Diez MC, Tortella GR, Rubilar O. Extracellular biosynthesis of copper and copper oxide nanoparticles by *Stereum hirsutum*, a native white-rot fungus from chilean forests. J Nanomater 2015; 2015: 1-7.
       [http://dx.doi.org/10.1155/2015/789089]

[108] Salvadori MR, Lepre LF, Ando RA, Oller do Nascimento CA, Corrêa B. Biosynthesis and uptake of copper nanoparticles by dead biomass of *Hypocrea lixii* isolated from the metal mine in the Brazilian Amazon region. PLoS One 2013; 8(11): e80519.
       [http://dx.doi.org/10.1371/journal.pone.0080519]

[109] Majumder DR. Bioremediation: copper nanoparticles from electronic-waste. Int J Eng Sci Technol 2012; 4(10): 4380-9.

[110] Honary S, Barabadi H, Gharaei-Fathabad E, Naghibi F. Green synthesis of copper oxide nanoparticles using *Penicillium aurantiogriseum*, *Penicillium citrinum* and *Penicillium waksmanii*. Dig J Nanomater Biostruct 2012; 7(3): 999-1005.

[111] Shamsuzzaman , Mashrai A, Khanam H, Aljawfi RN. Biological synthesis of ZnO nanoparticles using C. albicans and studying their catalytic performance in the synthesis of steroidal pyrazolines. Arab J Chem 2017; 10: S1530-6.
       [http://dx.doi.org/10.1016/j.arabjc.2013.05.004]

[112] Sandana Mala JG, Rose C. Facile production of ZnS quantum dot nanoparticles by *Saccharomyces cerevisiae* MTCC 2918. J Biotechnol 2014; 170: 73-8.
       [http://dx.doi.org/10.1016/j.jbiotec.2013.11.017]

[113] Shankar S, Jaiswal L, Aparna RSL, Prasad RGSV. Synthesis, characterization, *in vitro* biocompatibility, and antimicrobial activity of gold, silver and gold silver alloy nanoparticles prepared from *Lansium domesticum* fruit peel extract. Mater Lett 2014; 137: 75-8.
       [http://dx.doi.org/10.1016/j.matlet.2014.08.122]

[114] Akilandaeaswari B, Muthu K. Green method for synthesis and characterization of gold nanoparticles using *Lawsonia inermis* seed extract and their photocatalytic activity. Mater Lett 2020; 277: 128344.

[http://dx.doi.org/10.1016/j.matlet.2020.128344]

[115]   Song JY, Jang HK, Kim BS. Biological synthesis of gold nanoparticles using *Magnolia kobus* and *Diopyros kaki* leaf extracts. Process Biochem 2009; 44(10): 1133-8.
        [http://dx.doi.org/10.1016/j.procbio.2009.06.005]

[116]   Keshavarzi M, Davoodi D, Pourseyedi S, Taghizadeh S. The effects of three types of alfalfa plants (*Medicago sativa*) on the biosynthesis of gold nanoparticles: an insight into phytomining. Gold Bull 2018; 51(3): 99-110.
        [http://dx.doi.org/10.1007/s13404-018-0237-0]

[117]   Vankar PS, Bajpai D. Preparation of gold nanoparticles from *Mirabilis jalapa* flowers. Indian J Biochem Biophys 2010; 47(3): 157-60.

[118]   Bindhu MR, Vijaya Rekha P, Umamaheswari T, Umadevi M. Antibacterial activities of *Hibiscus cannabinus* stem-assisted silver and gold nanoparticles. Mater Lett 2014; 131: 194-7.
        [http://dx.doi.org/10.1016/j.matlet.2014.05.172]

[119]   Abdel-Raouf N, Al-Enazi NM, Ibraheem IBM. Green biosynthesis of gold nanoparticles using *Galaxaura elongata* and characterization of their antibacterial activity. Arab J Chem 2017; 10: S3029-39.
        [http://dx.doi.org/10.1016/j.arabjc.2013.11.044]

[120]   Arunachalam K, Annamalai S, Shanmugasundaram Hari . One-step green synthesis and characterization of leaf extract-mediated biocompatible silver and gold nanoparticles from *Memecylon umbellatum*. Int J Nanomedicine 2013; 8: 1307-15.
        [http://dx.doi.org/10.2147/IJN.S36670]

[121]   Parashar UK, Saxena PS, Srivastava A. Bioinspired synthesis of silver nanoparticles. Dig J Nanomater Biostruct 2009; 4(1): 159-66. [DJNB].

[122]   Gopu C, Chirumamilla P, Kagithoju S, Taduri S. Green synthesis of silver nanoparticles using *Momordica cymbalaria* aqueous leaf extracts and screening of their antimicrobial activity. Proc Natl Acad Sci, India, Sect B Biol Sci 2022; 92(4): 771-82.
        [http://dx.doi.org/10.1007/s40011-022-01367-x]

[123]   Menon S, Agarwal H, Rajeshkumar S, Kumar SV. Anticancer assessment of biosynthesized silver nanoparticles using *Mucuna pruriens* seed extract on Lung Cancer Treatment. Res J Pharm Techno 2018; 11(9): 3887-91.
        [http://dx.doi.org/10.5958/0974-360X.2018.00712.6]

[124]   Shilpha J, Meyappan V, Sakthivel N. Bioinspired synthesis of gold nanoparticles from *Hemidesmus indicus* L. root extract and their antibiofilm efficacy against *Pseudomonas aeruginosa*. Process Biochem 2022; 122: 224-37.
        [http://dx.doi.org/10.1016/j.procbio.2022.10.018]

[125]   Devi GK, Kumar KS, Parthiban R, Kalishwaralal K. An insight study on HPTLC fingerprinting of Mukia maderaspatna : Mechanism of bioactive constituents in metal nanoparticle synthesis and its activity against human pathogens. Microb Pathog 2017; 102: 120-32.
        [http://dx.doi.org/10.1016/j.micpath.2016.11.026]

[126]   Miri A, Sarani M, Bazaz M R, Darroudi M. Plant-mediated biosynthesis of silver nanoparticles using *Prosopis farcta* extract and its antibacterial properties. Spectrochim Acta A Mol Biomol Spectrosc 2015; 15(141): 287-91.
        [http://dx.doi.org/10.1016/j.saa.2015.01.024]

[127]   Gopalakrishnan K, Ramesh C, Ragunathan V, Thamilselvan M. Antibacterial activity of Cu_2O nanoparticles on E. coli synthesized from *Tridax procumbens* leaf extract and surface coating with polyaniline. Dig J Nanomater Biostruct 2012; 7(2): 833-9.

[128]   Chan Y, Selvanathan V, Tey LH, *et al.* Effect of Calcination Temperature on Structural, Morphological and Optical Properties of Copper Oxide Nanostructures Derived from Garcinia

mangostana L. Leaf Extract. Nanomaterials 2022; 12(20): 3589.
[http://dx.doi.org/10.3390/nano12203589]

[129] Suresh D, Shobharani RM, Nethravathi PC, Pavan Kumar MA, Nagabhushana H, Sharma SC. Artocarpus gomezianus aided green synthesis of ZnO nanoparticles: Luminescence, photocatalytic and antioxidant properties. Spectrochim Acta A Mol Biomol Spectrosc 2015; 141: 128-34.
[http://dx.doi.org/10.1016/j.saa.2015.01.048]

[130] Sangeetha G, Rajeshwari S, Venckatesh R. Green synthesis of zinc oxide nanoparticles by aloe barbadensis miller leaf extract: Structure and optical properties. Mater Res Bull 2011; 46(12): 2560-6.
[http://dx.doi.org/10.1016/j.materresbull.2011.07.046]

[131] Chandra H, Patel D, Kumari P, Jangwan JS, Yadav S. Phyto-mediated synthesis of zinc oxide nanoparticles of *Berberis aristata*: Characterization, antioxidant activity and antibacterial activity with special reference to urinary tract pathogens. Mater Sci Eng C 2019; 102: 212-20.
[http://dx.doi.org/10.1016/j.msec.2019.04.035]

[132] Vidya C, Hiremath S, Chandraprabha MN, *et al.* Green synthesis of ZnO nanoparticles by *Calotropis gigantea*. Int J Curr Eng Technol 2013; 1(1): 118-20.

[133] Sana SS, Badineni VR, Arla SK, Naidu Boya VK. Eco-friendly synthesis of silver nanoparticles using leaf extract of *Grewia flaviscences* and study of their antimicrobial activity. Mater Lett 2015; 145: 347-50.
[http://dx.doi.org/10.1016/j.matlet.2015.01.096]

[134] Wade EA, Massie CJ, Harris DF. Synthesis and characterization of ZnO nanoparticles from extracts of allium sativum and *hydrastis canadensis*. Idaho Confe on Underg Res. 2020.

[135] Velmurugan P, Cho M, Lim SS, *et al.* Phytosynthesis of silver nanoparticles by *Prunus yedoensis* leaf extract and their antimicrobial activity. Mater Lett 2015; 138: 272-5.
[http://dx.doi.org/10.1016/j.matlet.2014.09.136]

[136] Yousefzadi M, Rahimi Z, Ghafori V. The green synthesis, characterization and antimicrobial activities of silver nanoparticles synthesized from green alga *Enteromorpha flexuosa* (wulfen) J. Agardh. Mater Lett 2014; 137: 1-4.
[http://dx.doi.org/10.1016/j.matlet.2014.08.110]

[137] Iyer RI, Panda T. Biosynthesis of gold and silver nanoparticles using extracts of callus cultures of pumpkin (*Cucurbita maxima*). J Nanosci Nanotechnol 2018; 18(8): 5341-53.
[http://dx.doi.org/10.1166/jnn.2018.15378]

[138] Sulaiman GM, Mohammed WH, Marzoog TR, Al-Amiery AAA, Kadhum AAH, Mohamad AB. Green synthesis, antimicrobial and cytotoxic effects of silver nanoparticles using *Eucalyptus chapmaniana* leaves extract. Asian Pac J Trop Biomed 2013; 3(1): 58-63.
[http://dx.doi.org/10.1016/S2221-1691(13)60024-6]

[139] Ahmad W, Kumar Jaiswal K, Amjad M. Euphorbia herita leaf extract as a reducing agent in a facile green synthesis of iron oxide nanoparticles and antimicrobial activity evaluation. Inorg and Nano-Metal Chem 2021; 51(9): 1147-54.

[140] Singh BK, Walker A. Microbial degradation of organophosphorus compounds. FEMS Microbiol Rev 2006; 30(3): 428-71.
[http://dx.doi.org/10.1111/j.1574-6976.2006.00018.x]

[141] Castro L, Blázquez ML, González F, Muñoz JA, Ballester A. Heavy metal adsorption using biogenic iron compounds. Hydrometallurgy 2018; 179: 44-51.
[http://dx.doi.org/10.1016/j.hydromet.2018.05.029]

[142] Aziz N, Faraz M, Pandey R, *et al.* Facile algae-derived route to biogenic silver nanoparticles: synthesis, antibacterial, and photocatalytic properties. Langmuir 2015; 31(42): 11605-12.
[http://dx.doi.org/10.1021/acs.langmuir.5b03081]

[143] Furgal KM, Meyer RL, Bester K. Removing selected steroid hormones, biocides and pharmaceuticals

from water by means of biogenic manganese oxide nanoparticles *in situ* at ppb levels. Chemosphere 2015; 136: 321-6.
[http://dx.doi.org/10.1016/j.chemosphere.2014.11.059]

[144] Kümmerer K. Antibiotics in the aquatic environment – A review – Part I. Chemosphere 2009; 75(4): 417-34.
[http://dx.doi.org/10.1016/j.chemosphere.2008.11.086]

[145] Forrez I, Carballa M, Fink G, *et al.* Biogenic metals for the oxidative and reductive removal of pharmaceuticals, biocides and iodinated contrast media in a polishing membrane bioreactor. Water Res 2011; 45(4): 1763-73.
[http://dx.doi.org/10.1016/j.watres.2010.11.031]

[146] Yogalakshmi K N, Das A, Rani G, Jaswal V, Randhawa J S. Nano-bioremediation: A new age technology for the treatment of dyes in textile effluents. Bioremed of Indust Waste for Environ Saf 2019; 313-47.
[http://dx.doi.org/10.1007/978-981-13-1891-7_15]

[147] Wang P, Song Y, Fan H, Yu L. Bioreduction of azo dyes was enhanced by *in-situ* biogenic palladium nanoparticles. Bioresour Technol 2018; 266: 176-80.
[http://dx.doi.org/10.1016/j.biortech.2018.06.079]

[148] Johnson A, Merilis G, Hastings J, Elizabeth Palmer M, Fitts JP, Chidambaram D. Reductive degradation of organic compounds using microbial nanotechnology. J Electrochem Soc 2013; 160(1): G27-31.
[http://dx.doi.org/10.1149/2.053301jes]

[149] Hennebel T, Verhagen P, Simoen H, *et al.* Remediation of trichloroethylene by bio-precipitated and encapsulated palladium nanoparticles in a fixed bed reactor. Chemosphere 2009; 76(9): 1221-5.
[http://dx.doi.org/10.1016/j.chemosphere.2009.05.046]

[150] Resgalla C Jr, Noldin JA, Tamanaha MS, Deschamps FC, Eberhardt DS, Rörig LR. Risk analysis of herbicide quinclorac residues in irrigated rice areas, Santa Catarina, Brazil. Ecotoxicology 2007; 16(8): 565-71.
[http://dx.doi.org/10.1007/s10646-007-0165-x]

[151] Shi G, Li Y, Xi G, *et al.* Rapid green synthesis of gold nanocatalyst for high-efficiency degradation of quinclorac. J Hazard Mater 2017; 335: 170-7.
[http://dx.doi.org/10.1016/j.jhazmat.2017.04.042]

[152] Sengupta S. Nano-Remediation: Carving solution to pesticide pollution Agrobios Newsl 2019; 18: 134-5.

[153] Das SK, Das AR, Guha AK. Gold nanoparticles: Microbial synthesis and application in water hygiene management. Langmuir 2009; 25(14): 8192-9.
[http://dx.doi.org/10.1021/la900585p]

[154] Lee MK, Saunders JA, Wilson T, *et al.* Field-scale bioremediation of arsenic-contaminated groundwater using sulfate-reducing bacteria and biogenic pyrite. Bioremediat J 2019; 23(1): 1-21.
[http://dx.doi.org/10.1080/10889868.2018.1516617]

[155] Davolos D, Pietrangeli B. A molecular study on bacterial resistance to arsenic-toxicity in surface and underground waters of Latium (Italy). Ecotoxicol Environ Saf 2013; 96: 1-9.
[http://dx.doi.org/10.1016/j.ecoenv.2013.05.039]

[156] Bai Y, Yang T, Liang J, Qu J. The role of biogenic Fe-Mn oxides formed in situ for arsenic oxidation and adsorption in aquatic ecosystems. Water Res 2016; 98: 119-27.
[http://dx.doi.org/10.1016/j.watres.2016.03.068]

[157] Zhou D, Kim DG, Ko SO. Heavy metal adsorption with biogenic manganese oxides generated by *Pseudomonas putida* strain MnB1. J Ind Eng Chem 2015; 24: 132-9.
[http://dx.doi.org/10.1016/j.jiec.2014.09.020]

[158] Xu J, Bravo AG, Lagerkvist A, Bertilsson S, Sjöblom R, Kumpiene J. Sources and remediation techniques for mercury contaminated soil. Environ Int 2015; 74: 42-53.
[http://dx.doi.org/10.1016/j.envint.2014.09.007]

[159] Wang X, Zhang D, Pan X, *et al.* Aerobic and anaerobic biosynthesis of nano-selenium for remediation of mercury contaminated soil. Chemosphere 2017; 170: 266-73. b
[http://dx.doi.org/10.1016/j.chemosphere.2016.12.020]

[160] Dave D, Ghaly AE. Remediation technologies for marine oil spills: a critical review and comparative analysis. Am J Environ Sci 2011; 7(5): 424-40.
[http://dx.doi.org/10.3844/ajessp.2011.424.440]

[161] Agarwal N, Solanki VS, Gacem A, *et al.* Bacterial laccases as biocatalysts for the remediation of environmental toxic pollutants: a green and eco-friendly approach—a review. Water 2022; 14(24): 4068.
[http://dx.doi.org/10.3390/w14244068]

[162] Omokhagbor Adams G, Tawari Fufeyin P, Eruke Okoro S, Ehinomen I. Bioremediation, biostimulation and bioaugmention: a review. Int J Environ Bioremediat Biodegrad 2020; 3(1): 28-39.
[http://dx.doi.org/10.12691/ijebb-3-1-5]

[163] Meng L, Liu H, Bao M, Sun P. Microbial community structure shifts are associated with temperature, dispersants and nutrients in crude oil-contaminated seawaters. Mar Pollut Bull 2016; 111(1-2): 203-12.
[http://dx.doi.org/10.1016/j.marpolbul.2016.07.010]

[164] Sadeghpour A, Pirolt F, Glatter O. Submicrometer-sized Pickering emulsions stabilized by silica nanoparticles with adsorbed oleic acid. Langmuir 2013; 29(20): 6004-12.
[http://dx.doi.org/10.1021/la4008685]

[165] Mehta D, Mazumdar S, Singh SK. Magnetic adsorbents for the treatment of water/wastewater—A review. J Water Process Eng 2015; 7: 244-65.
[http://dx.doi.org/10.1016/j.jwpe.2015.07.001]

[166] Saunders JA, Lee MK, Dhakal P, *et al.* Bioremediation of arsenic-contaminated groundwater by sequestration of arsenic in biogenic pyrite. Appl Geochem 2018; 96: 233-43.
[http://dx.doi.org/10.1016/j.apgeochem.2018.07.007]

[167] Verma A, Dua R, Singh A, Bishnoi NR. Biogenic sulfides for sequestration of Cr (VI), COD and sulfate from synthetic wastewater. Water Science 2015; 29(1): 19-25.
[http://dx.doi.org/10.1016/j.wsj.2015.03.001]

[168] Jencarova J, Luptakova A. The elimination of heavy metal ions from waters by biogenic iron sulphides. Chem Eng Trans 2012; 28: 205-10.

[169] Wang X, Gai Z, Yu B, *et al.* Degradation of carbazole by microbial cells immobilized in magnetic gellan gum gel beads. Appl Environ Microbiol 2007; 73(20): 6421-8.
[http://dx.doi.org/10.1128/AEM.01051-07]

[170] Bokare V, Murugesan K, Kim YM, Jeon JR, Kim EJ, Chang YS. Degradation of triclosan by an integrated nano-bio redox process. Bioresour Technol 2010; 101(16): 6354-60.
[http://dx.doi.org/10.1016/j.biortech.2010.03.062]

[171] Bokare V, Murugesan K, Kim JH, Kim EJ, Chang YS. Integrated hybrid treatment for the remediation of 2,3,7,8-tetrachlorodibenzo-p-dioxin. Sci Total Environ 2012; 435-436: 563-6.
[http://dx.doi.org/10.1016/j.scitotenv.2012.07.079]

[172] Le TT, Nguyen KH, Jeon JR, Francis AJ, Chang YS. Nano/bio treatment of polychlorinated biphenyls with evaluation of comparative toxicity. J Hazard Mater 2015; 287: 335-41.
[http://dx.doi.org/10.1016/j.jhazmat.2015.02.001]

[173] Wu R, Wu H, Jiang X, *et al.* The key role of biogenic manganese oxides in enhanced removal of highly recalcitrant 1,2,4-triazole from bio-treated chemical industrial wastewater. Environ Sci Pollut

Res Int 2017; 24(11): 10570-83.
[http://dx.doi.org/10.1007/s11356-017-8641-1]

[174] Wang R, Wang S, Tai Y, *et al.* Biogenic manganese oxides generated by green algae Desmodesmus sp. WR1 to improve bisphenol A removal. J Hazard Mater 2017; 339: 310-9.
[http://dx.doi.org/10.1016/j.jhazmat.2017.06.026]

[175] Chidambaram D, Hennebel T, Taghavi S, *et al.* Concomitant microbial generation of palladium nanoparticles and hydrogen to immobilize chromate. Environ Sci Technol 2010; 44(19): 7635-40.
[http://dx.doi.org/10.1021/es101559r]

[176] Windt WD, Aelterman P, Verstraete W. Bioreductive deposition of palladium (0) nanoparticles on *Shewanella oneidensis* with catalytic activity towards reductive dechlorination of polychlorinated biphenyls. Environ Microbiol 2005; 7(3): 314-25.
[http://dx.doi.org/10.1111/j.1462-2920.2005.00696.x]

[177] De Corte S, Sabbe T, Hennebel T, *et al.* Doping of biogenic Pd catalysts with Au enables dechlorination of diclofenac at environmental conditions. Water Res 2012; 46(8): 2718-26.
[http://dx.doi.org/10.1016/j.watres.2012.02.036]

[178] Yan FF, Wu C, Cheng YY, He YR, Li WW, Yu HQ. Carbon nanotubes promote Cr(VI) reduction by alginate-immobilized *Shewanella oneidensis* MR-1. Biochem Eng J 2013; 77: 183-9.
[http://dx.doi.org/10.1016/j.bej.2013.06.009]

[179] Fischer S, Krause T, Lederer F, *et al.* Bacillus safensis JG-B5T affects the fate of selenium by extracellular production of colloidally less stable selenium nanoparticles. J Hazard Mater 2020; 384: 121146.
[http://dx.doi.org/10.1016/j.jhazmat.2019.121146]

[180] Fernández PM, Viñarta SC, Bernal AR, Cruz EL, Figueroa LIC. Bioremediation strategies for chromium removal: Current research, scale-up approach and future perspectives. Chemosphere 2018; 208: 139-48.
[http://dx.doi.org/10.1016/j.chemosphere.2018.05.166]

[181] Ramezani M, Rad FA, Ghahari S, *et al.* Nano-bioremediation application for environment contamination by microorganism. Microbial rejuvenation of polluted environment, microorganisms for sustainability. 26th ed. Singapore: Springer 2021; pp. 349-78.
[http://dx.doi.org/10.1007/978-981-15-7455-9_14]

Phytoremediation/Phytoextraction: A Sustainable Approach to the Restoration of Chromium-Contaminated Soil

Pankaj Kumar[1,*], Deepankshi Shah[1], Manoj Kumar[2], Snigdha Singh[1], Virendra Kumar Yadav[3], Mohd. Tariq[4], Ramesh Kumar[5], Nakul Kumar[6], Shivraj Gangadhar Wanale[7] and Shipra Choudhary[8]

[1] *Department of Environmental Science, Parul Institute of Applied Sciences, Parul University, Vadodara, Gujarat, India*

[2] *Department of Hydro and Renewable Energy, Indian Institute of Technology Roorkee, Roorkee, Uttarakhand, India*

[3] *Department of Life Sciences, Hemchandracharya North Gujarat University, Matarvadi Part, Gujarat, India*

[4] *Department of Life Science, Parul Institute of Applied Sciences, Parul University, Vadodara, Gujarat, India*

[5] *Department of Environmental Science, School of Earth Sciences, Central University of Rajasthan, Ajmer, Rajasthan, India*

[6] *Gandhinagar Institute of Science, Gandhinagar University, Gandhinagar, Gujarat, India*

[7] *School of Chemical Sciences, Swami Ramanand Teerth Marathwada University, Nanded, Maharashtra, India*

[8] *Department of Microbiology and Biotechnology, Meerut Institute of Engineering & Technology, Meerut, Uttar Pradesh, India*

Abstract: Chromium is a major component that is responsible for environmental stress. It also has profound effects on the health of living beings because trace amounts of chromium in the environment have been linked to serious health problems in humans and plants. The dangers to human health, bioavailability, plant response to chromium toxicity, and phytoextraction storage in plants are issues of major concern. Understanding and optimizing the phytoextraction process would be immensely beneficial to know about metabolic pool changes that occur in plants in response to Cr toxicity. Therefore, the removal of chromium from the environment is necessary due to its toxic nature. However, the removal of chromium from the environment is a daunting task. Physico-chemical and biological techniques are either too expensive or inefficient to be widely implemented to eradicate chromium from the contaminated soil and environment. The challenges of widespread implementation can be met by adopting

---

[*] **Corresponding author Pankaj Kumar:** Department of Environmental Science, Parul Institute of Applied Sciences, Parul University, Vadodara, Gujarat, India; Tel: +918460571814; E-mail: pankaj.kumar25135@paruluniversity.ac.in

integrated approaches, which are currently under consideration. The removal of chromium from the environment can be more economical and sustainable using phytoremediation technology.

In this chapter, we have discussed the phytoremediation technique as a green solution for the removal of chromium from polluted soil because phytoremediation, and especially phytoextraction, is a viable and sustainable solution to restore chromium-polluted soil.

Keywords: Chromium, Health hazard, Phytoremediation technology, Phytoextraction, Sustainable approaches.

INTRODUCTION

Soil is one of humanity's most valuable and important natural resources. Healthy soil is necessary for the continued success of agriculture and the well-being of civilization [1]. However, heavy metal-polluted soil poses severe health hazards to humans and is a major issue around the world that causes serious illnesses [2 - 4]. The harmful impacts of metals on humans have been observed since ancient times but acquaintance with the harmful effects of these chemicals is still insufficient. Metals can have devastating health impacts and can be lethal to humans. Anthropogenic activities are responsible for the release and accumulation of toxic chemicals into the environment. Heavy metal pollution also results from human activities as well as industrial operations, as mentioned in Fig. (**1**) [5 - 7].

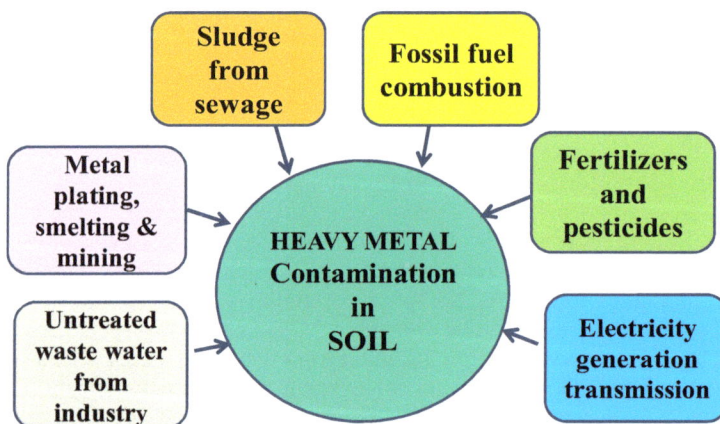

Fig. (1). Possible anthropogenic activities responsible for metal contamination in soil.

Heavy metals are metals with densities greater than 5 g/cm^3 and atomic numbers greater than 20 [8]. Toxic metals such as Arsenic (As), Nickel (Ni), Mercury (Hg), Zinc (Zn), Cadmium (Cd), Chromium (Cr), and Lead (Pb) are widely distributed in the soil and water. Trace elements are metals that account for less than 0.1% of a rock's total mass [9]. Some metals are necessary for the survival of human beings, whereas others are not. The important heavy metals are considered micronutrients but become toxic in high doses. Pb, Cd, and Hg are poisonous metals that are not required to live and are categorized as nonessential [10]. As the seventh most common element in the earth's crust, Cr is a major pollutant that affects ecosystems and human health when it continues to leak into the environment [11]. Sources of Cr include metal extraction, electroplating, leather industry, fertilizer use, and other anthropogenic and natural processes (Fig. **2**) [12]. Chromium and its derivatives have multiple practical applications; for instance, the metallurgical industries employ 90% of the world's chrome ore output to make steel, alloys, and non-ferrous alloys, the chemical (leather tanning and plating) and refractory (iron & steel, cement, and glass) sectors have utilized about 5% of each [13]. Approximately 0.13% of the country's arable land is polluted with Cr and is therefore not suitable for cultivation, but 1.26% is in high danger of Cr contamination. This data suggests that a systematic exploration is required to alleviate unfavorable impacts from the persistence of Cr pollution in soil [14]. Traditional ways of getting rid of metal are usually expensive, harmful, time-consuming, and cause more problems. Alternatively, phytoremediation is a cutting-edge, low-cost, environmentally-friendly remediation approach that is particularly well-suited to developing countries [15, 16]. Phytoremediation is among the most proficient approaches to cleaning metal-contaminated soil since it uses plants to lessen the pollutant's concentration [16, 17].

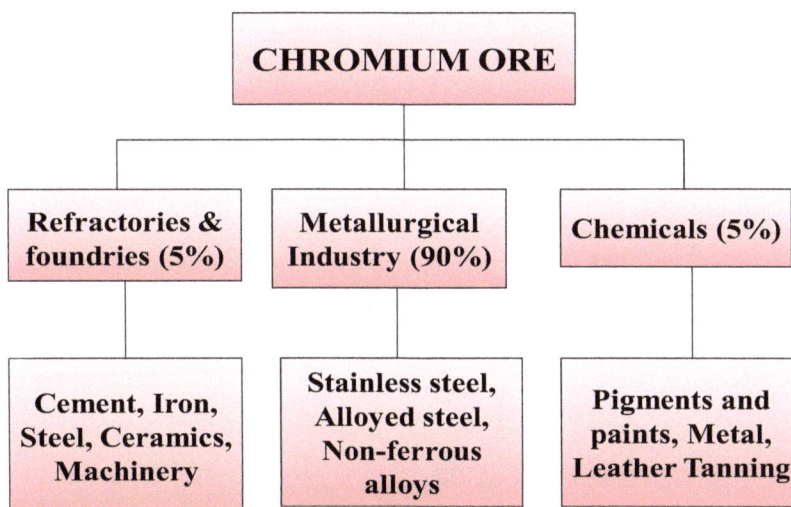

Fig. (2). Usage of chromium in different industries [13].

CHROMIUM: HEALTH HAZARDS

Metal soil contamination is constantly rising as a result of emissions from ongoing manufacturing activities in the industrial sector [18]. These contaminants (such as heavy metals) from soil accumulate in cereals that are cultivated in contaminated soil and pose serious health risks to humans who consume these grains (Fig. **3**) [19 - 21]. Heavy metal pollution thereby disrupts ecological harmony, degrades soil quality, reduces agricultural output, and endangers human health in one way or another *via* the food chain [22, 23]. Cr exposure can cause severe symptoms such as nausea, abdominal pain, diarrhea, kidney failure, irritated and ulcerated gastrointestinal tract, lung cancer, and even death from cardiovascular events [24]. Though Cr (III) is required for the breakdown of glucose, protein, and lipids in the human body, human health hazards have also been described for Cr (VI) form, particularly from acute and chronic inhalation exposures that exacerbate respiratory tract problems [25, 26]. According to research, the risk of developing lung cancer increases due to inhalation of Cr (VI). Animal studies have presented that chronic inhalation of the Cr (VI) form can lead to malignancies in the lungs [27]. It has also been shown that Cr (VI) is chronically harmful to invertebrates. Cr (VI) shortens the lifespan of fish, damages DNA, and accumulates on the gills, all of which contribute to the fish dying. The poison travels through the fish's circulatory system and kills organs like the liver and kidneys [28, 29].

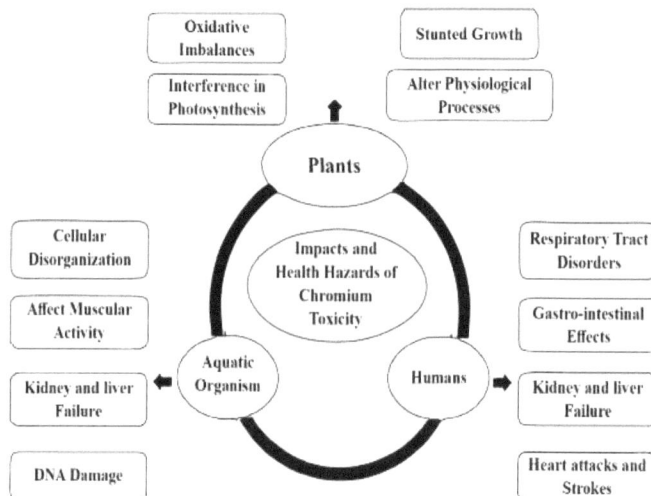

Fig. (3). Chromium toxicity-associated impacts and health hazards.

SOURCES OF CHROMIUM POLLUTION

Industrial development and combustion processes in different industries, such as the metal, chemical, and tanning industries, contribute to the steady increase in Cr content [30]. Anthropogenic sources of Cr include leather processing, chrome plating, pigment manufacture, wood preservation, and corrosion inhibition in cooling towers [31]. Electroplating industries, the creation of resistant alloy goods, aviation, nuclear reactors, electronics, and cement manufacturing sectors all use Cr as a protective coating due to its resistance to corrosive chemicals. It is also used in pyrotechnics, glass, ceramics, dye synthesis, wood preservation, tanneries, the manufacturing of negative film, and the preservation of wood [32].

Geological processes, aforementioned human activities, spillage and dumping of wastes containing Cr, and other anthropogenic activities all contribute to the introduction of chromium and its compounds into soils. The weathering of rocks by wind and rain and industrial and municipal trash are additional potential sources of Cr emissions into the environment [33, 34]. Even though Cr pollution is not a worldwide issue, it can have an impact on local biogeochemical cycles due to the metal's penetration into soil, water, and air [35]. Furthermore, Cr is a byproduct of the mining industry and can be found in the environment in the trivalent (Cr^{3+}) and hexavalent (Cr^{6+}) forms, which are stable forms of Cr. Trivalent Cr[Cr (III)] occurs naturally in soils, where it is used as a nutrient by organisms for healthy growth and development despite its poor solubility and greater inclination to adsorb on soil particles. Many enzymes require trivalent chromium as a cofactor or as a trace element [27, 36]. It can remain in contaminated soil for a long time because of its inert and non-biodegradable nature. Soil can store heavy metals in a different oxidation state but cannot break them [37, 38]. Different industrial sources of Cr and its deposition pathways are depicted in Fig. (**4**).

PHYSICOCHEMICAL APPROACHES FOR CHROMIUM REMEDIATION

Cr cannot be destroyed; hence, it is typically contained during decontamination, unlike organic molecules, which are generally biodegradable. Physicochemical and biological processes are both being researched as potential removal strategies for protecting our soil, sewage, and groundwater. Biological processes have the potential to deliver a more effective and cost-effective technical solution [39]. Chemical precipitation, electrochemical, ion exchange, reverse osmosis, and adsorption are all examples of traditional physico-chemical remediation methods [40, 41]. However, they are either prohibitively expensive or produce poisonous sludge [42]. In addition, waste products of treated effluents rise as a result of these approaches, leading to an increase in secondary pollution. These remediation strategies put a load on the ecosystem by eliminating biotic consortia, which in turn reduces soil fertility. As a result, environmental engineers face a formidable challenge when tasked with reducing Cr (VI) concentrations in effluents to below the Maximum Achievable Control Technology (MACT) standard [43]. Although physicochemical technologies are effective at mitigating the harmful effects of metals, they have significant drawbacks. Consequently, the improvement of more cost-efficient, secure, and ecologically acceptable ways for the remediation of Cr from soil necessitates the search for cheaper and more effective solutions [44].

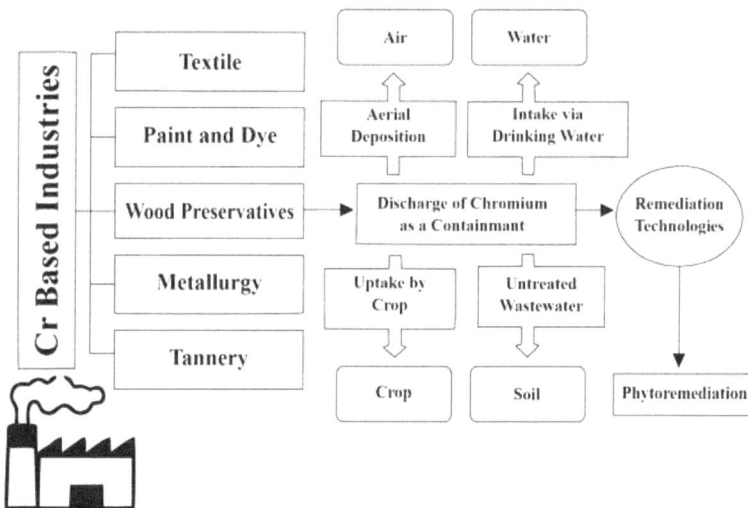

Fig. (4). Industrial sources of Cr and its deposition pathway.

IMPACTS OF CHROMIUM CONTAMINATION

Effects of Chromium Contamination on Properties of Soil

Cr is a common mineral, soil, and rock constituent. Higher amounts of Cr can accumulate in soils due to human activities like industrial operations and incorrect disposal of industrial waste [45]. Cr contamination can have negative effects on soil properties such as pH, making it more acidic. Cr has the potential to alter the plant nutrients' solubility and availability [46]. Cr pollution can cause soil aggregate breakdown, which in turn can enhance soil erosion and degrade soil quality. Soil microbial activity decreases, and the nutrient cycle is disrupted when Cr concentrations are too high. Soil organic matter content can be negatively impacted by Cr pollution, which in turn may affect soil fertility and water-holding ability [47].

Effects of Chromium Toxicity on Plant Growth

Despite the lack of evidence, several studies suggest that Cr may promote plant growth at low doses and hinder higher ones [48]. Seed germination, photosynthetic competence, root impairment, and, ultimately, plant growth are all adversely affected by high concentrations of Cr [49]. The interaction of Cr with soil can disrupt the plant's absorption pattern of critical elements, including calcium (Ca^{2+}) and magnesium (Mg^{2+}) [50]. In addition, agricultural soils reduce crop output due to Cr toxicity [51, 52]. Cr also plays a major part in root growth and development. Due to their dominant role in nutrient intake, plants' roots are also sensitive to Cr toxicity [53]. High concentrations of Cr (III) may inhibit plant growth, but Cr (VI) is devastating to plants as it interferes with a wide variety of physiological and metabolic processes and can lead to complete plant death [54]. Research has shown that when agricultural plants on land take up Cr from the soil, it can be transferred and eventually stored in the plant's aerial sections [55]. Seed germination, photosynthetic rate, suppression of critical enzyme activity, and nutrient uptake are all said to suffer as a result of its presence. Cr (VI) can cause oxidative imbalances and mutagenesis, as well as decrease plant biomass, hinder blooming and fruit set, diminish crop yields, and degrade food-grain quality [55, 56].

Effects of Chromium Toxicity on Germination

The presence of Cr in the growing medium can hinder several plant activities, including germination. Consequently, germination success may be an indicator of Cr tolerance [30]. In tannery sludge soil (with 4000 mg Cr kg^{-1}), the germination of oats was drastically reduced, while in tannery sludge soil (with 8000 mg Cr kg^{-1}), both oat and sorghum germination was inhibited [57]. *T. aestivum* seed germination was likewise inhibited by 100 mg L^{-1} of Cr (VI) [58]. Despite these results, it has been hypothesized that seed germination is unaffected by Cr (VI) treatment, but radicle growth is inhibited upon contact with Cr solution [59].

Effects of Chromium Toxicity on Root Growth

Heavy metals often inhibit root development as well as germination; for example, paddy (*Oriza sativa L.*) growth was found to be stunted by Cr (VI) concentrations as high as 200 mg L^{-1} [60]. Root development was likewise stunted in Cr (VI)-exposed sensitive mung bean cultivars (VI) [61]. *Zea mays* L. roots exposed to Cr (VI) also shrank in size, darkened in color, and displayed a decreased quantity of root hairs [56]. Soil Cr values of 100 mg Cr (VI) kg^{-1} were observed to inhibit the root development of oat and sorghum [62]. Cr (VI) causes a shallower root surface. There may be a correlation between plant stress and a reduced ability to find water in the soil [30].

Effects of Chromium Toxicity on Stem Growth

Cr exposure is also harmful to stem development [63]. *Zea mays* L. showed a significant reduction in shoot length after 7 days when exposed to 9 gmL^{-1} Cr (VI) [64]. It is also described that Cr exposure causes sensitive mung bean plants to grow shorter and produce fewer shoots [65]. For 7 days, *T. aestivum* L. seedlings were treated to 100 mgL^{-1} of Cr (VI), and the reductions in root and shoot length were 63% and 44%, respectively [66]. The shorter plant could be the result of stunted root development and a lack of resources being brought to the upper levels of the plant. Additionally, Cr translocation to the shoots can affect cellular metabolism, which in turn contributes to the plant's reduced stature [56].

Effects of Chromium Toxicity on Leaf Growth

Increasing Cr concentration also often reduces leaf area [67]. Leaves on cauliflower grew in sand containing 0.5mM Cr (III) and were smaller in comparison to the normal growth [68]. When grown in the presence of Cr (VI), watermelon plants had fewer and smaller leaves; these leaves also became yellow, wilted, and hung limply from their petioles due to a loss of turgor [56].

NECESSITY FOR CHROMIUM REMEDIATION

The risks associated with Cr exposure are extensively acknowledged by the WHO. The United States Environmental Protection Agency has designated chromium (VI) as one of the top seventeen contaminants that pose the greatest hazard to humans [69]. The EPA has designated it as a pollutant of extremely high concern. Due to its mutagenic and carcinogenic qualities, Cr in its various forms is often regarded as extremely toxic to virtually all forms of life [70]. Since Cr has a high positive redox potential, it can easily permeate cell membranes and destroy cellular and molecular components in all living organisms, such as damage to membranes, protein degradation, and DNA changes. Single-strand breaks and mutations are the results of Cr (VI) interference with DNA protein cross-links [71]. The risk of lung cancer is also connected with Cr (VI) exposure above the acceptable level. Kidneys and liver may also be damaged, and there is a risk of nausea, allergy, stomach ulcers, and bleeding [72].

PHYTOREMEDIATION TECHNOLOGIES

Soil remediation practices like washing, vitrification, electro-kinetic extraction, bioremediation, and phytoremediation are being applied to restore the metal-polluted soil. Among these, phytoremediation is considered a green technology and the most effective method for achieving the desired results [73]. Phytoremediation is the application of plants to clean polluted areas. Grass, bushes, and trees, along with microorganisms, can degrade, accumulate, and stabilize pollutants in the environment [74]. This "green technology" can reduce secondary wastes along with toxins from the soil [75]. This technology is easy to use, economically viable, aesthetically beautiful, and widely accepted by the general population. It is a green technique for cleaning polluted ground without compromising the area's natural fertility or biodiversity [76, 77]. Phytoremediation also does not need any specialized workers or high-priced machinery to be put into action [73].

Different techniques of phytoremediations of soil are phytoextraction (uptake of metals), phytoaccumulation (accretion and translocation release to air), phytovolatilization, and photostabilization (stabilization in the root zone) [78]. These techniques are described below and illustrated in Fig. (**5**).

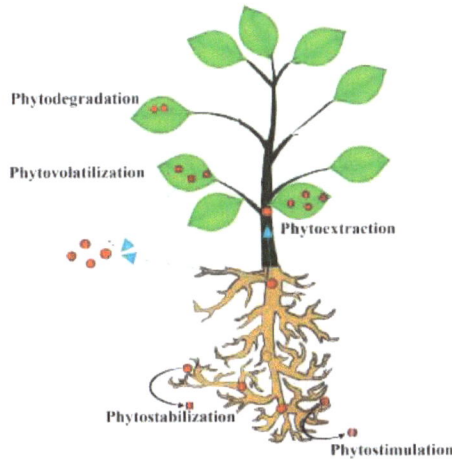

Fig. (5). Schematic representation of different phytoremediation techniques.

Phytostabilization

In-situ immobilization of metals, also known as phytostabilization, decreases heavy metal bioavailability and stops their movement to other areas [79, 80]. To prevent metal from leaching into groundwater or entering the food chain, plants can be used for phytostabilization or phyto-immobilization. This is accomplished through a variety of mechanisms, such as adsorption, precipitation, and complexation in the root zone [81]. With this method, metals are absorbed by plant roots and rendered immobile. Phytostabilization differs from conventional phytoremediation approaches in its primary goal of not removing hazardous toxins but alleviating them, thereby reducing the threat. Phytostabilization refers to a plant-based inactivation technique for dealing with metal-contaminated soil.

This strategy reduces heavy metals' bioavailability and mobility, which in turn reduces their leaching and entry into groundwater and their influence on the food chain [82]. This method comprises the physical and chemical immobilization of metal impurities by dissolving them on the roots and fixing them with various soil additives [83]. Plants' main functions are pollution mitigation and pollution control, which include lowering soil erosion, water percolation, pollutant interaction, and contaminant migration [84]. Phytostabilization is more of a management strategy because metal contaminants will always be present in the soil. Phytostabilization is favored over competing strategies due to its low price tag and simple implementation [85]. Soils polluted with zinc, cadmium, Cr, lead, and copper are typical candidates for this method of cleanup [86]. Multiple metals such as Cr, zinc, and cadmium can be phytostabilized by the plant species *Vossia cuspidata*. Heavy metal precipitation by various soil additions might enhance stabilization [87].

Phytoextraction

Most metals and metalloids can be removed from polluted soils, water, and biosolids by the phytoremediation technique known as phytoextraction. It is superior to other phytoremediation methods in terms of suitability for commercial use [88]. Heavy metals can be extracted using fast-growing plants in a process called phytoextraction. Natural and chemically induced phytoextraction are the two methods of phytoextraction [89]. Heavy metals can be extracted continuously through a system of plant roots that are guided upwards into the plant's top tissues in a process known as "phytoextraction" [90]. Plant biomass that has been harvested can be utilized to create biogas or burned as fuel. Plant biomass that has been burned has multiple potential uses, including metal recovery, brick-making, and land clearing.

Plants with metal-restoring properties are worth harvesting [91]. This new environmentally friendly solution would be ten times less expensive than conventional remedial methods [92]. Phytoextraction efficiency is affected by a wide range of parameters, such as soil qualities, metal availability to plant, metal speciation, and features of the relevant plant. Phytoaccumulation is the process by which the metal is taken up by the plant, precipitated, and stored in its upper portions such as leaves and stems [93]. Plants that may store Cr, cadmium, lead, and zinc include many examples, such as *Brassica juncea*, *Arabidopsis thaliana*, *Cynodon dactylon*, *Helianthus annuus*, and *Calendula officinalis* [94 - 96].

Phytovolatilization

Another technique called phytovolatilization involves the release of metal into the atmosphere *via* stomata after the metal is converted into a volatile form. Metals are taken up by plants, transformed into volatile forms, and then exhaled by stomata into the air through a process called phytovolatilization [97].

Phytofiltration

The use of plant roots to clean polluted water with lower pollutant concentrations is called phytofiltration [98]. Firstly, the polluted water is used to acclimatize the plants, and then the acclimated plants are moved to the polluted area. Harvesting occurs when the roots have reached saturation. This method decreases the leaching of pollutants into groundwater by causing them to be absorbed, adsorbed, or precipitated [99]. Plant metals can be precipitated onto their roots if the pH of the rhizospheric zone is altered by root exudates [100]. Cadmium, lead, nickel, zinc, copper, and Cr can all be extracted *via* rhizofiltration. Because of their longer and more fibrous root systems, terrestrial plants are typically utilized [101].

MECHANISMS FOR CHROMIUM PHYTOREMEDIATION

The mechanism by which plants absorb Cr is still poorly understood. In-depth analyses of metabolic and absorption trails from soil and aquatic medium have indicated that certain plant species bioaccumulate significant amounts of Cr. Plants use several processes, including bioaccumulation, biosorption, and precipitation, to resist Cr [102]. Plants show the ability to transform Cr (VI) to the safer Cr (III). In addition, the chromate efflux process is used by plants, and this can be a useful strategy for lowering Cr pollution [103]. The uptake of Cr (III) by plants is a passive procedure; the plant makes no active effort to do so. Carriers are hypothesized to actively transport Cr (VI) in exchange for necessary elements like sulfate. For carrier binding, chromium competes with Fe, S, and P [104].

Not much research has been done on the plant's Cr uptake pathway. One of the major elements influencing Cr transport within a plant cell is the valence state of Cr. Cr is a non-essential constituent thus, plants do not have a specialized absorption system for it [105]. There is evidence that Cr (VI) is converted to the less toxic Cr (III) on the root surface before it can enter the plant cell. In plants, the reduction process involves many different cellular components, including NAD(P)H, glutathione, cyanocobalamin, cytochrome P-450, and the mitochondrial respiratory chain [55]. On the other hand, other researchers claim that Cr (III) becomes impermeable to biomembranes because it generates water-insoluble composites in non-acidic solutions. Some investigations have even shown that plants may uptake Cr (VI) without it being reduced first. Because Cr (VI) molecules have the same chemical structure as SO_4^{2-} ions, these vital plant nutrients are taken up by the cell *via* transporters for sulfate and phosphate. As a result, it is believed that chromatin is actively transported across biological membranes [106].

Tolerance to Cr toxicity has been developed by the molecular inspection of various Cr detoxifying systems. Cr stress triggers a cascade of events that ultimately contribute to the plants' bioaccumulation capacity, including (i) reactive oxygen species signaling, (ii) antioxidant reactions, (iii) activation of defense proteins like phytochelatins, metallothionine, and Glutathione-S transferases, and (iv) phyto-sequestration and classification. To improve the accumulation rate for Cr, a different method may be developed by modifying the genes involved in the absorption, transportation, and sequestration of Cr [53].

It concludes that the oxidation state of Cr determines whether it has an active or passive role in the transportation pathway. For instance, the uptake of Cr (III) by plants through cation exchange sites in the cell wall is a simple and passive process. In contrast, plant-based sulfate transporters actively transfer Cr (VI). The roots transform Cr (VI) to Cr (III) after absorption [107]. Fig. (**6**) shows the speciation and phytoremediation process of Cr-contaminated soil.

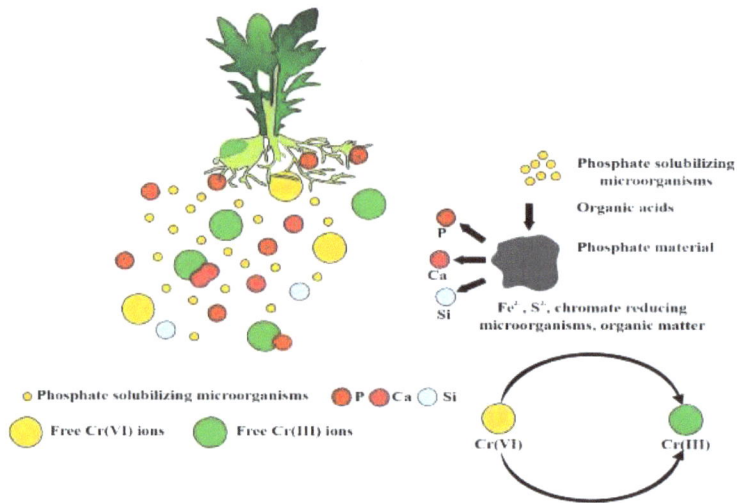

Fig. (6). Speciation and phytoremediation of Cr-contaminated soil [108].

PHYTOREMEDIATION OF CHROMIUM: CASE STUDIES

Phytoremediation relies heavily on plants' inherent capacity to absorb and eliminate toxic metals [109]. In reaction to environmental heavy metal stress, plants develop chelators and organic acids that can bind with metal ions and protect the plant from damage. Sequestration of the metal-chelators complex by the cell leads to metal-ion inactivation through compartmentalization in plant cells [110]. In the last few decades, researchers have identified many tolerant and hyperaccumulator plants for studying their mechanism and use in the phytoremediation process. Nearly 500 plant species from 45 different families have been discovered so far. When exposed to harmful metals, many of the tolerant hyperaccumulator plants convert them into less dangerous and immobile forms [111]. Numerous investigations revealed that different plants are proficient in accumulating chromium, making them potentially excellent candidates for phytoremediation (Table 1). Hyper accumulators of Cr and the microflora that live with them are effective at cleaning polluted or contaminated areas with excess Cr and other organic waste. Plant-microbe interaction is highly effective, inexpensive, and environmentally friendly in Cr detoxification [112].

Table 1. Phytoremediation studies conducted on Chromium removal from the environmental component.

Plant	Condition	Chromium Concentration	Duration	Chromium Removal Efficiency	Refs.
Pongamia pinnata (L.) Pierre	Pots containing chromium-treated soil	50-500 µg/g.	45 days	up to 200 µg/g.	[113]
(a)*Brassica juncea*, (b) *Oryza sativa* (c) *Triticum aestivum*	Earthen pots with polluted soil	5.9871±0.9032 mg/kg dry weight) (46.4739± 2.2920 mg/kg dry weight) 22.5622±2.9256 mg/kg dry weight)	30 days	-	[114]
Portulaca oleracea	Simulated test soils	200ppm	21 days	-	[115]
Cirsium vulgare	Pots containing Cr-contaminated soil	30 mgkg^{-1}	60 days	8.237 mgkg^{-1}	[116]
(a) *Sansevieria trifasciata* var. *hahnii*, (b) *Canna indica* (L.) (c)*Nephrolepis exaltata* (L.)	Pot culture experiment	(250, 500, 750 mg Cr kg^{-1}soil)	30 days and 60 days	-	[117]
Ricinus communis	Mixed tannery sludge containing chromium	Sludge Drying Bed - 9.640 ppm Sludge Pit - 14.936 ppm	30-75 days	-	[118]
Phragmites australis and *Helianthus annuus*	Pots containing Cr-contaminated soil	10 mg L^{-1} Cr(VI)	90 days	-	[119]

(Table 1) cont.....

Plant	Condition	Chromium Concentration	Duration	Chromium Removal Efficiency	Refs.
Pennisetum sinese	Pots and Hoagland nutrient solution	Tolerance: 0, 50, 250, 500, 1000, and 2000 µM as (K2Cr$_2$O$_7$). Absrption:400 and 800 µM	3 days (0-72 hours)	150.99 (shoots) and 979.03 (roots) mgkg^{-1}	[120]
Spirogyra, Eichhornia crassipes Pistia stratiotes	wastewater	1, 3, 5 mg/l	15 days	*Spirogyra* 86% *Eichhornia crassipes* 92% *Pistia stratiotes* 99%	[121]
Azolla filiculoids	Sewage	27.2 mg/kg	28 days	126%	
Ipomoea aquatica	Microcosm Mesocosm experiment synthetic wastewater	Microcosm 0.01, 0.1, 0.9, and 4.4 mg/L Mesocosm 10 mg/L	4 days	82.8% in microcosm 90.4% mesocosm	[122]
Eichhornia crassipes Pistia stratiotes	4% Batik wastewater	596mg/l	15 days	*E. crassipes* 63.76% *P. stratiotes.* 83.39%	[123]
Chrysopogon zizanoides	Artificial electroplating wastewater	Different levels of chromium concentration	28 days	61.10%	[124]
Azolla pinnata Salvinia molesta	Hoagland's nutrient solution with various chromium concentrations	0.07, 0.09, 0.11, and 0.13 mg/l along with control	15 days	*Azolla pinnata* 72.86 - 97.69% *Salvinia molesta* 75.71 - 90%	[125]
Brachiaria mutica, Canna indica, Cyperus laevigatus, Leptochloa fusca, Typha domingensis	Tannery effluent	Cr6$^-$ (mg/L) 0.48 ± 0.15 Cr3+ (mg/L) 133 ± 1.4	42 days	*Leptochloa fusca* 55% *T.domingensis* 48% *B.mutica* 35%	[126]
Chrysopogon zizanioides	Laboratory condition	2L Cr (VI) to 5, 10, 30, and 70 ppm	49 days	87% reduction in Cr (VI) 5 ppm container 51% in 10 ppm container	[127]
Salvinia molesta Lemna gibba	Wastewater	1.58±0.08 mg/l	7 days	*S. molesta* 81.66% *L. gibba* 86.99%	[128]

(Table 1) cont.....

Plant	Condition	Chromium Concentration	Duration	Chromium Removal Efficiency	Refs.
Pistia stratiotes Eichhornia crassipes	Solution of various concentration of metal	2, 4, 6 and 8 mg/L	30 days	*Pistia stratiotes* 77.3% *Eichhornia crassipes* 80.9%	[129]
Pistia stratiotes	Wastewater from the Nickel industry	0.5, 1, 2, 5, and 7 ppm	20 days	89%	[130]
Phyllostachys pubescens	Pot culture experiment	100 mg/L	84 days	49.2% to 61.7%	[131]
Brachiaria mutica Leptochloa fusca	Pot culture experiment	25, 50, and 100 mg/Kg	3 months	-	[132]
Typha angustifolia L. *Canna indica* L. *Hydrocotyle umbellatae* L.	Andosol soil	10, 30 and 50 mg/L	9 days	99.78% 99.67% 86.36%	[133]

ADVANTAGES OF PHYTOREMEDIATION

In phytoremediation, plants' roots are used for their exceptional selective and natural absorption capacities, while the plant itself is put to use for its translocation, bioaccumulation, and pollutant storage/degradation properties. The procedure is carried out on-site, is solar-powered, requires very little upkeep once it is set up, and is very cost-effective. It is roughly ten times cheaper than alternative physical, chemical, or thermal remediation approaches [134]. The procedure can also be enjoyable from a leisure standpoint and a design perspective. When compared to other methods of Cr detoxification, phytoremediation has many advantages [16] (Fig. 7). These are described as:

i. It is a solar-powered autotrophic system that is easy to maintain and cheap to set up.
ii. Safe for the ecology and the environment, meaning that it reduces the amount of pollution released into the air.
iii. Usability, *i.e.*, its potential for widespread implementation and straightforward disposal.
iv. It reduces the potential for contamination spread by stabilizing heavy metals, which stops erosion and metal leaching.
v. It can increase soil fertility by allowing a variety of organic materials to be returned to the ground.

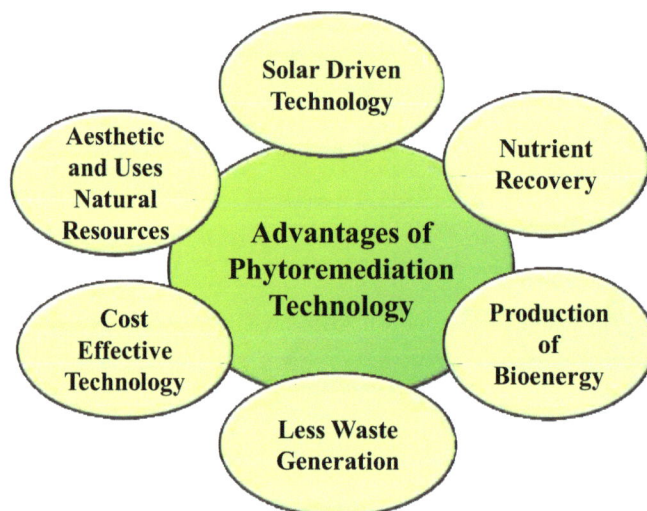

Fig. (7). Advantages of Phytoremediation Technology.

LIMITATIONS OF PHYTOREMEDIATION

The amount of pollutants and excess poisons present should not exceed what the plants utilized. Selecting plants that can successfully remove different types of pollutants is a difficult task. These limitations (Table **2**) and the likelihood that these contaminants will reach the food chain should be considered while implementing this technology [135].

Table 2. Limitations of phytoremediation.

Implementation	• It mainly applies to mine tailings and the top layer of soil. • Its applicability to a wide range of waste types is limited, mainly when dealing with toxic wastes.
Time and cost factors	• Extremely hazardous environments may cause plant deaths, which could raise the process's price. • Most often, pollutants are removed partially and with long-term poor performance. • To avoid incidents, good maintenance and cultivation techniques are necessary.
Performance	• Seasonal variations have an impact on this rehabilitation procedure's efficacy. • It is necessary to consider the physiological alterations and plant performance in response to various waste types.
Impacts on the population and the environment	• Pollutants have the potential to bioaccumulate in the food chain. • It is feasible for undesirable invasive plant species to be introduced and spread. • Plant matter must be disposed of properly, and risks must be assessed.

CONCLUSION AND FUTURE DIRECTIONS

Heavy metals, particularly Cr, enter the environment *via* several routes and are considered one of the most significant dangers to soil and human health. Traditional cleanup methods are very expensive and harmful to the ecosystem. As a result, remediating soils polluted with metals requires the use of methods that are both inexpensive and gentle on the environment. Metal phytoremediation is the most efficient plant-based method for cleaning polluted regions without damaging the soil. Phytoremediation is a more economical and eco-friendly substitute to traditional physicochemical procedures. More research on the benefits and drawbacks of phytoremediation will help in the advancement of this technique for cleaning the soil. Phytoremediation can be utilized for a long time, requires little specialized expertise, and has a high success rate. Improvements in plant genetics have allowed for greater accumulation and tolerance capability in phytoremediation. Large-scale decontamination of polluted soils may also make use of more modern phytoremediation methods like chemical-assisted phytoextraction and microbial-assisted phytoremediation. To make these techniques more efficient, less time-consuming, and economical, additional study on genetic engineering is required to enhance the capacities of transgenic plants and to recognize the operation and efficiency of phytoremediation.

ACKNOWLEDGEMENT

The authors are grateful to the Department of Environmental Science, Parul Institute of Applied Sciences, Parul University, Vadodara, for providing facilities.

REFERENCES

[1]   Lone MI, He Z, Stoffella PJ, Yang X. Phytoremediation of heavy metal polluted soils and water: Progresses and perspectives. J Zhejiang Univ Sci B 2008; 9(3): 210-20.
[http://dx.doi.org/10.1631/jzus.B0710633] [PMID: 18357623]

[2]   Briffa J, Sinagra E, Blundell R. Heavy metal pollution in the environment and their toxicological effects on humans. Heliyon 2020; 6(9): e04691.
[http://dx.doi.org/10.1016/j.heliyon.2020.e04691] [PMID: 32964150]

[3]   Singh JK, Kumar P, Kumar R. Ecological risk assessment of heavy metal contamination in mangrove forest sediment of Gulf of Khambhat region, West Coast of India. SN Appl Sci 2020; 2(12): 2027.
[http://dx.doi.org/10.1007/s42452-020-03890-w]

[4]   Kumar P, Fulekar MH. Multivariate and statistical approaches for the evaluation of heavy metals pollution at e-waste dumping sites. SN Appl Sci 2019; 1(11): 1506.
[http://dx.doi.org/10.1007/s42452-019-1559-0]

[5]   Adnan M, Xiao B, Xiao P, Zhao P, Li R, Bibi S. Research progress on heavy metals pollution in the soil of smelting sites in china. Toxics 2022; 10(5): 231.
[http://dx.doi.org/10.3390/toxics10050231] [PMID: 35622644]

[6]   Ali H, Khan E, Sajad MA. Phytoremediation of heavy metals—Concepts and applications. Chemosphere 2013; 91(7): 869-81.
[http://dx.doi.org/10.1016/j.chemosphere.2013.01.075] [PMID: 23466085]

[7]   Kumar P. Electronic waste - hazards, management and available green technologies for remediation - a review. Int Res J Environ Sci 2018; 7(5): 57-68.

[8]   Malik B. Occurrence and impact of heavy metals on environment. 2023, Available from: https://www.sciencedirect.com/science/article/pii/S2214785323004078

[9]   Alengebawy A, Abdelkhalek ST, Qureshi SR, Wang MQ. Heavy metals and pesticides toxicity in agricultural soil and plants: ecological risks and human health implications. Toxics 2021; 9(3): 42.
[http://dx.doi.org/10.3390/toxics9030042] [PMID: 33668829]

[10]   Gupta N, Yadav KK, Kumar V, Prasad S, Cabral-Pinto MMS, Jeon BH, *et al.* Investigation of heavy metal accumulation in vegetables and health risk to humans from their consumption. Front Environ Sci https://www.frontiersin.org/articles/10.3389/fenvs.2022.7910522022; 10.

[11]   Antoniadis V, Polyzois T, Golia EE, Petropoulos SA. Hexavalent chromium availability and phytoremediation potential of *Cichorium spinosum* as affect by manure, zeolite and soil ageing. Chemosphere 2017; 171: 729-34.
[http://dx.doi.org/10.1016/j.chemosphere.2016.11.146] [PMID: 27939668]

[12]   Xia S, Song Z, Jeyakumar P, *et al.* A critical review on bioremediation technologies for Cr(VI)-contaminated soils and wastewater. Crit Rev Environ Sci Technol 2019; 49(12): 1027-78.
[http://dx.doi.org/10.1080/10643389.2018.1564526]

[13]   Dhal B, Thatoi HN, Das NN, Pandey BD. Chemical and microbial remediation of hexavalent chromium from contaminated soil and mining/metallurgical solid waste: A review. J Hazard Mater 2013; 250-251: 272-91.
[http://dx.doi.org/10.1016/j.jhazmat.2013.01.048] [PMID: 23467183]

[14]   Zhang X, Gai X, Zhong Z, *et al.* Understanding variations in soil properties and microbial communities in bamboo plantation soils along a chromium pollution gradient. Ecotoxicol Environ Saf 2021; 222: 112507.
[http://dx.doi.org/10.1016/j.ecoenv.2021.112507] [PMID: 34265530]

[15]   Bhat SA, Bashir O, Ul Haq SA, *et al.* Phytoremediation of heavy metals in soil and water: An eco-friendly, sustainable and multidisciplinary approach. Chemosphere 2022; 303(Pt 1): 134788.
[http://dx.doi.org/10.1016/j.chemosphere.2022.134788] [PMID: 35504464]

[16]   Kumar P, Gacem A, Ahmad MT, Yadav VK, Singh S, Yadav KK, *et al.* Environmental and human health implications of metal(loid)s: Source identification, contamination, toxicity, and sustainable clean-up technologies. Front Environ Sci 2022; 10: 949581.

[17]   Kumar P, Fulekar MH. Cadmium phytoremediation potential of Deenanath grass (*Pennisetum pedicellatum*) and the assessment of bacterial communities in the rhizospheric soil. Environ Sci Pollut Res Int 2021; 29(2): 2936-53.
[http://dx.doi.org/10.1007/s11356-021-15667-8] [PMID: 34382164]

[18]   Zhou XY, Wang XR. Impact of industrial activities on heavy metal contamination in soils in three major urban agglomerations of China. J Clean Prod 2019; 230: 1-10.

[19]   O'Connor D, Hou D, Ok YS, Lanphear BP. The effects of iniquitous lead exposure on health. Nat Sustain 2020; 3(2): 77-9.
[http://dx.doi.org/10.1038/s41893-020-0475-z]

[20]   Pizarro I, Gómez-Gómez M, León J, Román D, Palacios MA. Bioaccessibility and arsenic speciation in carrots, beets and quinoa from a contaminated area of Chile. Sci Total Environ 2016; 565: 557-63.
[http://dx.doi.org/10.1016/j.scitotenv.2016.04.199] [PMID: 27196992]

[21]   Liu Q, Li X, He L. Health risk assessment of heavy metals in soils and food crops from a coexist area of heavily industrialized and intensively cropping in Chengdu Plain, Sichuan, China. Front Chem 2022, 10, 988587. https://www.frontiersin.org/articles/10.3389/fchem.2022.988587

[22]   Mao X, Jiang R, Xiao W, Yu J. Use of surfactants for the remediation of contaminated soils: A review.

J Hazard Mater 2015; 285: 419-35.
[http://dx.doi.org/10.1016/j.jhazmat.2014.12.009] [PMID: 25528485]

[23]   Ahmad R, Tehsin Z, Malik ST, *et al.* Phytoremediation Potential of Hemp (*Cannabis sativa* L.):
       Identification and Characterization of Heavy Metals Responsive Genes. Clean (Weinh) 2016; 44(2):
       195-201.
       [http://dx.doi.org/10.1002/clen.201500117]

[24]   Hessel EVS, Staal YCM, Piersma AH, den Braver-Sewradj SP, Ezendam J. Occupational exposure to
       hexavalent chromium. Part I. Hazard assessment of non-cancer health effects. Regul Toxicol
       Pharmacol 2021; 126: 105048.
       [http://dx.doi.org/10.1016/j.yrtph.2021.105048] [PMID: 34563613]

[25]   Cabral-Pinto MMS, Inácio M, Neves O, *et al.* Human health risk assessment due to agricultural
       activities and crop consumption in the surroundings of an industrial area. Expo Health 2020; 12(4):
       629-40.
       [http://dx.doi.org/10.1007/s12403-019-00323-x]

[26]   Cabral Pinto MMS, Ferreira da Silva EA. Heavy metals of santiago island (Cape Verde) alluvial
       deposits: Baseline value maps and human health risk assessment. Int J Environ Res Public Health
       2019; 16.

[27]   Lewicki S, Zdanowski R, Krzyżowska M, *et al.* The role of Chromium III in the organism and its
       possible use in diabetes and obesity treatment. Ann Agric Environ Med 2014; 21(2): 331-5.
       [http://dx.doi.org/10.5604/1232-1966.1108599] [PMID: 24959784]

[28]   Tumolo M, Ancona V, De Paola D, *et al.* Chromium pollution in european water, sources, health risk,
       and remediation strategies: an overview. Int J Environ Res Public Health 2020; 17(15): 5438.
       [http://dx.doi.org/10.3390/ijerph17155438] [PMID: 32731582]

[29]   Gupta N, Yadav KK, Kumar V, Kumar S, Chadd RP, Kumar A. Trace elements in soil-vegetables
       interface: Translocation, bioaccumulation, toxicity and amelioration - A review. Sci Total Environ
       2019; 651(2): 2927-42.
       [http://dx.doi.org/10.1016/j.scitotenv.2018.10.047] [PMID: 30463144]

[30]   Sharma A, Kapoor D, Wang J, *et al.* Chromium bioaccumulation and its impacts on plants: An
       overview. Vol. 9. Plants 2020; 9(1): 100.
       [http://dx.doi.org/10.3390/plants9010100]

[31]   Kumar P, Kumar R, Reddy MV. Assessment of sewage treatment plant effluent and its impact on the
       surface water and sediment quality of river Ganga at Kanpur. Int J Sci Eng Res 2018; 8(1): 1315-24.
       [http://dx.doi.org/10.14299/ijser.2018.01.003]

[32]   Saha B, Amine A, Verpoort F. Special Issue: Hexavalent Chromium: Sources, Toxicity, and
       Remediation. Chemistry Africa 2022; 5(6): 1779-80.
       [http://dx.doi.org/10.1007/s42250-022-00443-z]

[33]   Müller A, Österlund H, Marsalek J, Viklander M. The pollution conveyed by urban runoff: A review
       of sources. Sci Total Environ 2020; 709: 136125.
       [http://dx.doi.org/10.1016/j.scitotenv.2019.136125] [PMID: 31905584]

[34]   Li Z, Wang Q, Xiao Z, Fan L, Wang D, Li X, *et al.* Behaviors of chromium in coal-fired power plants
       and associated atmospheric emissions in Guizhou, Southwest China. Atmosphere 2020; 11.

[35]   Yang Y, Huang J, Sun Q, *et al.* microRNAs: Key Players in Plant Response to Metal Toxicity. Int J
       Mol Sci 2022; 23(15): 8642.
       [http://dx.doi.org/10.3390/ijms23158642]

[36]   Ahemad M. Enhancing phytoremediation of chromium-stressed soils through plant-growth-promoting
       bacteria. J Genet Eng Biotechnol 2015; 13(1): 51-8.
       [http://dx.doi.org/10.1016/j.jgeb.2015.02.001] [PMID: 30647566]

[37]   Mandal P. An insight of environmental contamination of arsenic on animal health. Emerg Contam

2017; 3(1): 17-22.
[http://dx.doi.org/10.1016/j.emcon.2017.01.004]

[38]   Coetzee JJ, Bansal N, Chirwa EMN. Chromium in environment, its toxic effect from chromite-mining and ferrochrome industries, and its possible bioremediation. Expo Health 2020; 12(1): 51-62.
[http://dx.doi.org/10.1007/s12403-018-0284-z]

[39]   Ranieri E, Gikas P. Effects of plants for reduction and removal of hexavalent chromium from a contaminated soil. Water Air Soil Pollut 2014; 225(6): 1981.
[http://dx.doi.org/10.1007/s11270-014-1981-2]

[40]   Fu F, Wang Q. Removal of heavy metal ions from wastewaters: A review. J Environ Manage 2011; 92(3): 407-18.
[http://dx.doi.org/10.1016/j.jenvman.2010.11.011] [PMID: 21138785]

[41]   Heidmann I, Calmano W. Removal of Cr(VI) from model wastewaters by electrocoagulation with Fe electrodes. Separ Purific Techno 2008; 61(1,6): 15-21.https://www.sciencedirect.com/science/article/pii/S1383586607004285

[42]   Kurniawan TA, Chan GYS, Lo WH, Babel S. Physico–chemical treatment techniques for wastewater laden with heavy metals. Chem Eng J 2006; 118(1-2): 83-98.
[http://dx.doi.org/10.1016/j.cej.2006.01.015]

[43]   Ryskie S, Neculita C, Rosa E, Coudert L, Couture P. Active treatment of contaminants of emerging concern in cold mine water using advanced oxidation and membrane-related processes: A review. Minerals 2021; 11(3): 259.
[http://dx.doi.org/10.3390/min11030259]

[44]   Fernández PM, Viñarta SC, Bernal AR, Cruz EL, Figueroa LIC. Bioremediation strategies for chromium removal: Current research, scale-up approach and future perspectives. Chemosphere 2018; 208: 139-48.
[http://dx.doi.org/10.1016/j.chemosphere.2018.05.166] [PMID: 29864705]

[45]   Sharma P, Singh SP, Parakh SK, Tong YW. Health hazards of hexavalent chromium (Cr (VI)) and its microbial reduction. Bioengineered 2022; 13(3): 4923-38.
[http://dx.doi.org/10.1080/21655979.2022.2037273] [PMID: 35164635]

[46]   Xu S, Yu C, Wang Q, Liao J, Liu C, Huang L, *et al.* Chromium contamination and health risk assessment of soil and agricultural products in a rural area in southern china. Toxics 2023; 11.

[47]   Zulfiqar U, Haider FU, Ahmad M, Hussain S, Maqsood MF, Ishfaq M, *et al.* Chromium toxicity, speciation, and remediation strategies in soil-plant interface: A critical review. Front Plant Sci 2023; 13: 1081624.

[48]   Ao M, Chen X, Deng T, *et al.* Chromium biogeochemical behaviour in soil-plant systems and remediation strategies: A critical review. J Hazard Mater 2022; 424(Pt A): 127233.
[http://dx.doi.org/10.1016/j.jhazmat.2021.127233] [PMID: 34592592]

[49]   Zaheer IE, Ali S, Rizwan M, *et al.* Citric acid assisted phytoremediation of copper by *Brassica napus* L. Ecotoxicol Environ Saf 2015; 120: 310-7.
[http://dx.doi.org/10.1016/j.ecoenv.2015.06.020] [PMID: 26099461]

[50]   Wakeel A, Xu M. Chromium Morpho-Phytotoxicity. Plants 2020; 9(5): 564.
[http://dx.doi.org/10.3390/plants9050564]

[51]   Adrees M, Ali S, Iqbal M, *et al.* Mannitol alleviates chromium toxicity in wheat plants in relation to growth, yield, stimulation of anti-oxidative enzymes, oxidative stress and Cr uptake in sand and soil media. Ecotoxicol Environ Saf 2015; 122: 1-8.
[http://dx.doi.org/10.1016/j.ecoenv.2015.07.003] [PMID: 26164268]

[52]   Kanwal U, Ali S, Shakoor MB, *et al.* EDTA ameliorates phytoextraction of lead and plant growth by reducing morphological and biochemical injuries in *Brassica napus* L. under lead stress. Environ Sci Pollut Res Int 2014; 21(16): 9899-910.

[http://dx.doi.org/10.1007/s11356-014-3001-x] [PMID: 24854501]

[53]   Srivastava D, Tiwari M, Dutta P, Singh P, Chawda K, Kumari M, *et al.* Chromium stress in plants: Toxicity, tolerance and phytoremediation. Switzerland: Sustainability 2021; 13..

[54]   Ali S, Mir RA, Tyagi A, *et al.* Chromium toxicity in plants: signaling, mitigation, and future perspectives. Plants 2023; 12(7): 1502.
[http://dx.doi.org/10.3390/plants12071502]

[55]   Stambulska UY, Bayliak MM, Lushchak VI. Chromium (VI) toxicity in legume plants: Modulation effects of rhizobial symbiosis. Biomed Res Int 2018; 14(2018): 8031213.
[http://dx.doi.org/10.1155/2018/8031213]

[56]   Oliveira H. Chromium as an Environmental Pollutant: Insights on Induced Plant Toxicity Deckert J. 2012; pp. 1-8.
[http://dx.doi.org/10.1155/2012/375843]

[57]   Mubeen S, Ni W, He C, Yang Z. Agricultural Strategies to Reduce Cadmium Accumulation in Crops for Food Safety. Switzerland: Agriculture 2023; 13.

[58]   Medda S, Mondal NK. Chromium toxicity and ultrastructural deformation of *Cicer arietinum* with special reference of root elongation and coleoptile growth. Ann Agrar Sci 2017; 15(3): 396-401.
[http://dx.doi.org/10.1016/j.aasci.2017.05.022]

[59]   Kundu D, Dey S, Raychaudhuri SS. Chromium (VI) – induced stress response in the plant *Plantago ovata* Forsk *in vitro*. Genes Environ 2018; 40(1): 21.
[http://dx.doi.org/10.1186/s41021-018-0109-0] [PMID: 30349616]

[60]   Nagarajan M, Ganesh KS. Effect of chromium on growth, biochemicals and nutrient accumulation of paddy (*Oryza sativa* L.) Intern Let Nat Sci 2014; 23: 63-71 .https://www.academicoa.com/ILNS.23.63

[61]   Husain T, Suhel M, Prasad SM, Singh VP. Ethylene needs endogenous hydrogen sulfide for alleviating hexavalent chromium stress in *Vigna mungo* L. and *Vigna radiata* L. Environ Pollut 2021; 290: 117968.
[http://dx.doi.org/10.1016/j.envpol.2021.117968] [PMID: 34523532]

[62]   López-Luna J, González-Chávez MC, Esparza-García FJ, Rodríguez-Vázquez R. Toxicity assessment of soil amended with tannery sludge, trivalent chromium and hexavalent chromium, using wheat, oat and sorghum plants. J Hazard Mater 2009; 163(2-3): 829-34.
[http://dx.doi.org/10.1016/j.jhazmat.2008.07.034] [PMID: 18814962]

[63]   Singh D, Sharma NL, Singh CK, Yerramilli V, Narayan R, Sarkar SK, *et al.* Chromium (VI)-induced alterations in physio-chemical parameters, yield, and yield characteristics in two cultivars of mungbean (*Vigna radiata* L.) Front Plant Sci 2021; 12.

[64]   Mallick S, Sinam G, Kumar Mishra R, Sinha S. Interactive effects of Cr and Fe treatments on plants growth, nutrition and oxidative status in *Zea mays* L. Ecotoxicol Environ Saf 2010; 73(5): 987-95.
[http://dx.doi.org/10.1016/j.ecoenv.2010.03.004] [PMID: 20363501]

[65]   Rout GR, Samantaray S, Das P. Differential chromium tolerance among eight mungbean cultivars grown in nutrient culture. J Plant Nutr 1997; 20(4-5): 473-83.
[http://dx.doi.org/10.1080/01904169709365268]

[66]   Dey SK, Jena PP, Kundu S. Antioxidative efficiency of *Triticum aestivum* L. exposed to chromium stress. J Environ Biol 2009; 30(4): 539-44.
[PMID: 20120493]

[67]   Sundaramoorthy P, Chidambaram A, Ganesh KS, Unnikannan P, Baskaran L. Chromium stress in paddy: (i) Nutrient status of paddy under chromium stress; (ii) Phytoremediation of chromium by aquatic and terrestrial weeds. C R Biol 2010; 333(8): 597-607.
[http://dx.doi.org/10.1016/j.crvi.2010.03.002] [PMID: 20688280]

[68]   Chatterjee J, Chatterjee C. Phytotoxicity of cobalt, chromium and copper in cauliflower. Environ

Pollut 2000; 109(1): 69-74.
[http://dx.doi.org/10.1016/S0269-7491(99)00238-9] [PMID: 15092914]

[69] Sun H, Brocato J, Costa M. Oral chromium exposure and toxicity. Curr Environ Health Rep 2015; 2(3): 295-303.
[http://dx.doi.org/10.1007/s40572-015-0054-z]

[70] Tchounwou PB, Yedjou CG, Patlolla AK, Sutton DJ. Heavy metal toxicity and the environment.Molecular, Clinical and Environmental Toxicology. Basel: Springer Basel 2012; pp. 133-64. [Internet]
[http://dx.doi.org/10.1007/978-3-7643-8340-4_6]

[71] Jan AT, Azam M, Siddiqui K, Ali A, Choi I, Haq QMR. Heavy metals and human health: mechanistic insight into toxicity and counter defense system of antioxidants. Int J Mol Sci 2015; 16: 29592-630.

[72] McCarroll N, Keshava N, Chen J, Akerman G, Kligerman A, Rinde E. An evaluation of the mode of action framework for mutagenic carcinogens case study II: Chromium (VI). Environ Mol Mutagen 2010; 51(2): 89-111.
[http://dx.doi.org/10.1002/em.20525] [PMID: 19708067]

[73] Liu L, Li W, Song W, Guo M. Remediation techniques for heavy metal-contaminated soils: Principles and applicability. Sci Total Environ 2018; 633: 206-19.
[http://dx.doi.org/10.1016/j.scitotenv.2018.03.161] [PMID: 29573687]

[74] Cameselle C, Gouveia S. Phytoremediation of mixed contaminated soil enhanced with electric current. J Hazard Mater 2019; 361: 95-102.
[http://dx.doi.org/10.1016/j.jhazmat.2018.08.062] [PMID: 30176420]

[75] Derakhshan Nejad Z, Jung MC, Kim KH. Remediation of soils contaminated with heavy metals with an emphasis on immobilization technology. Environ Geochem Health 2018; 40(3): 927-53.
[http://dx.doi.org/10.1007/s10653-017-9964-z] [PMID: 28447234]

[76] Złoch M, Kowalkowski T, Tyburski J, Hrynkiewicz K. Modeling of phytoextraction efficiency of microbially stimulated *Salix dasyclados* L. in the soils with different speciation of heavy metals. Int J Phytoremediation 2017; 19(12): 1150-64.
[http://dx.doi.org/10.1080/15226514.2017.1328396] [PMID: 28532161]

[77] Xiao R, Ali A, Wang P, Li R, Tian X, Zhang Z. Comparison of the feasibility of different washing solutions for combined soil washing and phytoremediation for the detoxification of cadmium (Cd) and zinc (Zn) in contaminated soil. Chemosphere 2019; 230: 510-8.
[http://dx.doi.org/10.1016/j.chemosphere.2019.05.121] [PMID: 31125879]

[78] Yan A, Wang Y, Tan SN, Mohd Yusof ML, Ghosh S, Chen Z. Phytoremediation: a promising approach for revegetation of heavy metal-polluted land. Front Plant Sci 2020; 11: 359.
[http://dx.doi.org/10.3389/fpls.2020.00359] [PMID: 32425957]

[79] Mahar A, Wang P, Ali A, *et al.* Challenges and opportunities in the phytoremediation of heavy metals contaminated soils: A review. Ecotoxicol Environ Saf 2016; 126: 111-21.
[http://dx.doi.org/10.1016/j.ecoenv.2015.12.023] [PMID: 26741880]

[80] Radziemska M, Koda E, Bilgin A, Vaverková MD. Concept of aided phytostabilization of contaminated soils in postindustrial areas. Int J Environ Res Public Health 2018; 15.

[81] Bakshe P, Jugade R. Phytostabilization and rhizofiltration of toxic heavy metals by heavy metal accumulator plants for sustainable management of contaminated industrial sites: A comprehensive review. J Hazard Mater Adv 2023; 10: 100293.
[http://dx.doi.org/10.1016/j.hazadv.2023.100293]

[82] Khalid S, Shahid M, Niazi NK, Murtaza B, Bibi I, Dumat C. A comparison of technologies for remediation of heavy metal contaminated soils. J Geochem Explor 2017; 182: 247-68.
[http://dx.doi.org/10.1016/j.gexplo.2016.11.021]

[83] Wuana RA, Okieimen FE. Heavy metals in contaminated soils: A review of sources, chemistry, risks

and best available strategies for remediation. Intern Schol Res Not. 2011.

[84] Akhtar MS, Chali B, Azam T. Bioremediation of arsenic and lead by plants and microbes from contaminated soil. Res Plant Sci 2013; 1(3): 68-73.
[http://dx.doi.org/10.12691/plant-1-3-4]

[85] Babu SMOF, Hossain MB, Rahman MS, *et al.* Phytoremediation of toxic metals: a sustainable green solution for clean environment. Appl Sci 2021; 11(21): 10348.
[http://dx.doi.org/10.3390/app112110348]

[86] Haq S, Bhatti AA, Dar ZA, Bhat SA. Phytoremediation of Heavy Metals: An Eco-Friendly and Sustainable Approach BT - Bioremediation and Biotechnology: Sustainable Approaches to Pollution Degradation. Bioremediation and Biotechnology. Cham: Springer International Publishing 2020; pp. 215-31.
[http://dx.doi.org/10.1007/978-3-030-35691-0_10]

[87] Galal TM, Gharib FA, Ghazi SM, Mansour KH. Phytostabilization of heavy metals by the emergent macrophyte *Vossia cuspidata* (Roxb.) Griff.: A phytoremediation approach. Int J Phytoremediation 2017; 19(11): 992-9.
[http://dx.doi.org/10.1080/15226514.2017.1303816] [PMID: 28323451]

[88] Sabreena HS, Hassan S, Bhat SA, Kumar V, Ganai BA, Ameen F. Phytoremediation of heavy metals: an indispensable contrivance in green remediation technology. Plants 2022; 11(9): 1255.
[http://dx.doi.org/10.3390/plants11091255] [PMID: 35567256]

[89] Pajević S, Borišev M, Nikolić N, Arsenov DD, Orlović S, Župunski M. Phytoextraction of Heavy Metals by Fast-Growing Trees: A Review BT - Phytoremediation: Management of Environmental Contaminants.Phytoremediation. Cham: Springer International Publishing 2016; Vol. 3: pp. 29-64.
[http://dx.doi.org/10.1007/978-3-319-40148-5_2]

[90] DalCorso G, Fasani E, Manara A, Visioli G, Furini A. Heavy Metal Pollutions: State of the Art and Innovation in Phytoremediation. Int J Mol Sci 2019; 20(14): 3412.
[http://dx.doi.org/10.3390/ijms20143412] [PMID: 31336773]

[91] Dinh T, Dobo Z, Kovacs H. Phytomining of noble metals – A review. Chemosphere 2022; 286(Pt 3): 131805.
[http://dx.doi.org/10.1016/j.chemosphere.2021.131805] [PMID: 34391113]

[92] Wan X, Lei M, Chen T. Cost–benefit calculation of phytoremediation technology for heavy-metal contaminated soil. Sci Total Environ 2016; 563-564: 796-802.
[http://dx.doi.org/10.1016/j.scitotenv.2015.12.080] [PMID: 26765508]

[93] Kafle A, Timilsina A, Gautam A, Adhikari K, Bhattarai A, Aryal N. Phytoremediation: Mechanisms, plant selection and enhancement by natural and synthetic agents. Environ Adv 2022; 8: 100203.
[http://dx.doi.org/10.1016/j.envadv.2022.100203]

[94] Alaboudi KA, Ahmed B, Brodie G. Phytoremediation of Pb and Cd contaminated soils by using sunflower (*Helianthus annuus*) plant. Ann Agric Sci 2018; 63(1): 123-7.
[http://dx.doi.org/10.1016/j.aoas.2018.05.007]

[95] Goswami S, Das S. Copper phytoremediation potential of *Calandula officinalis* L. and the role of antioxidant enzymes in metal tolerance. Ecotoxicol Environ Saf 2016; 126: 211-8.
[http://dx.doi.org/10.1016/j.ecoenv.2015.12.030] [PMID: 26773830]

[96] Wei H, Huang M, Quan G, Zhang J, Liu Z, Ma R. Turn bane into a boon: Application of invasive plant species to remedy soil cadmium contamination. Chemosphere 2018; 210: 1013-20.
[http://dx.doi.org/10.1016/j.chemosphere.2018.07.129] [PMID: 30208525]

[97] Limmer M, Burken J. Phytovolatilization of organic contaminants. Environ Sci Technol 2016; 50(13): 6632-43.
[http://dx.doi.org/10.1021/acs.est.5b04113] [PMID: 27249664]

[98] Nedjimi B. Phytoremediation: a sustainable environmental technology for heavy metals

decontamination. SN Appl Sci 2021; 3(3): 286.
[http://dx.doi.org/10.1007/s42452-021-04301-4]

[99]   Kumar P, Fulekar MH. Rhizosphere Bioremediation of Heavy Metals (Copper and Lead) by *Cenchrus ciliaris.* Res J Environ Sci 2018; 12(4): 166-76.
[http://dx.doi.org/10.3923/rjes.2018.166.176]

[100] Javed MT, Tanwir K, Akram MS, Shahid M, Niazi NK, Lindberg S. Chapter 20 - phytoremediation of cadmium-polluted water/sediment by aquatic macrophytes: Role of plant-induced ph changes. In: Hasanuzzaman M, Prasad MNV, Fujita MBT-CT, T. P, Eds. Cadmium Toxic and Toler in Plants. Academic Press 2019; pp. 495-529.

[101] Dhanwal P, Kumar A, Dudeja S, Chhokar V, Beniwal V. Recent Advances in Phytoremediation Technology BT - Advances in Environmental Biotechnology. Advances in Environmental Biotechnology. Singapore: Springer Singapore 2017; pp. 227-41.
[http://dx.doi.org/10.1007/978-981-10-4041-2_14]

[102] Ranieri E, Moustakas K, Barbafieri M, *et al.* Phytoextraction technologies for mercury- and chromium-contaminated soil: a review. J Chem Technol Biotechnol 2020; 95(2): 317-27.
[http://dx.doi.org/10.1002/jctb.6008]

[103] Mushtaq Z, Liaquat M, Nazir A, *et al.* Potential of plant growth promoting rhizobacteria to mitigate chromium contamination. Environ Technol Innov 2022; 28: 102826.
[http://dx.doi.org/10.1016/j.eti.2022.102826]

[104] Jeyakumar P, Debnath C, Vijayaraghavan R, Muthuraj M. Trends in bioremediation of heavy metal contaminations. Environ Eng Res 2023; 28(4): 220631.
[http://dx.doi.org/10.4491/eer.2021.631]

[105] Wani KI, Naeem M, Aftab T. Chromium in plant-soil nexus: Speciation, uptake, transport and sustainable remediation techniques. Environ Pollut 2022; 315: 120350.
[http://dx.doi.org/10.1016/j.envpol.2022.120350] [PMID: 36209933]

[106] Marieschi M, Gorbi G, Zanni C, Sardella A, Torelli A. Increase of chromium tolerance in Scenedesmus acutus after sulfur starvation: Chromium uptake and compartmentalization in two strains with different sensitivities to Cr(VI). Aquat Toxicol 2015; 167: 124-33.
[http://dx.doi.org/10.1016/j.aquatox.2015.08.001] [PMID: 26281774]

[107] Schiavon M, Galla G, Wirtz M, *et al.* Transcriptome profiling of genes differentially modulated by sulfur and chromium identifies potential targets for phytoremediation and reveals a complex S–Cr interplay on sulfate transport regulation in *B. juncea.* J Hazard Mater 2012; 239-240: 192-205.
[http://dx.doi.org/10.1016/j.jhazmat.2012.08.060] [PMID: 22995205]

[108] Guo S, Xiao C, Zhou N, Chi R. Speciation, toxicity, microbial remediation and phytoremediation of soil chromium contamination. Environ Chem Lett 2021; 19(2): 1413-31.
[http://dx.doi.org/10.1007/s10311-020-01114-6]

[109] Shmaefsky BR. Principles of Phytoremediation BT - Phytoremediation: *In-situ* Applications. In: Shmaefsky BR, Ed. Phytoremediation Concepts and Strategies in Plant Sciences. Cham: Springer International Publishing 2020; pp. 1-26.
[http://dx.doi.org/10.1007/978-3-030-00099-8_1]

[110] Hossain MA, Piyatida P, da Silva JAT, Fujita M. Molecular mechanism of heavy metal toxicity and tolerance in plants: Central role of glutathione in detoxification of reactive oxygen species and methylglyoxal and in heavy metal chelation In: Polle A, Ed. J Bot. : 2012; 2012: p. 872875.
[http://dx.doi.org/10.1155/2012/872875]

[111] Buendía-González L, Orozco-Villafuerte J, Cruz-Sosa F, Barrera-Díaz CE, Vernon-Carter EJ. *Prosopis laevigata* a potential chromium (VI) and cadmium (II) hyperaccumulator desert plant. Bioresour Technol 2010; 101(15): 5862-7.
[http://dx.doi.org/10.1016/j.biortech.2010.03.027] [PMID: 20347590]

[112] Siyar R, Doulati Ardejani F, Norouzi P, *et al.* Phytoremediation Potential of Native Hyperaccumulator Plants Growing on Heavy Metal-Contaminated Soil of Khatunabad Copper Smelter and Refinery, Iran. Water 2022; 14(22): 3597.
[http://dx.doi.org/10.3390/w14223597]

[113] Das PK, Das BK, Das BP, Dash P. Evaluation Of Remediation Ability of *Pongamia pinnata* (L.) Pierre Under Hexavalent Chromium Stress Soil Conditions. Pollut Res 2022; 41: 989-96.
[http://dx.doi.org/10.53550/PR.2022.v41i03.033]

[114] Halder S, Anirban A. Phytoremediation of EMS induced *Brassica juncea* heavy metal hyperaccumulator genotypes. bioRxiv 2021.

[115] Alyazouri A, Jewsbury R, Tayim H, Humphreys P, Al-Sayah MH. Uptake of chromium by portulaca oleracea from soil: Effects of organic content, pH, and sulphate concentration. In: Nayak AK, Ed. Appl Environ Soil Sci. 2020; 2020: p. 3620726.
[http://dx.doi.org/10.1155/2020/3620726]

[116] Dökmeci AH, Adiloğlu S. The phytoremediation of chromium from soil using *Cirsium Vulgare* and the health effects. Biosci Biotechnol Res Asia 2020; 17(3): 535-41.
[http://dx.doi.org/10.13005/bbra/2857]

[117] Sehrawat G, Singh R, Kaushik A. Tolerance of three ornamental plant species to chromium contamination in soil and their potential for phytoextraction and phytostabilization of the toxic metal. Curr World Environ 2021; 16(2): 386-98.
[http://dx.doi.org/10.12944/CWE.16.2.06]

[118] Nyangeri EK, Oonge Z, Nyakundi E. Phytoremediation of Hexavalent Chromium in Effluent Tannery Sludge and Chromium Contaminated Soils using *Ricinus Communis*. Int J Eng Res Technol (Ahmedabad) 2022; 11(08): 226-33.

[119] Ranieri E, Fratino U, Petruzzelli D, Borges AC. A comparison between *phragmites australis* and *Helianthus annuus* in chromium phytoextraction. Water Air Soil Pollut 2013; 224(3): 1465.
[http://dx.doi.org/10.1007/s11270-013-1465-9]

[120] Chen X, Tong J, Su Y, Xiao L. Pennisetum sinese: A potential phytoremediation plant for chromium deletion from soil. Switzerland: Sustainability 2020; Vol. 12.

[121] Nasiru Alhaji S, Asmau Umar S, Abdullahi Muhammad S, Kasimu S, Aliyu S. Cadmium, iron and chromium removal from simulated waste water using algae, water hyacinth and water lettuce. Amer J App Chem 2021; 9(1): 36.
[http://dx.doi.org/10.11648/j.ajac.20210901.15]

[122] Haokip N, Gupta A. Phytoremediation of chromium and manganese by *Ipomoea aquatica* Forssk. from aqueous medium containing chromium-manganese mixtures in microcosms and mesocosms. Water Environ J 2021; 35(3): 884-91.
[http://dx.doi.org/10.1111/wej.12676]

[123] Muchtasjar B, Hadiyanto H, Izzati M, V.Gaile Z, Hendroko R. The ability of water hyacinth (*Eichhornia crasipes Mart.*) and water lettuce (*Pistia stratiotes Linn.*) for reducing pollutants in batik wastewater E3S Web Conf 2021; 1: 226-10.

[124] Nugroho AP, Butar ESB, Priantoro EA, Sriwuryandari L, Pratiwi ZB, Sembiring T. Phytoremediation of electroplating wastewater by vetiver grass (*Chrysopogon zizanoides* L.). Sci Rep 2021; 11(1): 14482.
[http://dx.doi.org/10.1038/s41598-021-93923-0] [PMID: 34262111]

[125] Parida P, Satapathy KB, Mohapatra A. Phytoremediation potential of aquatic macrophyte *Azolla pinnata* R. BR. and *Salvinia molesta Mitchell* to remove chromium from wastewater. Plant Arch 2020; 20(1): 6063.

[126] Ashraf S, Naveed M, Afzal M, *et al.* Unveiling the potential of novel macrophytes for the treatment of tannery effluent in vertical flow pilot constructed wetlands. Water 2020; 12(2): 549.

[http://dx.doi.org/10.3390/w12020549]

[127] Masinire F, Adenuga D, Tichapondwa S, Chirwa EMN. Remediation of chromium (vi) containing wastewater using *chrysopogon zizanioides* (Vetiver grass). Chem Eng Trans 2020; 1(79): 385-90.https://www.cetjournal.it/index.php/cet/article/view/CET2079065

[128] Abhayawardhana MLDD, Bandara NJGJ, Rupasinge SKLS. Removal of Heavy Metals and Nutrients from Municipal Wastewater using *Salvinia molesta* and *Lemna gibba.* J Trop For Environ 2019; 9(2): 9.
[http://dx.doi.org/10.31357/jtfe.v9i2.4469]

[129] Tabinda AB, Irfan R, Yasar A, Iqbal A, Mahmood A. Phytoremediation potential of *Pistia stratiotes* and *Eichhornia crassipes* to remove chromium and copper. Environ Technol 2020; 41(12): 1514-9.
[http://dx.doi.org/10.1080/09593330.2018.1540662] [PMID: 30355050]

[130] Maurizka E, Moersidik S, Saria L. Control of chromium hexavalent (Cr -VI) pollution on waste water in nickel ore extraction industry with phytoremediation technology E3S Web Conf 2018; 1: 68-3011.

[131] Ranieri E, D'Onghia G, Ranieri F, Petrella A, Spagnolo V, Ranieri AC. Phytoextraction of Cr(VI)-Contaminated Soil by *Phyllostachys pubescens*: A Case Study. Toxics 2021; 9(11): 312.
[http://dx.doi.org/10.3390/toxics9110312] [PMID: 34822703]

[132] Ullah S, Mahmood S, Ali R, Khan MR, Akhtar K, Depar N. Comparing chromium phyto-assessment in *Brachiaria mutica* and *Leptochloa fusca* growing on chromium polluted soil. Chemosphere 2021; 269: 128728.
[http://dx.doi.org/10.1016/j.chemosphere.2020.128728] [PMID: 33143883]

[133] Taufikurahman T, Pradisa M A S, Amalia SG, Hutahaean GEM. Phytoremediation of chromium (Cr) using Typha angustifolia L., *Canna indica* L. and *Hydrocotyle umbellata* L. in surface flow system of constructed wetland. IOP Conf Ser Earth Environ Sci 2019; 308(1): 012020.
[http://dx.doi.org/10.1088/1755-1315/308/1/012020]

[134] Alsafran M, Usman K, Ahmed B, Rizwan M, Saleem MH, Al Jabri H. Understanding the phytoremediation mechanisms of potentially toxic elements: A proteomic overview of recent advances Frontiers in Plant Science 2022; 13.https://www.frontiersin.org/articles/10.3389/fpls.2022.881242

[135] Asante-Badu B, Kgorutla LE, Li SS, Danso PO, Xue Z, Qiang G. Phytoremediation of organic and inorganic compounds in a natural and an agricultural environment: A review. Appl Ecol & Environ Res 2020; 18(5).
[http://dx.doi.org/10.15666/aeer/1805_68756904]

CHAPTER 10

Significance of Nanobioremediation for the Removal of Contaminants from Water: Challenges and Future Prospects

Shraddha Bais[1,*] and **Ruchi Shrivastava**[1]

[1] *Department of Chemistry, Institute of Science and Research, IPS Academy, Indore, Madhya Pradesh, India*

Abstract: Water is the most crucial natural resource required for the survival of humankind, but chemical industries, household activities, foul practices, *etc.*, are responsible for polluting it. This leads us to work on the purification and bioremediation of water contaminants. Diverse techniques have been developed globally for decontamination/purification of water, but owing to their high cost, tediousness and time consumption, it has become necessary to work on those methods that are comparatively cheaper, techno-feasible and employ a green process. In the contemporary world, the use of nanotechnology in the bioremediation of water pollutants revolutionarily provides a way to incorporate functional chemicals in notably reduced quantities to fulfil the desired purpose. Globally, water pollution is primarily caused by the rainwater containing the pollutants present in the air. Industrial and domestic wastewater, which is a major source of toxic heavy metals, industrial dyes, pesticides and insecticides, is used in agriculture. These water pollutants adversely affect the health of human beings as well as the whole ecosystem of the affected region. The employment of nano-materials (NMs) degrades the pollutants from the water source to its standard permissible level. The current chapter comprises a detailed discussion on the immense potential of NMs in the bioremediation of polluted water using different NMs and also provides a comprehensive comparison with other conventional bioremediation methods to make water environmentally non-hazardous.

Keywords: Bioremediation, Contaminants, Conventional treatment techniques, Green methods, Heavy Metals, Metal oxides, Nanotechnology, Nanomaterials, Toxic pollutants, Wastewater, Water pollutants, Zero-valent metals.

* Corresponding author Shraddha Bais: Department of Chemistry, Institute of Science and Research, IPS Academy, Indore, Madhya Pradesh, India; E-mail: shraddhabais@ipsacademy.org

Neha Agarwal, Vijendra Singh Solanki and Sreekantha B. Jonnalagadda (Eds.)

INTRODUCTION

Toxic pollutants are of serious public health concern because they flow with effluents like heavy metals, pesticide residues, antibiotics, chemicals and hydrogen. There is an urgent need to address the environmental issues caused by these effluents. Bioremediation is one of the major water treatment techniques that involve the use of microorganisms in the degradation of hazardous pollutants. By this technique, the pollutants get degraded either to their permissible limit or to less harmful substances [1]. This method can be mostly employed in situ by creating microbial consortia under required conditions like humidity, temperature, and nutrition to promote the growth of required microorganisms. Microorganisms that are used to remove contaminants from water have different impacts on different pollutants and are capable of mineralizing the contaminant into mineral acids [2 - 5]. Different categories of wastewater effluents and their bioremediation outcomes are mentioned in Table **1**, while some of the bioremediation agents and their effects are listed in Table **2** [6, 7].

Table 1. Different categories of wastewater effluents and their bioremediation outcomes.

Effluent	Bioremediation Outcomes
OMWs-Olive Mill Wastewater	Saprophytes are widely employed in the removal of heavy metals like Hg, Pb, Fe, U
Dyes, Colorants, Textile Industries	Bacilli are frequently used in the degradation of heavy metals like Hg, Pb, Cd, Zn, dyes, pesticides
Pulp, paper, Textile	Saccharomyces cerevisiae decreases COD and also removes heavy metals and organic matter.
Sewage water	Algae remove toxic metals and reduce BOD and COD to optimum levels.
Pharmaceutical Industries' wastewater	Enzymes and microorganisms eliminate and reduce active pharmaceutical ingredients, drugs, hormones, detergents, disinfectants, antioxidants, *etc.*
Battery industries' wastewater	Lysinibacillus, Paenibacillus, Bacillus, and Acidithiobacillus are employed to remove heavy metals and lithium from the battery wastewater.
Plastic industries' wastewater	Enzymes mediated removal of by-products or intermediate products of polymerization process released in water *i.e.*, resin, organic acids, tetrahydric alcohol, pentaerythritol, formaldehyde, sodium formate, phenols, urea, benzene, *etc.*

CONVENTIONAL TECHNIQUES TO TREAT WATER POLLUTANTS

Water pollution is primarily caused by microbial infections and the presence of hazardous organic as well as inorganic chemical compounds. The most ancient method to disinfect water was introduced by Holmes in 1835 in Boston and Semmelweis in Vienna in 1837 [10]. They used chlorine as chlorinated lime for water disinfection and after that, a solution of chlorine of 1-3 ppm concentration was used as a major treatment to disinfect drinking water. However, the major flaw in the chlorination of water was that it did not work against most of the chemical pollutants. Later on, oxidation by various oxidizing reagents like H_2O_2 was done to treat polluted water containing chemical pollutants [11]. Other conventional techniques that are used to treat water pollutants are listed in Table **3**. Most of the chemicals and heavy metals cannot be removed easily by normal physical, chemical or biological methods. Therefore, electrochemical treatment, ion exchange chemical redox process, membrane filtration, reverse osmosis, photocatalytic degradation, microbial treatment, NMs-based treatments and their combined approaches are some common techniques that are under the area of research to treat water pollutants commercially in the near future [12, 13].

Table 2. Different types of bioremediation techniques, their significance and deficits [8, 9].

Bioremediation Techniques	Significance	Deficits
Mycoremediation	• This involves the use of fungi, which act as decomposers. They feed on dead organic matter, degrade oils, hydrocarbons, and aromatic compounds • Some examples like mushrooms *Agaricus, Amanita, Cortinarius, saprotrophs, hyphae, Suillus,* and *Phellinus* are used for mobilization/complexation of different heavy metals in soil	Incapable of complete removal of pollutants.
Phytoremediation	• Phytoextraction and phytostabilization are remediation techniques that absorb contaminants from water.	Only relocate the toxic pollutants and applicable only on the surface of the wastewater.
Phycoremediation	• This is the most effective method for aquatic ecosystems. Microalgae present in water grow as an algal bloom and easily absorb pollutants through bioassimilation and biosorption.	Low toxin tolerance capacity and slow toxin removal rate.
Microorganisms	• Genetically modified microorganisms absorb pollutants and heavy metals like mercury and aromatic hydrocarbons • Especial microorganisms are effective in remediating oil spills, which is a major cause of aquatic pollution • Bioaugmentation is another advanced method that degrades pollutants.	Some microorganisms produce more toxic products during bioremediation; pollutants with higher molecular weight (PAN, PAH *etc.*) are difficult to be removed by microorganisms.

(Table 2) cont.....

Bioremediation Techniques	Significance	Deficits
Nematodes	• The Nematode is specifically used as a bioremediator. It is also helpful in cleaning, nutrient mobilization, nitrification, enzyme activation and in the removal of heavy metals from wastewater. • Some examples of productive bioremediators are Caenorhabditis elegans, Plectus acuminatus, Heterocephalobus pauciannulatus.	Many nematodes are toxic and infectious to humans and other members of an ecosystem.

Table 3. Demerits of conventional techniques for wastewater treatment.

Conventional Method	Demerits	Refs.
Adsorption	*Acidic effluents increase the desorption rate. *High mobility and the addition of more active functional groups are required to apply these adsorbents in large-scale applications in water treatment. * Expensive and complex synthesis process * Inadequate selectivity of heavy metal ions. * Low-quality yield	[14 - 16]
Chemical oxidation	* Expensive * Toxic byproducts	[17, 18]
Ion exchange	* Harmful effect of nitric acid on ion exchange capacity of anion resin. * Dilution of wastewater before treatment must be done.	[19, 20]
Ion Exchange by membrane	* Expensive *Toxic byproducts	[21]

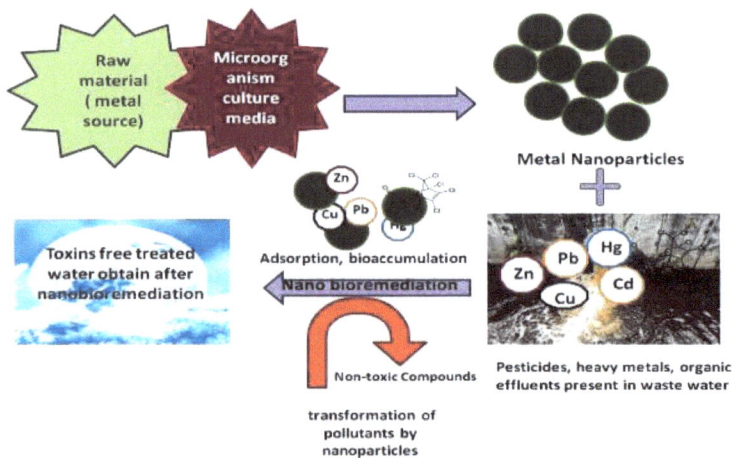

Fig. (1). Bioremediation by NPs synthesized by microorganisms.

BIOREMEDIATION OF WATER CONTAMINANTS

Bioremediation is a technique to remove pollutants from the affected site using living microorganisms (Fig. **1**). The remediation process conducted by microorganisms induces the breakdown of the pollutants into non-toxic, biodegradable materials *via* the microbiological process. The bioremediation proceeds by microorganisms either by biostimulation or by bio augmentation. The use of microorganisms facilitates the *in-situ* removal of contaminants from underground sites and is a cost-efficient, eco-friendly technique. Some of the recent investigations for wastewater management using bioremediation to understand the role of microorganisms in the remediation process are given in Table **4**. The investigations on bioremediation using microorganisms indicated its higher efficiency toward wastewater remediation; only the selection and identification of appropriate microbe is needed. When bioremediation is combined with other techniques such as nanotechnology, it provides a more effective, cost-efficient and environmentally safe approach to wastewater management.

Table 4. Bioremediation activity of different microorganisms on various water pollutants.

Types of Water Pollutants Treated	Micro-organisms used for Bioremediation	Sources of Micro-organism Isolates	Outcomes	Refs.
Petroleum (hydrocarbon pollutants)	Acinetobacter faecalis WD2, Staphylococcus. sp DD3, Neisseria elongate TDA4.2, Pseudomonas putida TAM4.4.	groundwater and wastewater resources of Terengganu oil refinery, Malaysia	*Peptone is a good source of N_2 for the growth of tested bacterial strains that help in the biodegradation of petroleum wastes. *Around 98-100% degradation of hydrocarbons present as pollutants is observed in 15 days of incubation.	[22]
Oil contaminants	Alcanivorax borkumensis SK2, Nocardia sp. SoB	Preexisting collections from IAMC-CNR, Messina and the University of Palermo, Italy	*Porous Polycaprolactone sponge as carrier material for oil adsorbent bacteria is a successful approach to increase the sorption capacity and bacterial activity to remediate oil pollutants from water samples.	[23]

(Table 4) cont.....

Types of Water Pollutants Treated	Micro-organisms used for Bioremediation	Sources of Micro-organism Isolates	Outcomes	Refs.
Crude oil contaminants	Alcanivorax borkumensis SK2, Oleibacter marinus 5, Nocardia sp., Gordonia sp. SoCg	Pre-existing cultures	*A combined approach by immobilizing selected bacterial species on the polylactic acid and polycaprolactone electrospun oil-absorbing membranes increased the biodegradation of oil up to 23%. *Over 66% biodegradation of oil is observed with 100% absorption of present crude oil.	[24]
Endosulfan	Acinetobacter, Klebsiella, Alcaligenes, Bacillus, Flavobacterium	Kor River surrounding Fars Province, Iran	*The highest bacterial count was observed in samples collected in summer, while it was lowest in the samples of autumn. * All the bacterial cultures present in the sample had shown 80-95% degradation of both alpha and beta Endosulfan contaminants.	[25]
PAH compounds: Acenaphthene& Fluorene	Raoultella ornithinolytica, Bacillus megaterium, Serratia marcescens, Aeromonas hydrophila	Diep and Plankenburg Rivers, Western Cape, South Africa	*All the tested isolates had shown more than 95% removal tendency toward both the pollutants except Bacillus megaterium for Acenaphthene. *Bacillus megaterium had shown 90.2% removal of Acenaphthene.	[26]
Pollutants from the food industry, dye industry, pharmaceutical industry, Phosphates, Chromium and lead	Enzymes produced by Bacillus sp.	Sampling water and sediments from 5 different hot spring locations in Limpopo, South Africa	*The cultured bacteria from the selected location produced a variety of enzymes *i.e.*, oxidoreductases, amylase, protease, phosphatase and ribonuclease, which can further be used for wastewater remediation. *1-pyrroline-5-carboxylase dehydrogenase and delta-aminolevulinic dehydratase enzymes can be used to monitor lead in polluted water. *Enzymes have shown excellent properties in reducing the phenol in the laboratory environment.	[27]

Types of Water Pollutants Treated	Micro-organisms used for Bioremediation	Sources of Micro-organism Isolates	Outcomes	Refs.
Odorous Sulphate contaminants	Paracoccus sp., Pseudomonas sp. N2, N5,S3 & S4, Gordonia sp. N4, Bacillus sp. N6, Gordonia sp. N8,	Odorous river of Panyu District (Guangzhou), China And laboratory prepared odorous water	* All the tested organisms are mixtopic sulfate oxidizing Bacteria and show good absorption and removal capacity towards sulfide and sulfate ions.	[28]
Organophosphorus compounds present in pesticides	Brevundimonas sp., Sphingomonas sp.	Soil samples from Daireaux, Buenos Aires province, Argentina treated with chlorpyrifos and coumaphos. .	* Both the tested species have phosphotriesterase activity (PTE) which is responsible for the degradation of Organophosphorus compounds present in agricultural wastes. * The PTE was found to be maximum at pH 6 and temperature between 30-60^{0}C.	[29]
Heavy metals	Metal-resistant gram-positive and gram-negative bacteria	-	* Mechanisms of heavy metal accumulation by the bacterial cell have been analyzed. * Gram-positive bacteria are better in bioremediation of wastewater because their heavy metal accumulation is much more than gram-negative bacteria.	[30]
Arsenic present in groundwater	Gram-positive bacteria: Bacillus sp. and Aneurinibacillus aneurinilyticus	12 villages of Purbasthali, West bangal, India	* The isolated bacteria oxidize the arsenite to the less toxic arsenate. * Both the isolates cannot reduce the arsenate further.	[31]
Arsenic from high arsenic acid mine water of Aguas Calientes Area, Northern Chile	Sulfate reducing bacteria	Bacteria-enriched culture obtained from Winogradsky column	* SRB has a high potential to remediate rich acid mine water. * Low pH and high metal concentration are favorable for bacterial growth. Increased pH and sulfide concentration lead to iron and Arsenic removal, probably by the adsorption of iron ores.	[32]

(Table 4) cont.....

Types of Water Pollutants Treated	Micro-organisms used for Bioremediation	Sources of Micro-organism Isolates	Outcomes	Refs.
Arsenic	Micrococcus luteus strain AS2	Industrial wastewater of District Sheikhupura, Pakistan.	* Genomic analysis revealed the presence of the aioB gene, which encodes arsenite oxidase smaller subunit. The presence of arsC1, arsC2, and ACR3 was also identified. * Genes responsible for other heavy metals (Cd, Zn, Hg, Ni, Co) removal are also identified during genomic analysis, which indicated the broad coverage of HM bioremediation using the analyzed Bacterial strain.	[33]
Arsenic	Klebsiella pneumoniae RnASA11	Soil samples of Chhattisgarh, India	*The mechanism of arsenic bioremediation is a function of two genes, arsC, which reduced arsenate into arsenite, and arsB gene, which extruded arsenite from the cell, making the bacterial cells arsenic-resistant.	[34]
Inorganic pollutants, heavy metals, and Coli bacteria in sewage, sea and well water	Chlorella vulgaris, Chlorella Salina	C. vulgaris from El-Salhya sewage station, Qena, Egypand C. salina from Lake Marriott, Alexandria, Egypt	*C. vulgaris is superior in treating pathogens like Coli. *Both the tested algae had shown moderate to excellent heavy metals adsorption ability.	[35]
Nitrogen content, organic and inorganic wastes containing $KMnO_4$	bacterial algae	-	*The bacterial algae provide good removal efficiency for ammonia and $KMnO_4$ from the Boezem water of Kalidami, Indonesia *Surabaya tested in the laboratory environment. The presence of a 3% potassium and carbon source decreased the removal efficiency of algae.	[36]

NANOTECHNOLOGY AND EXCLUSIVITY OF NPS IN THE TREATMENT OF WATER POLLUTANTS

Nanotechnology has emerged as one of the most expedient technologies for wastewater treatment owing to its reduced size, definite morphology, filtration, and high surface-to-volume ratio [37]. It is an efficient, cost-effective and recent technique employed for the treatment of industrial effluents [38 - 42]. In the commercial world, a hub of industries has been developed that inculcate battery manufacturing, alloy, mining, pharmaceuticals, dyes, pesticides, and electroplating. The contaminants released from these industries affect human health and the environment. Hazardous components include organic compounds, heavy metals, saline and turbid effluents that must be treated. NMs are used for the treatment of wastewater for the following reasons:

- Nanocatalysts are widely prepared to carry out photocatalytic reactions and can further be employed for the treatment of wastewater.
- Nanoadsorbents are metallic and magnetic nano-structured particles that possess a high affinity to adsorb organic or inorganic contaminants.
- Nanomembranes are broadly used to remove heavy metals, dyes and other contaminants. These may be in the form of carbon nanotubes (CNTs), nanoribbons, and nanofibers. CNTs are electrically conducting membranes specifically employed for the biological treatments of industrial wastewater and desalination of industrial brine. Table **5** explains the predominance of metal-based nanoparticles (NPs) in the remediation of a variety of pollutants.

Table 5. Predominance of metal-based NPs in remediation of pollutants [43 - 46].

Fe-NPs	**Fe-NPs have a small size and can be directly injected into soil, which forms a suspension or matrix with sediments and can be used further to treat water to form a slurry.**
Au-NPs	Gold metal-based NPs can be utilized to eliminate toxic metals like Hg^{2+}. These are found to be cost-effective, efficient and reliable.
Ag-NPs	Silver metal can easily form NPs that have high absorption capacity and are efficient in removing mercury and cadmium.
Ti-NPs	TiO_2 forms a highly crystalline and the most effective adsorption medium for the bioremediation process. It can reduce the toxicity of Copper, Lead and Arsenic.
Ce-NPs	These can easily eliminate chromium from organic contaminants.
Cu-NPs	These metal NPs possess high porosity and high surface region, which can eliminate Cd, Ni, and Pb.
Polymer-based	Polymeric NPs act as surfactants with hydrophobic and hydrophilic ends. They can be utilized to form stable particles of crosslink polymer chains. Amphiphilic polyurethane (APU) is a known example of a remediation agent. Some others are polyurethane acrylate anionomer (UAA) and poly(ethylene glycol)-modified urethane acrylate (PMUA)

GREEN BIOREMEDIATION OF WASTE WATER WITH NPS

Bioremediation with NMs Synthesized/embedded with Microorganisms

The production of NMs using different microorganisms like bacteria, algae, fungi or yeast is an eco-friendly and cost-effective technique due to minimal or no use of harmful chemicals. Microorganisms are also termed as bio-factory to fabricate NPs [47]. Salvadori MR *et al.* used the dead biomass of *Trichoderma koningiopsis* to synthesize Copper NPs extracellularly from the wastewater and concluded that the adsorption process followed Langmuir isotherm [48]. The Fe_3O_4 NPs synthesized from three different seaweeds *Petalonia fascia, Colpomenia sinuosa and Padina pavonica* were used to control the growth of hazardous algae development in seawater by adsorbing the Nitrogen and Phosphorus from marine water [49]. Fe_3O_4 NPs synthesized from *Aspergillus tubingensis* had also shown excellent adsorption properties toward heavy metals from wastewater, and the revival and reusability of Fe_3O_4 NPs were found in up to five adsorption/desorption practices. Application-based nanobioremediation of municipal wastewater was carried out by N. Khan *et al.* They used the Ag-Au NPs and plant growth-promoting Rizobectaria to see their effect on Maize plant growth by municipal wastewater. They concluded that before irrigating the maize plant with municipal water, the Ag-Au NPs and rhizobacteria treatment of the wastewater is needed to promote plant growth and heavy metal remediation from the wastewater [50]. Similarly, for the rice plant's growth, antioxidative properties and osmolyte contents were investigated in heavy metals contaminated water, and the joint effect of Zn NPs and *Bacillus cereus* (PMBL-3) and *L. macroides* (PMBL-7) bacterial strains on the analyzed properties were also examined in the same water source. The investigation revealed the adverse effects of heavy metal contaminants on the seedling, germination, and osmolyte contents. Heavy metal-induced stress affects the antioxidative properties, but the addition of Zn NPs with bacterial strain promotes the growth factors of plants even under heavy metal stress.

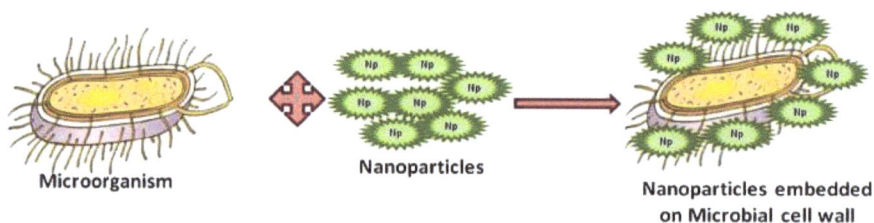

Fig. (2). Incorporation of NMs on microorganism's cell wall to enhance adsorption of pollutants from wastewater.

The microorganisms combined with NPs show enhanced adsorption and bioaccumulation of heavy metals and other micro-pollutants from wastewater. Fig. (**2**) shows the diagrammatic representation of the incorporation of NMs with microorganisms. M. Shakya *et al.* (2018) analyzed the fungus-based nanobioremadiation of heavy metals and concluded that the biosorption of heavy metals on NPs is based on the special cellular function of fungus and the properties of NMs [51]. Likewise, another fungus *Fusarium solani* YMM20 and chitosan/tripiolyphosphate NPs (CS/TPPNP) were used to remove heavy metal particles from wastewater by reducing them into heavy metal NPs. These were later adsorbed on CS/TPP-FF NPs and removed from the water by centrifugation process. This method is suitable for the removal of metal ion NPs and the maximum adsorption affinity of fungus-embedded NPs for Pb^{2+} ions and minimum adsorption affinity for Co^{2+} ions [52]. The *Alcaligenes faecalis* isolated from sewage wastewater can synthesize Ag^{2+} NPs. Another group of researchers obtained nine different isolates that possess heavy metal-resistant properties from sewage water of Taif province, Saudi Arabia. They characterized the isolate by 16S rRNA and found that four of them were highly resistant toward heavy metals [53].

Cadmium is a toxic heavy metal that can easily accumulate in the biotic environment and later shows hazardous effects on the ecosystem. The removal of Cadmium ions from an aqueous solution is an important task that was performed by C.P. Devatha *et al.* (2020) by using the Magamite NPs coating on *Bacillus Subtilis* bacterial strain. They found an 85% removal tendency of Magamite NPs coated B. Subtilis at 30°C temperature and a pH of 4 and an optimum time of one hour by biosorption of cadmium ions by the NP-coated bacterial strain. They also found the excellent revival rate of applied NP-coated bacterial strain to reuse it for more cadmium removal [54]. *Phanerochaete chrysosporium* is a fungus that has a good removal capacity for Pb and Cd from contaminated water by absorption mechanism. A ZnS NP coating on *Phanerochaete chrysosporium* cell increased its adsorption and removal tendency of heavy metal ions up to 1.4 to 1.6 folds than the uncoated cells. The detailed kinetic analysis revealed the adsorption of Cd^{2+} and Pb^{2+} ions on the ZnS- *Phanerochaete chrysosporium* cell has been completed through chemisorption and diffusion mechanism [55]. In the mixed fermentative bacterial culture, when incorporated with α-Fe_2O_3, NiO, and ZnO NPs, the enzymatic activity of enzymes involved in the metabolism of mono-ethylene glycol (*i.e.*, alcohol dehydrogenase, aldehyde dehydrogenase and hydrogenase) was increased. This indicated the incorporation of fermentative bacterial culture with NPs having a positive impact on hydrogen production from wastewater having mono-ethylene glycol as a pollutant [56]. *E. Priyadarshini et al.* (2021) assessed the metal resistance mechanism in fungal cells and found the biosorption and bioaccumulation and sometimes biotransformation of toxic metals by the fungal cells to resist heavy metals toxicity. They also found that due to biosorption and biotransformation, some fungal cells convert the heavy metals to NPs, which can further be used in wastewater management [57].

Bioremediation of Wastewater using Enzyme-embedded NMs

The enzymes are superior adsorbents of heavy metals and organic and inorganic pollutants. Microorganisms are the main host of enzymes involved in the bioremediation of wastewater. Generally, the enzymes catalyze the adsorption, bioaccumulation or biotransformation of heavy metals and inorganic/organic pollutants to increase the rate of bioremediation of wastewater. The current research confirms that when the enzymes are loaded with NMs, it usually enhances their capacity to adsorb pollutants from wastewater. Fig. (**3**) contains the schematic representation of enzyme immobilization by NPs to ease wastewater management by remediating it.

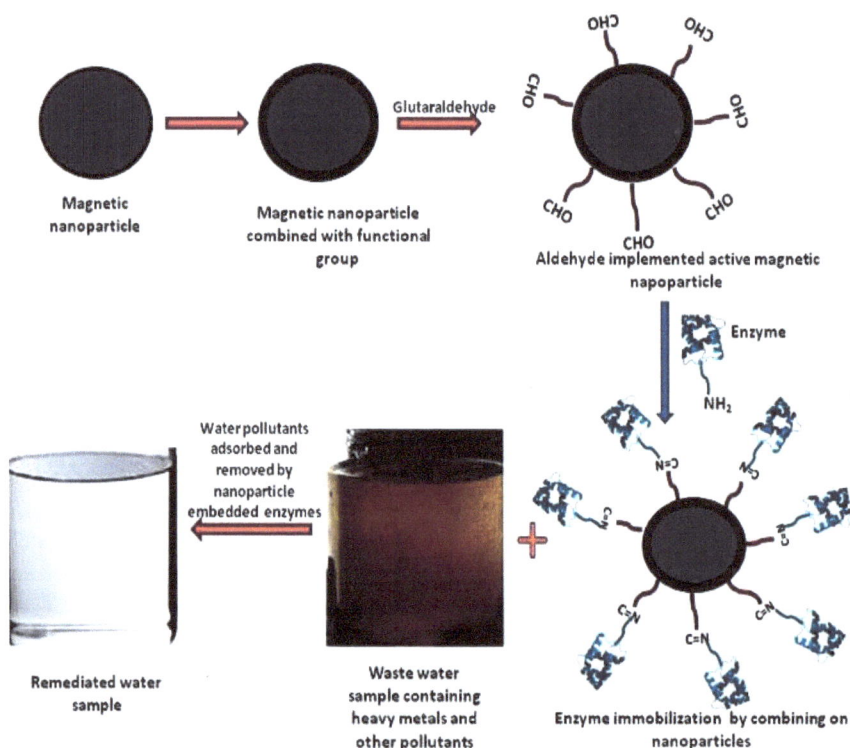

Fig. (3). Schematic diagram of enzyme immobilization by NPs for bioremediation of wastewater.

Peroxidase enzymes are excellent absorbers of azo dye, and when it is immobilized on Fe_3O_4 NPs, they give excellent results by decolorizing the water containing various chemicals (dyes) in a bioreactor established in a laboratory environment [58]. Similarly, cyanate is a more toxic oxidative byproduct of cyanide. To achieve the effective removal of cyanate from the wastewater, a combination of cyanase and carbonic anhydrase immobilized on silanized Fe_3O_4 NPs was analyzed, and around 85% removal from industrial wastewater and 100% removal from the buffered sample was achieved. Likewise, the cyanate hydratase enzyme was immobilized on magnetic (Fe_3O_4 filled)-multiwall CNTs by B. Ranjan *et al.* (2019) to remediate the synthetic water containing cyanate contaminant [59]. The investigators found the alkaline pH for the most stable adhesion of cyanate hydratase to magnetic multiwalled CNTs which is stable for up to 30 days. The tested enzyme-CNT showed excellent cyanate removal capacity that could be reused for up to 10 consecutive cycles [60]. Chang *et al.* prepared graphene oxide/Fe_3O_4 NPs by an ultrasonic-assisted reverse co-precipitation method and immobilized the horseradish peroxidase enzyme onto it by using 1-ethyl-3-(3-dimethylaminopropyl) carbodiimide as a cross-linking agent. The GO/Fe_3O_4 -horseradish peroxidase composite showed a 95% phenol removal tendency from the aqueous environment and was easily separated by a magnetic separation technique. The composite was capable of catalyzing the reduction of H_2O_2 to remove phenol and 2,4-dichlorophenol from the aqueous sample [61]. Ali *et al.* (2017) applied an immobilization technique to the enzyme ginger peroxidase on amino-functionalized silica-coated titanium dioxide, which enhanced the catalytic activity toward dye decoloration. The authors claimed that the increased activity is an effect of conformational change in the secondary structure of the enzyme due to immobilization. Ali and coworkers found immobilized ginger peroxidase on amino-functionalized silica-coated titanium dioxide is 23% more effective against yellow 45 dye decoloration as compared to simple ginger peroxidase [62]. Peroxidase enzyme was isolated from Bacillus amyloliquefaciens and immobilized on a chitosan-coated, halloysite nanotube, which was modified by dopamine and found effective in more than 90% decoloration and detoxification of wastewater. The immobilized enzymes showed reusability in the range of more than 50% even after 6 reuse cycles [63]. The magnetic chitosan NPs embedded with laccase enzyme obtained from *Weissella viridescens* LB37 were investigated for biodegradation of azo dye from wastewater by H. Nadaroglu and coworkers. The cross-linking of NPs with laccase was done by 1% L-glutaraldehyde for one hour immobilization time at pH 6 and 30°C optimum temperature. The catalytic activity for degradation of azo dye was found to be doubled as compared to untreated laccase enzyme, and the reusability of the employed enzyme-NP system was also found satisfactory in a range of around 47% after 10 degradation cycles [64]. The immobilization of laccase enzyme on zeolitic imidazolate (ZI) framework-coated multi-walled carbon nanotubes was successfully achieved by P. Habimana and his fellow investigators at optimum pH 3 at 60°C. The investigators found the Laccase-ZI CNT composite was excellent against the decoloration of Eriochrome black T and Acid red 88 dyes [65]. More than 95% decoloration was observed in the case of dye in sample water, and as found by earlier investigators, immobilization of laccase increased the recovery and reusability rate of Laccase-ZI carbon nanotube composite [65].

The environmental dependence of enzyme stability is a major subject of concern. The bioremediation of wastewater using NMs in the treatment of wastewater containing dye was assessed by J.K. Hong Wong *et al.* (2019) [66]. They found that nanotechnology incorporation increases the enzyme catalytic activity towards dye decoloration. The authors concluded that more investigations are needed for the practical application of nanotechnology-based enzyme biodegradation of wastewater having dye as a potent pollutant [66]. Similarly, V. Karthik and coworkers also investigated the enzyme immobilization on NPs and found an increment in the degradation potential of enzymes for wastewater containing agricultural wastes, dyes, phenolic and pharmaceutical contaminants [67]. Another interesting research on the use of enzyme-embedded NPs for water treatment was reported by Chao *et al.* (2017) where they carried out enzymatic treatment over TiO_2 based NM. They used SDS-PAG as an enzyme extract and assessed its activity at different pH and temperature conditions [68]. Some researchers also accounted for the treatment of antibiotic residues obtained from pharmaceutical industries. Nguyen and his coworker synthesized Fe_3O_4 NPs and studied their effect on immobilized β-lactones [69]. Another study described the usage of new biocatalyst magnetite for the removal of some typical organic pollutants like phenol and azo dye [70]. They used laccases as enzyme catalysts owing to their low cost and efficiency as effective bioreactors [71]. The degradation was noted at the optimum pH range to evaluate its performance [70]. Another study by Agarwal *et al.* reported the role of bacterial laccases as biocatalysts for the degradation of toxic environmental pollutants [72].

Wang and his associates (2012) developed another technology that involves the use of enzyme-immobilized laccase over silica NMs, which reflected a strong potential to mineralize organic pollutants and also reported the recovery of such immobilized enzymes [73].

NEGATIVE ASPECTS OF NANO-BIOREMEDIATION

Apart from the enormous benefits of nano-bioremediation, there are some negative factors associated with the combined approach of nanotechnology with microorganisms for wastewater bioremediation. When an NM is introduced into the aquatic environment, it can easily accumulate and is transferred into the food chain and possibly generate harmful effects. Ecological providence and toxicity of NMs are the core issues in designing and utilizing NPs in polluted water cleaning treatments. A.S. Karakti *et al.* (2006) reviewed the toxic behavior of NPs when entered into living cells and found that the surface chemistry of NMs is mainly responsible for their toxicity. Particle size, zeta potential, surface area, surface charge, *etc.*, are some important factors that have to be taken care of to know the toxicity mechanism of NPs in living cells [74]. P. Ganguly *et al.* assessed and combined the various toxicity assaying techniques *i.e.*, genotoxicity, cell proliferation, and cytotoxicity, and discussed the recent advances in in-vivo, in-vitro and *in silico* study of NPs along with their cellular uptake mechanism [75].

Further research is required to know the impact of NPs on the ecosystem before applying the nano-bioremediation of wastewater commercially. The interaction of NMs with biotic and abiotic factors is still not clearly understood. The NMs used in bioremediation can transform into their oxidized or reduced form by photochemical reactions. This will lead to the adsorption and simultaneous desorption of pollutants and alter the pollutant removal capacity from the wastewater [76]. Another issue associated with NPs is self-toxicity. Carbon nanowires and CNTs can easily be used in the bioremediation of water because they are water-insoluble and can easily be separated from the water bodies after treating the pollutants. However, there are several toxic effects coupled with carbon NMs, such as central nervous system toxicity, kidney toxicity, cardiovascular toxicity, and sometimes oxidative stress too [77]. G.V.Cimbaluk *et al.* (2018) studied the effect of commercially available CNTs on two fish species, *Danio rerio and Astyanax altiparanae,* and found the possibility of DNA cross-linking with carbon nanotubes. They also concluded the subchronic neurotoxicity in *Daino Rerio* and the acute neurotoxicity in *Astyanax altiparanae* [78]. Furthermore, the toxic effects generated by Ag NPs in *Labeo rohita* fish were also studied for nuclear alteration and oxidative stress [79]. Several metal oxides also show toxicity and can become the cause of pulmonary disorders, oxidative stress, fibrosis, *etc.* [80]. Systematic assessment and management of the self-toxicity of NMs before using them in nano bioremediation is a gigantic challenge.

CONCLUSION AND FUTURE RECOMMENDATIONS

The bioremediation of wastewater using nanotechnology is one of the highly captivating areas of research at the current time due to the unique properties of NPs like smaller size, shape, chemical behavior, and sorption properties. This technique is mainly based on the adsorption power of NPs along with the digestive properties of microorganisms to treat a wide range of water pollutants like heavy metals, organic compounds, industrial effluents, domestic wastewater, *etc.* NMs have the knack to penetrate deep into the site of action so it is possible to apply this technique for *in-situ* wastewater treatment in the near future. Apart from the huge benefits of NMs in bioremediation of water, there are some toxic effects of NMs that can adversely affect the whole ecosystem. The main challenge before using nano bioremediation for commercial application is to study the additive effect of NMs on human health and living beings. The addition of NPs with microorganisms for bioremediation of polluted water can be a safer option as compared to the conventional techniques of bioremediation because this technique does not release hazardous chemicals after the remediation process. The use of biosynthesized NMs could lead nano-bioremediation toward a more eco-friendly, greener approach to remediate wastewater. The recovery and reprocessing of NPs after completion of bioremediation of water will provide a more cost-efficient approach, so efforts should be made to discover and establish such processes.

REFERENCES

[1]   Azubuike CC, Chikere CB, Okpokwasili GC. Bioremediation techniques–classification based on site of application: principles, advantages, limitations and prospects. World J Microbiol Biotechnol 2016; 32(11): 180.
[http://dx.doi.org/10.1007/s11274-016-2137-x] [PMID: 27638318]

[2]   Raychoudhury T, Prajapati SK. Bioremediation of pharmaceuticals in water and wastewater. Micro Bior Biod 2020; 425-46.
[http://dx.doi.org/10.1007/978-981-15-1812-6_16]

[3]   Wang Y, Tam NF. Microbial Remediation of organic pollutants. In World Seas: An Environ Eval 2019; 796: 283-303.
[http://dx.doi.org/10.1088/1755-1315/796/1/012012]

[4]   Zouboulis AI, Moussas PA, Psaltou SG. Groundwater and soil pollution: bioremediation. Encyclo of Environ Heal. 2019; pp. 1037-44.
[http://dx.doi.org/10.1016/B978-0-444-52272-6.00035-0]

[5]   Agarwal N, Yadav R S, Solanki V S. The Latest Trends In Bioremediation of Pharmaceutical Contaminants: Limitations and Future Prospects. In: Agarwal N Eds. Pharmaceuticals: Boon or Bane, Nova Science Publishers, Newyork, 2023, pp- 221-244.
[http://dx.doi.org/10.52305/GPMC4427]

[6]   Wasewar KL, Singh S, Kansal SK. Process intensification of treatment of inorganic water pollutants. Inorganic Pollutants in Water 2020; pp. 245-71.
[http://dx.doi.org/10.1016/B978-0-12-818965-8.00013-5]

[7]   Kumar M, Borah P, Devi P. Priority and emerging pollutants in water. Inorganic Pollutants in Water 2020; pp. 33-49.
[http://dx.doi.org/10.1016/B978-0-12-818965-8.00003-2]

[8]   Dar A. NaseerA Recent Applications of Bioremediation and Its Impact. Hazardous Waste Management 2022.

[9]   Kumar A, Bisht BS, Joshi VD, Dhewa T. Review on bioremediation of polluted environment: a management tool. Int J Environ Sci 2011; 1(6): 1079-93.
[http://dx.doi.org/10.4236/ijg.2015.63018]

[10]   Tsuchiya Y. Inorganic chemicals including radioactive materials in waterbodies. Water Quality Stand 2011; 2(7): 172.

[11]   Asghar A, Abdul Raman AA, Wan Daud WMA. Advanced oxidation processes for *in-situ* production of hydrogen peroxide/hydroxyl radical for textile wastewater treatment: a review. J Clean Prod 2015;

87: 826-38.
[http://dx.doi.org/10.1016/j.jclepro.2014.09.010]

[12]   Arola K, Van der Bruggen B, Mänttäri M, Kallioinen M. Treatment options for nanofiltration and reverse osmosis concentrates from municipal wastewater treatment: A review. Crit Rev Environ Sci Technol 2019; 49(22): 2049-116.
[http://dx.doi.org/10.1080/10643389.2019.1594519]

[13]   Vinardell S, Astals S, Mata-Alvarez J, Dosta J. Techno-economic analysis of combining forward osmosis-reverse osmosis and anaerobic membrane bioreactor technologies for municipal wastewater treatment and water production. Bioresour Technol 2020; 297: 122395.
[http://dx.doi.org/10.1016/j.biortech.2019.122395] [PMID: 31761630]

[14]   Galán J, Rodríguez A, Gómez JM, Allen SJ, Walker GM. Reactive dye adsorption onto a novel mesoporous carbon. Chem Eng J 2013; 219: 62-8.
[http://dx.doi.org/10.1016/j.cej.2012.12.073]

[15]   Gupta A, Sharma V, Sharma K, *et al.* A review of adsorbents for heavy metal decontamination: Growing approach to wastewater treatment. Materials 2021; 14(16): 4702.
[http://dx.doi.org/10.3390/ma14164702] [PMID: 34443225]

[16]   Velusamy S, Roy A, Sundaram S, Kumar Mallick T. A review on heavy metal ions and containing dyes removal through graphene oxide-based adsorption strategies for textile wastewater treatment. Chem Rec 2021; 21(7): 1570-610.
[http://dx.doi.org/10.1002/tcr.202000153] [PMID: 33539046]

[17]   Miralles-Cuevas S, Oller I, Agüera A, Llorca M, Sánchez Pérez JA, Malato S. Combination of nanofiltration and ozonation for the remediation of real municipal wastewater effluents: Acute and chronic toxicity assessment. J Hazard Mater 2017; 323(Pt A): 442-51.
[http://dx.doi.org/10.1016/j.jhazmat.2016.03.013] [PMID: 26988902]

[18]   Boczkaj G, Fernandes A. Wastewater treatment by means of advanced oxidation processes at basic pH conditions: A review. Chem Eng J 2017; 320: 608-33.
[http://dx.doi.org/10.1016/j.cej.2017.03.084]

[19]   Leaković S, Mijatović I, Cerjan-Stefanović Š, Hodžić E. Nitrogen removal from fertilizer wastewater by ion exchange. Water Res 2000; 34(1): 185-90.
[http://dx.doi.org/10.1016/S0043-1354(99)00122-0]

[20]   Rengaraj S, Yeon KH, Moon SH. Removal of chromium from water and wastewater by ion exchange resins. J Hazard Mater 2001; 87(1-3): 273-87.
[http://dx.doi.org/10.1016/S0304-3894(01)00291-6] [PMID: 11566415]

[21]   Ibrahim Y, Abdulkarem E, Naddeo V, Banat F, Hasan SW. Synthesis of super hydrophilic cellulose-alpha zirconium phosphate ion exchange membrane *via* surface coating for the removal of heavy metals from wastewater. Sci Total Environ 2019; 690: 167-80.
[http://dx.doi.org/10.1016/j.scitotenv.2019.07.009] [PMID: 31288108]

[22]   Mukred AM, Hamid AA, Hamzah A, Yusoff WMW. Development of three bacteria consortium for the bioremediation of crude petroleum-oil in contaminated water. Online J Biol Sci 2008; 8(4): 73-9.
[http://dx.doi.org/10.3844/ojbsci.2008.73.79]

[23]   Scaffaro R, Lopresti F, Catania V, *et al.* Polycaprolactone-based scaffold for oil-selective sorption and improvement of bacteria activity for bioremediation of polluted water. Eur Polym J 2017; 91: 260-73.
[http://dx.doi.org/10.1016/j.eurpolymj.2017.04.015]

[24]   Catania V, Lopresti F, Cappello S, Scaffaro R, Quatrini P. Innovative, ecofriendly biosorbent-biodegrading biofilms for bioremediation of oil- contaminated water. N Biotechnol 2020; 58: 25-31.
[http://dx.doi.org/10.1016/j.nbt.2020.04.001] [PMID: 32485241]

[25]   Kafilzadeh F, Ebrahimnezhad M, Tahery Y. Isolation and identification of endosulfan-degrading bacteria and evaluation of their bioremediation in kor river, iran. Osong Public Health Res Perspect

2015; 6(1): 39-46.
[http://dx.doi.org/10.1016/j.phrp.2014.12.003] [PMID: 25737830]

[26]   Alegbeleye OO, Opeolu BO, Jackson V. Bioremediation of polycyclic aromatic hydrocarbon (PAH) compounds: (acenaphthene and fluorene) in water using indigenous bacterial species isolated from the Diep and Plankenburg rivers, Western Cape, South Africa. Braz J Microbiol 2017; 48(2): 314-25.
[http://dx.doi.org/10.1016/j.bjm.2016.07.027] [PMID: 27956015]

[27]   Jardine JL, Stoychev S, Mavumengwana V, Ubomba-Jaswa E. Screening of potential bioremediation enzymes from hot spring bacteria using conventional plate assays and liquid chromatography - Tandem mass spectrometry (Lc-Ms/Ms). J Environ Manage 2018; 223: 787-96.
[http://dx.doi.org/10.1016/j.jenvman.2018.06.089] [PMID: 29986326]

[28]   Sun Z, Pang B, Xi J, Hu HY. Screening and characterization of mixotrophic sulfide oxidizing bacteria for odorous surface water bioremediation. Bioresour Technol 2019; 290: 121721.
[http://dx.doi.org/10.1016/j.biortech.2019.121721] [PMID: 31301572]

[29]   Santillan JY, Rojas NL, Ghiringhelli PD, Nóbile ML, Lewkowicz ES, Iribarren AM. Organophosphorus compounds biodegradation by novel bacterial isolates and their potential application in bioremediation of contaminated water. Bioresour Technol 2020; 317: 124003.
[http://dx.doi.org/10.1016/j.biortech.2020.124003] [PMID: 32810733]

[30]   Nanda M, Kumar V, Sharma DK. Multimetal tolerance mechanisms in bacteria: The resistance strategies acquired by bacteria that can be exploited to 'clean-up' heavy metal contaminants from water. Aquat Toxicol 2019; 212: 1-10.
[http://dx.doi.org/10.1016/j.aquatox.2019.04.011] [PMID: 31022608]

[31]   Dey U, Chatterjee S, Mondal NK. Isolation and characterization of arsenic-resistant bacteria and possible application in bioremediation. Biotechnol Rep 2016; 10: 1-7.
[http://dx.doi.org/10.1016/j.btre.2016.02.002] [PMID: 28352518]

[32]   Serrano J, Leiva E. Removal of arsenic using acid/metal-tolerant sulfate reducing bacteria: A new approach for bioremediation of high-arsenic acid mine waters. Water 2017; 9(12): 994.
[http://dx.doi.org/10.3390/w9120994]

[33]   Sher S, Hussain SZ, Rehman A. Phenotypic and genomic analysis of multiple heavy metal–resistant Micrococcus luteus strain AS2 isolated from industrial waste water and its potential use in arsenic bioremediation. Appl Microbiol Biotechnol 2020; 104(5): 2243-54.
[http://dx.doi.org/10.1007/s00253-020-10351-2] [PMID: 31927763]

[34]   Kumar P, Dash B, Suyal DC, *et al.* Characterization of arsenic-resistant klebsiella pneumoniae RnASA11 from contaminated soil and water samples and its bioremediation potential. Curr Microbiol 2021; 78(8): 3258-67.
[http://dx.doi.org/10.1007/s00284-021-02602-w] [PMID: 34230990]

[35]   El-Sheekh MM, Farghl AA, Galal HR, Bayoumi HS. Bioremediation of different types of polluted water using microalgae. Rend Lincei Sci Fis Nat 2016; 27(2): 401-10.
[http://dx.doi.org/10.1007/s12210-015-0495-1]

[36]   Nurhayati I, Ratnawati R, Sugito . Effects of potassium and carbon addition on bacterial algae bioremediation of boezem water. Environ Eng Res 2019; 24(3): 495-500.
[http://dx.doi.org/10.4491/eer.2018.270]

[37]   Thangavelu L, Veeragavan GR. A survey on Nanotechnology-Based Bioremediation of waste water. Bioinorg Chem Appl 2022; 5063177
[http://dx.doi.org/10.1155/2022/5063177]

[38]   Gehrke I, Geiser A, Somborn-Schulz A. Innovations in nanotechnology for water treatment. Nanotechnol Sci Appl 2015; 8: 1-17.
[http://dx.doi.org/10.2147/NSA.S43773] [PMID: 25609931]

[39]   Singh R, Kumar S, Behara M. Nano-bioremediation: An innovative remediation technology for

treatment and management of contaminated sites. Bioremed Indust Waste for Environ Saf 2020; 165-84.
[http://dx.doi.org/10.1007/978-981-13-3426-9_7]

[40]   Bhattacharya S, Saha I, Mukhopadhyay A, Chattopadhyay D, Chand U, Chatterjee D. Role of nanotechnology in water treatment and purification: Potential applications and implications. Int J Chem Sci Technol 2013; 3(3): 59-64.

[41]   Bala S, Garg D, Thirumalesh BV, *et al.* Recent strategies for bioremediation of emerging pollutants: A review for a green and sustainable environment. Toxics 2022; 10: 484.
[http://dx.doi.org/10.3390/toxics10080484]

[42]   Yunus IS, Harwin , Kurniawan A, Adityawarman D, Indarto A. Nanotechnologies in water and air pollution treatment. Environ Technol Rev 2012; 1(1): 136-48.
[http://dx.doi.org/10.1080/21622515.2012.733966]

[43]   Cecchin I, Reddy KR, Thomé A, Tessaro EF, Schnaid F. Nanobioremediation: Integration of nanoparticles and bioremediation for sustainable remediation of chlorinated organic contaminants in soils. Int Biodeterior Biodegradation 2017; 119: 419-28.
[http://dx.doi.org/10.1016/j.ibiod.2016.09.027]

[44]   Rathoure AK. Biostimulation remediation technologies for groundwater contaminants. Advan Environ Engin Green Techno. IGI Global: Pennsylvania, United States 2018.
[http://dx.doi.org/10.4018/978-1-5225-4162-2]

[45]   Kaul I, Sharma JG. Nanotechnology for the bioremediation of organic and inorganic compounds in aquatic ecosystem/marine ecosystem. J Appl Biol Biotechnol 2022; 10(6): 22-33.
[http://dx.doi.org/10.7324/JABB.2022.100603]

[46]   Kalimuthu K, Suresh Babu R, Venkataraman D, Bilal M, Gurunathan S. Biosynthesis of silver nanocrystals by *Bacillus licheniformis.* Colloids Surf B Biointerfaces 2008; 65(1): 150-3.
[http://dx.doi.org/10.1016/j.colsurfb.2008.02.018] [PMID: 18406112]

[47]   Iravani S. Bacteria in Nanoparticle Synthesis: Current Status and Future Prospects. Int Sch Res Notices 2014; 2014: 1-18.
[http://dx.doi.org/10.1155/2014/359316] [PMID: 27355054]

[48]   Salvadori MR, Ando RA, Oller Do Nascimento CA, Corrêa B. Bioremediation from wastewater and extracellular synthesis of copper nanoparticles by the fungus *Trichoderma koningiopsis.* J Environ Sci Health Part A Tox Hazard Subst Environ Eng 2014; 49(11): 1286-95.
[http://dx.doi.org/10.1080/10934529.2014.910067] [PMID: 24967562]

[49]   El-Sheekh MM, El-Kassas HY, Shams El-Din NG, Eissa DI, El-Sherbiny BA. Green synthesis, characterization applications of iron oxide nanoparticles for antialgal and wastewater bioremediation using three brown algae. Int J Phytoremediation 2021; 23(14): 1538-52.
[http://dx.doi.org/10.1080/15226514.2021.1915957] [PMID: 33899605]

[50]   Mahanty S, Chatterjee S, Ghosh S, *et al.* Synergistic approach towards the sustainable management of heavy metals in wastewater using mycosynthesized iron oxide nanoparticles: Biofabrication, adsorptive dynamics and chemometric modeling study. J Water Process Eng 2020; 37: 101426.
[http://dx.doi.org/10.1016/j.jwpe.2020.101426]

[51]   Shakya M, Rene ER, Nancharaiah YV, Lens PNL. Fungal-based nanotechnology for heavy metal removal. Environmental Chemistry for a Sustainable World 2018; 11: 229-53.
[http://dx.doi.org/10.1007/978-3-319-70166-0_7]

[52]   Mohammed YMM, Khedr YI. Applications of *Fusarium solani* YMM20 in bioremediation of heavy metals via enhancing extracellular green synthesis of nanoparticles. Water Environ Res 2021; 93(9): 1600-7.
[http://dx.doi.org/10.1002/wer.1542] [PMID: 33617697]

[53]   Abo-Amer AE, El-Shanshoury AERR, Alzahrani OM. Isolation and Molecular Characterization of

Heavy Metal-Resistant *Alcaligenes faecalis* from Sewage Wastewater and Synthesis of Silver Nanoparticles. Geomicrobiol J 2015; 32(9): 836-45.
[http://dx.doi.org/10.1080/01490451.2015.1010754]

[54]   Devatha CP, S S. Novel application of maghemite nanoparticles coated bacteria for the removal of cadmium from aqueous solution. J Environ Manage 2020; 258: 110038.
[http://dx.doi.org/10.1016/j.jenvman.2019.110038] [PMID: 31929071]

[55]   Qin H, Hu T, Zhai Y, Lu N, Aliyeva J. Sonochemical synthesis of ZnS nanolayers on the surface of microbial cells and their application in the removal of heavy metals. J Hazard Mater 2020; 400: 123161.
[http://dx.doi.org/10.1016/j.jhazmat.2020.123161] [PMID: 32574881]

[56]   Elreedy A, Fujii M, Koyama M, Nakasaki K, Tawfik A. Enhanced fermentative hydrogen production from industrial wastewater using mixed culture bacteria incorporated with iron, nickel, and zinc-based nanoparticles. Water Res 2019; 151: 349-61.
[http://dx.doi.org/10.1016/j.watres.2018.12.043] [PMID: 30616047]

[57]   Priyadarshini E, Priyadarshini SS, Cousins BG, Pradhan N. Metal-Fungus interaction: Review on cellular processes underlying heavy metal detoxification and synthesis of metal nanoparticles. Chemosphere 2021; 274: 129976.
[http://dx.doi.org/10.1016/j.chemosphere.2021.129976] [PMID: 33979913]

[58]   Darwesh OM, Matter IA, Eida MF. Development of peroxidase enzyme immobilized magnetic nanoparticles for bioremediation of textile wastewater dye. J Environ Chem Eng 2019; 7(1): 102805.
[http://dx.doi.org/10.1016/j.jece.2018.11.049]

[59]   Ranjan B, Pillai S, Permaul K, Singh S. Simultaneous removal of heavy metals and cyanate in a wastewater sample using immobilized cyanate hydratase on magnetic-multiwall carbon nanotubes. J Hazard Mater 2019; 363: 73-80.
[http://dx.doi.org/10.1016/j.jhazmat.2018.07.116] [PMID: 30308367]

[60]   Ranjan B, Pillai S, Permaul K, Singh S. A novel strategy for the efficient removal of toxic cyanate by the combinatorial use of recombinant enzymes immobilized on aminosilane modified magnetic nanoparticles. Bioresour Technol 2018; 253: 105-11.
[http://dx.doi.org/10.1016/j.biortech.2017.12.087] [PMID: 29331825]

[61]   Chang Q, Huang J, Ding Y, Tang H. Catalytic oxidation of phenol and 2,4-dichlorophenol by using horseradish peroxidase immobilized on graphene oxide/Fe_3O_4. Molecules 2016; 21(8): 1044.
[http://dx.doi.org/10.3390/molecules21081044] [PMID: 27517896]

[62]   Ali M, Husain Q, Alam N, Ahmad M. Enhanced catalytic activity and stability of ginger peroxidase immobilized on amino-functionalized silica-coated titanium dioxide nanocomposite: A cost-effective tool for bioremediation. Water Air Soil Pollut 2017; 228(1): 22.
[http://dx.doi.org/10.1007/s11270-016-3205-4]

[63]   Ren J, Huo J, Wang Q, *et al.* Characteristics of immobilized dye-decolorizing peroxidase from Bacillus amyloliquefaciens and application to the bioremediation of dyeing effluent. Biochem Eng J 2022; 182: 108430.
[http://dx.doi.org/10.1016/j.bej.2022.108430]

[64]   Nadaroglu H, Mosber G, Gungor AA, Adıguzel G, Adiguzel A. Biodegradation of some azo dyes from wastewater with laccase from *Weissella viridescens* LB37 immobilized on magnetic chitosan nanoparticles. J Water Process Eng 2019; 31: 100866.
[http://dx.doi.org/10.1016/j.jwpe.2019.100866]

[65]   Habimana P, Jiang Y, Gao J, *et al.* Enhancing laccase stability and activity for dyes decolorization using ZIF-8@MWCNT nanocomposite. Chin J Chem Eng 2022; 48: 66-75.
[http://dx.doi.org/10.1016/j.cjche.2021.05.044]

[66]   Wong JKH, Tan HK, Lau SY, Yap PS, Danquah MK. Potential and challenges of enzyme incorporated nanotechnology in dye wastewater treatment: A review. J Environ Chem Eng 2019; 7(4):

103261.
[http://dx.doi.org/10.1016/j.jece.2019.103261]

[67]   Karthik V, Senthil Kumar P, Vo DVN, *et al.* Enzyme-loaded nanoparticles for the degradation of wastewater contaminants: a review. Environ Chem Lett 2021; 19(3): 2331-50.
[http://dx.doi.org/10.1007/s10311-020-01158-8]

[68]   Gao XJ, Fan XJ, Chen XP. Immobilized b-lactamase on Fe$_3$O$_4$ magnetic nanoparticles for degradation of β-lactam antibiotics in wastewater. Int J Environ Sci Technol. 2017; pp. 1-10.
[http://dx.doi.org/10.1007/s13762-017-1596-4]

[69]   Ji C, Nguyen LN, Hou J, Hai FI, Chen V. Direct immobilization of laccase on titania nanoparticles from crude enzyme extracts of P ostreatus culture for micro-pollutant degradation. Seperat Purificat Technol 2017; pp. 1-39.
[http://dx.doi.org/10.1016/j.seppur.2017.01.043]

[70]   Penaranda PA, Noguera MJ, Florez SL, *et al.* Treatment of wastewater, phenols and dyes using novel magnetic torus microreactors and laccase immobilized on magnetite nanoparticles Nanomaterials 2022; 12(10): 1688.
[http://dx.doi.org/10.3390/nano12101688]

[71]   Bundschuh M, Seitz F, Rosenfeldt RR, Schulz R. Effects of nanoparticles in fresh waters: risks, mechanisms and interactions. Freshw Biol 2016; 61(12): 2185-96.
[http://dx.doi.org/10.1111/fwb.12701]

[72]   Agarwal N, Solanki VS, Gacem A, *et al.* Bacterial laccases as biocatalysts for the remediation of environmental toxic pollutants: A green and Eco-friendly approach—A review. Water 2022; 14(24): 4068.
[http://dx.doi.org/10.3390/w14244068]

[73]   Wang F, Hu Y, Guo C, Huang W, Liu CZ. Enhanced phenol degradation in coking wastewater by immobilized laccase on magnetic mesoporous silica nanoparticles in a magnetically stabilized fluidized bed. Bioresour Technol 2012; 110: 120-4.
[http://dx.doi.org/10.1016/j.biortech.2012.01.184] [PMID: 22382292]

[74]   Karakoti AS, Hench LL, Seal S. The potential toxicity of nanomaterials—The role of surfaces. J Miner Met Mater Soc 2006; 58(7): 77-82.
[http://dx.doi.org/10.1007/s11837-006-0147-0]

[75]   Ganguly P, Breen A, Pillai SC. Toxicity of nanomaterials: Exposure, pathways, assessment, and recent advances. ACS Biomater Sci Eng 2018; 4(7): 2237-75.
[http://dx.doi.org/10.1021/acsbiomaterials.8b00068] [PMID: 33435097]

[76]   Jain K, Patel AS, Pardhi VP, Flora SJS. Nanotechnology in wastewater management: A new paradigm towards wastewater treatment. Molecules 2021; 26(6): 1797.
[http://dx.doi.org/10.3390/molecules26061797] [PMID: 33806788]

[77]   Takahashi Y, Ishida I, Isoda K. Cerium (IV) oxide nanoparticles enhance hepatotoxic and nephrotoxic effects of paraquat, cisplatin, or acetaminophen in mice. BPB Reports 2023; 6(2): 33-6.
[http://dx.doi.org/10.1248/bpbreports.6.2_33] [PMID: 32878638]

[78]   Cimbaluk GV, Ramsdorf WA, Perussolo MC, *et al.* Evaluation of multiwalled carbon nanotubes toxicity in two fish species. Ecotoxicol Environ Saf 2018; 150: 215-23.
[http://dx.doi.org/10.1016/j.ecoenv.2017.12.034] [PMID: 29287268]

[79]   Khan MS, Qureshi NA, Jabeen F. Assessment of toxicity in fresh water fish *Labeo rohita* treated with silver nanoparticles. Appl Nanosci 2017; 7(5): 167-79.
[http://dx.doi.org/10.1007/s13204-017-0559-x]

[80]   Mohanta D, Patnaik S, Sood S, Das N. Carbon nanotubes: Evaluation of toxicity at biointerfaces. J Pharm Anal 2019; 9(5): 293-300.
[http://dx.doi.org/10.1016/j.jpha.2019.04.003] [PMID: 31929938]

CHAPTER 11

Current Trends in Biogenic Synthesis and Applications of Palladium Nanoparticles: A Sustainable Approach to Environmental Remediation

Gitanjali Arora[1], Anamika Srivastava[1,*], Manish Srivastava[2], Jaya Dwivedi[1], Shruti[1] and Rajendra[1]

[1] *Department of Chemistry, Banasthali Vidyapith, Banasthali, Rajasthan, India*

[2] *Department of Chemistry, University of Allahabad, Prayagraj, Uttar Pradesh, India*

Abstract: Nanotechnology is a multidisciplinary area with a wide range of applications. Recent developments in nanotechnology and nanoscience have also triggered the development of new nanomaterials (NMs), enhancing the hazards to human health and the environment. There has been a rise in interest in creating ecologically friendly techniques for producing metallic nanoparticles (MNPs). The aim is to reduce the harmful effects of synthetic technologies, the chemicals used in association with them, and the derivative products. A useful strategy in green nanotechnology is the utilization of various biomolecules for the fabrication of NPs. MNPs that are inexpensive, energy-efficient, nontoxic, and beneficial to the environment have been produced using biological resources, including bacteria, algae, fungi, and plants. Plant components are mainly employed as capping and reducing agents in green synthesis. MNPs of various sizes and forms have been created using bark, leaves, fruits, and flower extracts. In this chapter, we have addressed the green synthesis of palladium NPs to remove positive ions, negative ions, and dye from wastewater, their potential applications and the directions for future research.

Keywords: Application of nanoparticles, Dye removal, Green synthesis, Ions removal, Leaf extracts, Palladium nanoparticles, Wastewater treatment.

GREEN NANOTECHNOLOGY FOR SUSTAINABLE DEVELOPMENT

Water microbiological contamination and purification are major global issues, and the majority of commonly used treatments have a variety of drawbacks, such as being expensive and hazardous to the environment. Therefore, to solve the problem of water filtration, new methods and materials are urgently needed.

---

* **Corresponding author Anamika Srivastava:** Department of Chemistry, Banasthali Vidyapith, Banasthali, Rajasthan, India; E-mail: anamika.chemistry2011@gmail.com

Neha Agarwal, Vijendra Singh Solanki and Sreekantha B. Jonnalagadda (Eds.)

Numerous industries, including those in the leather, textile, agricultural, pharmaceutical, plastic, paper, wood, and cosmetics sectors, have drastically increased their production of organic pollutants and dye-containing wastewater.

According to claims, the textile sector is the biggest wastewater source containing wastewater that contains highly concentrated dyes (10–200 mg/L) [1 - 3]. Wastewater discharged into an aquatic environment containing these cancer-causing chromophores can limit light penetration in aqueous systems, hindering photosynthesis, affecting human health, and destroying aquatic living particles.

Therefore, more study is required to eliminate these toxins. Although several techniques for treating colored wastewater (such as activated carbon-based adsorption, biological techniques, coagulation-flocculation, chemical oxidation, *etc.*) have previously been industrialized, their shortcomings prevent their widespread use [4 - 8]. Chemical approaches, for instance, are expensive, ineffective at removing pollutants, and produce secondary pollutants in the form of sludge. Developing suitable, economical, and environmentally acceptable wastewater treatment methods that do not produce secondary pollutants is now necessary. Due to their strong reactivity, substantial surface area, and efficiency, MNPs have become a popular choice for use as catalysts [9 - 11].

Usually, MNPs are formed by employing a potent alkaline substance (reducing agent), such as sodium borohydride or sodium hydroxide, to chemically reduce metal ions found in salt solutions, followed by the addition of a stabilizing agent. However, the chemicals utilized as reducing agents and the solvents used to dissolve the stabilizers are frequently hazardous materials that might harm human health and the environment if residues are present in the finished nanosystems. As a result, this may give rise to several issues surrounding the safety of MNP applications as well as the search for creative new ways aimed at their synthesis. One of these options is the use of biological systems in the green synthesis of MNPs, which is based on the ideas of green chemistry (Fig. **1**) [12 - 16].

Prokaryotic or eukaryotic organisms (including microbes, plants, and animals) or components thereof may be used in the green synthesis of MNPs, which can take place through intracellular or extracellular routes. To encourage the reduction of a target metal ion and the creation of MNPs, the biological components function as agents. The surface of MNPs may develop a stabilizing layer (coating) from the same reducing substances or other nearby molecules, avoiding or at least minimizing the formation of agglomerates or disordered growth. Most MNPs made using green synthesis techniques exhibit traits that are desirable from a sustainability perspective, including being eco-friendly (using less toxic chemicals and solvents), quick and easy to make (fewer steps), biocompatible, biodegradable, low cost of production, and high yield [17 - 20]. Important examples of biosynthesis of MNPs by plants are summarized in Table **1**.

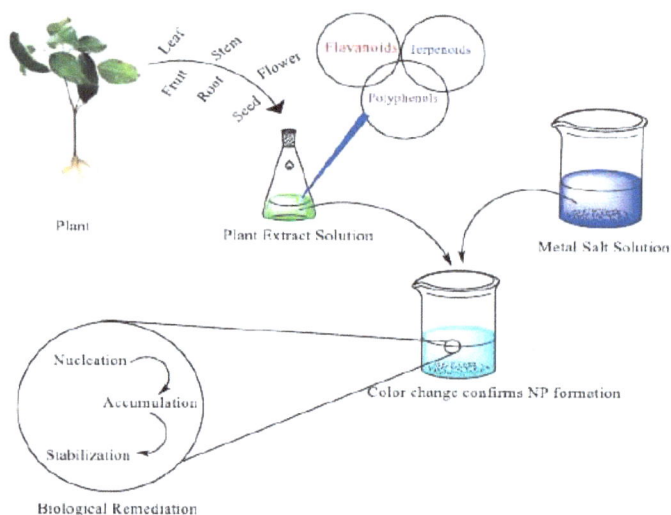

Fig. (1). Green chemistry involved in biological remediation.

Table 1. Important examples of biosynthesis of MNPs by Plants.

Plant Origin	NPs	Size(nm)	Morphology	Refs.
Aloe-vera	Au & Ag	—	Spherical, triangular	[28]
Acalypha indica	Ag	5-50	Spherical	[29]
Anogeissus latifolia Gum	Pd	4.8	Spherical	[30]
Apiin (from henna leaves)	Ag	39	Spherical, triangular	[31]
Apiin (from henna leaves)	Au	7.5-65	Quasi-spherical	[32]
Black tea leaf extract	Au & Ag	20	Spherical, prism	[33]
Brassica juncea (mustard)	Ag	2-35	Spherical	[34]
Camellia sinensis (green tea)	Au	40	Spherical, triangular, irregular	[35]
Carica papaya	Ag	60-80	Spherical	[36]
Chenopodium album	Au & Ag	10-30	Quasi-spherical	[37]
Cinnamomum camphora	Pd	3.2-6	—	[38]
Cinnamomum zeylanicum Bark	Pd	15-20	Crystalline	[39]
Citrus limon (lemon)	Ag	<50	Spherical, spheroidal	[40]
Coriandrum sativum (coriander)	Au	6.75-57.91	Spherical, triangular, truncated triangular, decahedral	[41]
Curcuma longa Tuber	Pd	10-15	Spherical	[42]
Cymbopogon flexuosus (lemongrass)	Au	200-500	Spherical, triangular	[43]

(Table 1) cont.....

Plant Origin	NPs	Size(nm)	Morphology	Refs.
Datura metel	Ag	16-40	Spherical, ellipsoidal	[44]
Eclipta sp.	Ag	2-6	Spherical	[45]
Emblica officinalis (Indian gooseberry)	Au &Ag	(10-20) &(15-25)	–	[46]
Eucalyptus citriodora (neelagiri)	Ag	~20	Spherical	[47]
Eucalyptus hybrida (safeda)	Ag	50-150	Crystalline, spherical	[48]
Euphorbia granulate leaves	Pd	25-35	–	[49]
Gardenia jasminoides Ellis	Pd	3-5	–	[50]
Glycine max leaves	Pd	15	Spherical	[51]
Hibiscus rosa sinensis	Au & Ag	14	Spherical, prism	[52]
Honey	Ag	4	Spherical	[53]
Medicago sativa (alfalfa)	Ti-Ni alloys	1-4	FCC-like geometry for the smallest clusters and complex arrays for the biggest one	[54]
Medicago sativa (alfalfa)	FeO	2-10	Crystalline	[55]
Mentha piperita (peppermint)	Ag	5-30	Spherical	[56]
Moringa oleifera	Ag	57	Spherical	[57]
Moringa oleifera Waste petal	Pd	10-50	Spherical	[58]
Moringa oleifera Peel extract	Pd	27±2	Spherical	[59]
Musa paradisiaca (banana peel extract)	Ag	–	Crystalline, irregular	[60]
Nelumbo nucifera (lotus)	Ag	25-80	Spherical, triangular, truncated triangular, decahedral	[61]
Ocimum sanctum (Tulsi; leaf extract)	Ag	10-20	Spherical	[62]
Pear fruit extract	Au	200-500	Triangular, hexagonal	[63]
Pelargonium graveolens (geranium)	Au	20-40	Decahedral, icosahedral	[64]
Physalis alkekengi	ZnO	72.5	Crystalline	[65]
Pinus resinosa Bark	Pd	16-20	Crystalline	[66]
Psidium guajava (guava)	Au	25-30	Mostly spherical	[67]
Pulicaria glutinosa Whole plant	Pd	20-25	Crystalline, spherical	[68]
Sedum alfredii Hance	ZnO	53.7	Hexagonal wurtzite and pseudo-spherical	[69]
Starch	Pd	1.5-4.5	Spherical	[70]
Syzygium aromaticum (clove)	Au	5-100	Crystalline, spherical, irregular, elliptical	[71]
Tamarindus indica (tamarind)	Au	20-40	Triangular	[72]

(Table 1) cont.....

Plant Origin	NPs	Size(nm)	Morphology	Refs.
Terminalia catappa (almond)	Au	10-35	Spherical	[73]

In comparison to other MNPs, palladium nanoparticle (PdNPs) catalysts have recently been used more frequently to reductively convert several refractory pollutants when there is an abundance of hydrogen/electron donors. PdNPs are used in both heterogeneous and homogeneous catalysis, and due to their substantial surface-to-volume proportion and high surface vitality, their importance in these applications has expanded significantly [21]. PdNPs are produced through electrochemical, chemical, laser pulse ablation, and sonochemical reduction techniques. New synthetic methods must be developed to create PdNPs with controlled size and shape for a wide range of future applications because chemical methods result in a harsh reaction and reduce the palladium's catalytic activity [22, 23].

Physical, chemical, enzymatic, and biological techniques can all be used to reduce metal ions into NPs. The high irradiation and concentrated reducing agents used in physical and chemical processes harm the environment and human health. The enzymatic method of producing NPs is safer but more expensive. Due to a growing desire to find environmentally acceptable methods for synthesizing nanomaterials, the employment of biological systems has emerged as an innovative and trustworthy technique for the manufacture of NPs in the most recent decade [24 - 26]. The manufacture of MNPs using microscopic organisms and plants has received a lot of attention. Plant-assisted NP synthesis has the benefit that its kinetics is much greater than those of other biosynthetic methods that are comparable to chemical NP creation [27].

GREEN SYNTHESIS OF PDNPS BY PLANTS

Plants are regarded as nature's efficient and affordable chemical production facilities. Because even tiny quantities of heavy metals can be harmful, even at extremely low levels, plants possess remarkable capabilities for both detoxifying and accumulating heavy metals. This potential of plants offers a promising solution to environmental pollution issues associated with heavy metals. Plants can reduce metal ions, which causes them to build up and deposit as NPs. As a result of variations in their penetration and localization of metal ions, entire plants develop NPs that vary in size and form, which alters their reducing property and prevents their usage in situations where selectivity and finely tuned structures are necessary. To get flexible control over the development and refinement of NPs with desired size and shape, this issue may be solved by using plant extracts rather than complete plants. The least expensive way to deal with the manufacture of environmentally acceptable NPs is a single-step bioreduction process employing plant extracts. Plant-based nanomaterial production has received a lot of attention due to its abundant biodiversity and accessibility. The reduction of metals in their ionic state into environmentally acceptable MNPs is known to be facilitated by unique primary as well as secondary metabolites found in plant crude extract [39, 32]. The use of plant extract-derived NPs in the food, beverage, wastewater treatment, and cosmetic sectors is widespread. Numerous studies have shown the effectiveness of plant-derived NPs as antibacterial, larvicidal, and cytotoxic agents. According to this research, a simple and efficient substitute for physical and chemical procedures is the biosynthetic process using plant sources. Even though recent research has reported on the production of PdNPs utilizing leaf extracts from a variety of plants, including *Prunus yedoensis, Hippophae rhamnoides, Sapium sebiferum, Camellia sinensis,* and *Catharanthus roseus*, the potential of plant sources has not yet been fully recognized. The evergreen tree *Filicium decipiens* is a member of the Sapindaceae family. It can grow in every soil type and is sometimes referred to as a fern tree. The plant's leaves are complex and have 12–16 leaflets per leaf. This tree's leaves are said to contain significant amounts of saponin and glycosides. Additionally, it functions as an anti-inflammatory, anti-microbial, and antioxidant. In this investigation, *F. decipiens* leaf extract was utilized as a bio-reducing agent for the manufacture of PdNPs [55, 59, 60].

CURRENT TRENDS IN GREEN SYNTHESIS

Researchers investigating MNPs can have a better understanding of the molecular mechanisms controlling nucleation, bioreduction, growth, and stabilization by employing green synthesis, a modern nanobiotechnology area that uses plants as a catalyst. Initial research efforts relied on the use of plant extracts chosen experimentally from local or worldwide biodiversity. MNPs were created in those experiments utilizing extracts from various plant species and components in the presence of metal salts, resulting in structures with various compositions, forms, sizes, and activities. On the reduction of noble metals like silver (Ag), gold (Au), palladium (Pd), copper (Cu), and platinum (Pt), research addressing the green synthesis of MNPs utilizing plants was conducted. Spheres and triangles are the MNP shapes that are typically seen in the green synthesis process employing plants.

Researchers can also change a variety of additional synthesis parameters in addition to the wide range of nanoscale properties that MNPs made utilizing various plant extracts can exhibit. Some of the adjustable factors that assist researchers in optimizing the effectiveness and speed of the green synthesis of MNPs while even resembling the usual aspects of traditional chemical synthesis include the concentration of extracts and/or metal ions, reaction time, pH, and temperature. It is important to note the significance of varying the synthesis parameters to produce MNPs with the appropriate properties [74]. Additionally, they provide insight into the MNPs' formation processes. In vivo, green synthesis of MNPs can also take place in plants. It is known that plants use phytochelatins and secondary metabolites to bind and complex with metal ions to protect themselves from the stress caused by their presence. When synthesizing MNPs, this method is more expensive and takes longer than using plant extracts [75, 76].

METHODS OF SYNTHESIS OF NPS

Compared to those generated by other organisms, the MNPs produced through plants are more stable and possess the ability to decrease metal ions more quickly than fungus or bacteria [74]. Plant extracts are essentially preferable to plant biomass or living plants for the industrial manufacture of well-dispersed MNPs because they are quick, safe, and environmentally friendly (Fig. **2**).

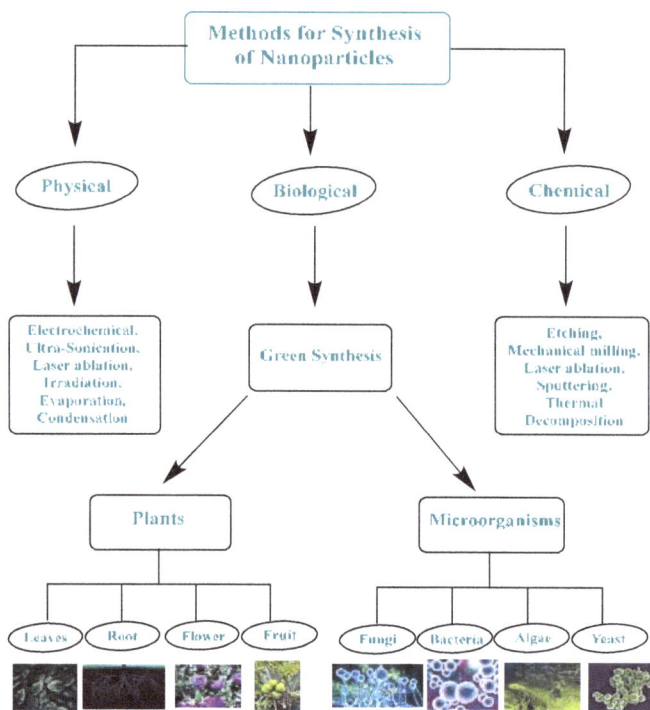

Fig. (2). Green synthesis of MNPs.

Researchers have focused their efforts on discovering and characterizing the biomolecules involved in the production of MNPs as well as studying the biochemical and enzymatic mechanisms of NP formation. In plants, a variety of biomolecules, including proteins and enzymes, amino acids, polysaccharides, alkaloids, alcoholic compounds, and vitamins, may play a role in the bioreduction, synthesis, and stabilization of MNPs. The synthesis of NPs is significantly influenced by the reduction potential of ions and the reducing capacity of plants, which depend on the availability of enzymes, polyphenols, and other chelating agents in plants. The improvement of reaction conditions and the development of recombinant organisms to produce large quantities of enzymes, proteins, and biomolecules involved in the biosynthesis and stabilization of NPs should be observed [72, 68].

Synthesis of PdNPs using Leaf Extracts

Several researchers have become intrigued by plant-derived NPs due to their environmentally friendly qualities, durability, and economical characteristics, all of which are reasons why palladium, an expensive metal with a high density, has captured their attention. Numerous plant extracts, including *Cinnamomum camphora, Gymnema Sylvestre, Gardenia jasminoides, Filicium decipiens, Pinus resinosa, Sepium sebiferum, Anogeissus latifolia, Ocimun sanctum, Glycine max, Doipyros kaki, Curcuma longa, Cinnamom zeylanicum, Musa paradisica, Pulicaria glutinosa*, and many more, have been used in the green synthesis of PdNPs [42 - 49]. Sequential steps involved in the creation, analysis, and applications of environmentally friendly PdNPs are mentioned in Fig. (**3**).

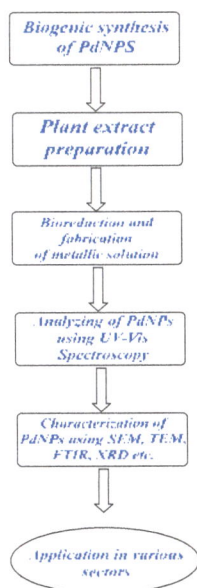

Fig. (3). A graphical representation of sequential steps involved in the creation, analysis, and uses of environmentally friendly PdNPs.

Synthesis of PdNPs using Sapium sebiferum

To extract phytoconstituents from *Sapium sebiferum* leaves, properly cleaned leaves are kept in deionized water to get rid of any dust. This plant's leaves are ground into powder using a grinder. Then, using 10 g of this powder, 100 mL of distilled water is heated at 70°C for two hours. Leaf extract is then filtered using Whatman Filter Paper No. 3. The extract is then centrifuged (at a speed of 1000 rpm) at 4°C to eliminate any remaining solid contaminants. Then, Sapium sebiferum leaf extract and 0.003 mM palladium chloride ($PdCl_2$) are combined in a 100 mL beaker, and the mixture is magnetically stirred until the color changes from light brown to dark brown. The Pd NPs are then produced, and the mixture is dried in a 6ES freeze dryer.

Application

Without utilizing any risky surfactants, the aqueous leaf extract of *Sapium sebiferum* is employed as a reducing cum stabilizing agent for the manufacture of PdNPs. Using varied concentrations of *Sapium sebiferum* leaf extract at various temperatures, the size and shape of the PdNPs can be successfully controlled. Significant antibacterial action had been demonstrated by the PdNPs synthesized under optimal circumstances against *S. aureus* and *B. subtilis*, but only modest antibacterial activity was seen against *P. aeruginosa*. The size-dependent photocatalytic reduction of Methylene blue using biologically synthesized PdNPs has also been examined. According to the research results, the PdNPs synthesized under optimal circumstances had substantial photocatalytic activity and only needed 70 minutes for the reduction of Methylene blue. Additionally, the greenly produced PdNPs can serve as a substitute in the future for biological and catalytic applications [77].

Synthesis of PdNPs using Filicium decipiens

Fresh *Filicium decipiens* leaves were obtained, thoroughly cleansed with tap water, and allowed to dry for ten days to eliminate water content. A household blender was used to finely grind the dried leaves. To make the leaf extract, 5 g of leaf powder was combined with 50 ml of double-distilled water and stored in a water bath for 10 minutes. The Whatman filter paper was used to filter the extract, which was subsequently preserved for future use. Then, 90 ml of 1 mM $PdCl_2$ solution was combined with 10 ml of water-based F. decipiens extract, and the mixture was left at room temperature for 4 days. The results from the UV spectrophotometer at the wavelength range of 400-900 nm were recorded to monitor the synthesis of PdNPs. To get PdNPs, it was centrifuged and dried after that.

Application

The conversion of Pd^{2+} ions to PdNPs is caused by the carboxyl, amine, carbonyl, and amide groups found in the leaf extract. Gram-positive and gram-negative bacteria are both significantly resistant to the antibacterial effects of PdNPs produced by F. decipiens. According to the study's findings, F. decipiens leaf extract is a low-cost, effective source for the synthesis of PdNPs. Additionally, produced PdNPs may be employed as a powerful antibacterial agent in biological applications [78].

Synthesis of PdNPs using Gymnema sylvestre

Gymnema Sylvestre (GSE) leaves that had been harvested are cleaned properly with distilled water to eliminate any surface dirt, and they are then air-dried in the shade at room temperature. The pestle and mortar are employed to grind leaves. 100 mL of deionized water and 10 g of powdered leaves are extracted for 5 minutes at 70 °C. Following the use of Whatman filter paper no. 1, muslin cloth is used to filter the mixture. For further usage, this extract (GSE) is stored in sterile vials and refrigerated at 3-5 °C. The 100 mL Pyrex beaker is utilized as the reactor for the PdNP synthesis. For the synthesis of PdNPs, a reaction system with a volume of 10 mL is set up. The necessary amount of $PdCl_2$ is added to the various GSE volumes (1, 2, 3, 4, and 5 mL) produced in deionized distilled water without the use of any other external reducing agent. For five minutes, the mixture is heated in a 350 W microwave. The resulting formation of PdNPs is confirmed by the reaction's shift in optical characteristics from yellow to dark brown. After boiling the resulting mixture consisting of GSE and $PdCl_2$ in the microwave for 5 minutes, precipitate starts to develop, which suggests the formation of GSE-PdNPs. By performing centrifugation at 10,000 rpm for 10 min, the synthesized GSE-PdNPs are purified. To get rid of any remaining unreacted GSE, the pellet is further cleaned with deionized water. Overnight drying of purified PdNPs prepares them for further characterization and catalytic usage. The synthesis of PdNPs using Gymnema Sylvestre leaves is shown in Fig. (**4**).

Application

The reduction of toxic chromium (VI) to its non-toxic chromium (III) state is achieved by utilizing formic acid as a reducing agent, biosynthesized GSE-PdNPs as a catalyst, and $K_2Cr_2O_7$ as a source of hazardous chromium. Various influential factors, including the quantity of HCOOH, initial concentration of Cr (VI), and GSE-PdNPs dosage, are investigated to optimize the catalytic efficiency of GSE-PdNPs in converting Cr (VI) to Cr (III) in a water-based solution. The reaction system is maintained at room temperature without agitation throughout the experiment. At 15-minute intervals, a 3 ml sample is extracted from the system and analyzed using UV-visible spectrophotometry in the wavelength range of 200-600 nm [79].

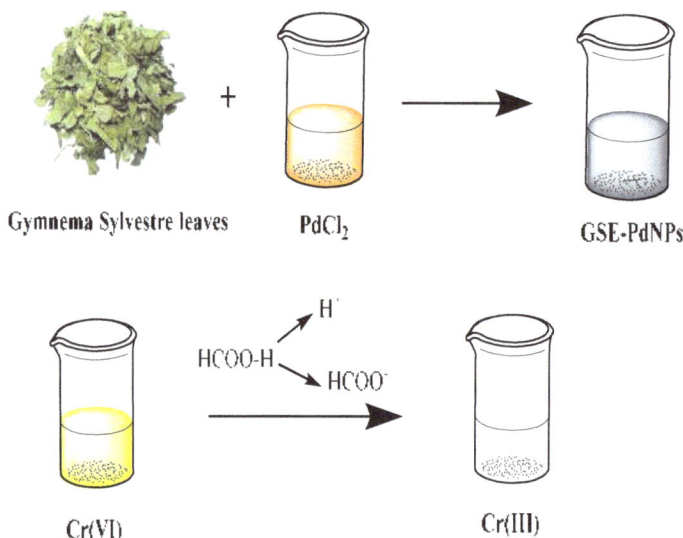

Fig. (4). Synthesis of PdNPs using *Gymnema Sylvestre* leaves.

Synthesis of PdNPs using Catharanthus roseus

When an aqueous solution of $[Pd(OAc)_2]$ is stirred for one hour at 60 °C with a methanol-based extract of Catharanthus roseus, which comprises a blend of eight compounds with -OH groups, is responsible for reducing the metal ion to MNPs. The color of the solution gets changed, which confirms the formation of Pd NPs and shows the absorption peak in the 360-400 nm range [80] (Fig. **5**).

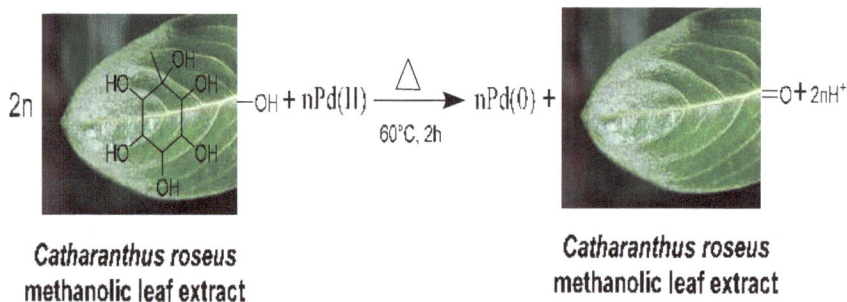

Fig. (5). Synthesis of PdNPs using *Catharanthus roseus* methanol-based extract and Pd ion.

Synthesis of PdNPs using Santalum album

The leaves of the santalum album are harvested, rinsed with water, and allowed to dry for a week. To prepare extracts, powdered dried leaves are employed. 5 g of leaf powder is dissolved in 50 mL of distilled water and maintained in a water bath at 60 °C for 10 minutes while being constantly stirred. The Whatman filter paper is used to filter the solution, which is then saved for later use. To synthesize PdNPs 90 ml of PdCl2 solution is mixed with 10 ml of aqueous S. album extract, and the mixture is left at room temperature for 4 days. A UV-spectrophotometer at wavelengths between 400 and 900 nm is used to monitor the PdNPs fabrication. To extract PdNPs, it is centrifuged and dried afterward.

Application

To determine the catalytic activity of phytosynthesized PdNPs, 4-nitrophenol (4-NP) was reduced catalytically to 4-aminophenol (4-AP) in the presence of excess NaBH$_4$. Due to the production of 4-nitrophenlate ions, the pale yellow color of 4-NP is transformed into vibrant yellow by the addition of NaBH$_4$ solution. For 30 minutes in the absence of the catalyst, the nitrophenolate ions displayed absorbance at 400 nm, but no color change was noticed. PdNPs, the catalyst, are then added to the solution, well mixed, and UV-Vis measurements are taken every two minutes. At 400 nm, 4-NP absorbance gradually decreased, and a new peak for 4-AP was developed at 300 nm. After 20 minutes, 96.8% of the bright yellow (4-NP) was entirely transformed into colorless (4-AP). PdNPs increase the rate of electron transport on its surface between donor (BH) and acceptor (4-NP). The kinetic barrier is reduced, and the reaction is promoted more quickly by PdNPs. PdNPs produced from S. album leaf extract have additionally demonstrated superior antibacterial properties against gram-negative bacteria compared to gram-positive bacteria. Greenly synthesized PdNPs' catalytic and antibacterial effects can be helpful in medical and pharmaceutical applications [81].

Synthesis of PdNPs using Chrysophyllum cainito

Chrysophyllum Cainito foliage essence is employed as a source of reducing species in a one-step nanopalladium production process. At normal temperatures, the process of making nanopalladium takes around 2-3 hours, while at higher temperatures (70–80 °C), the process of making colloidal nanopalladium is finished right away. The nanopalladium synthesized was employed as an eco-conscious nanocatalyst in aqueous settings, enabling C-C coupling reactions without the need for phosphine or oxygen-free conditions. Furthermore, the resulting nanopalladium is utilized as a sustainable nanocatalyst for the reduction of 3- and 4-nitrophenols using sodium borohydride. The nanopalladium exhibited enhanced and prolonged catalytic effectiveness, demonstrating its remarkable potential [82].

ENVIRONMENTAL APPLICATION OF GREEN SYNTHESIZED NPS

NPs are now in demand on the commercial market as a result of their broad spectrum of uses in industries, medical science disciplines, digital devices, the commercial sector, renewable resources, and particularly in chemistry. Both the developing multidisciplinary science of nanotechnology and biomedical fields are very much interested in NPs. PdNPs are often employed in several medical diagnoses without damaging the structure of deoxyribose nucleic acid (DNA). At pH 6, certain PdNPs are effective in photocatalytic activity for the breakdown of phenol red dye. Studies on dye degradation were conducted using different aliquots of PdNP solutions with pH values ranging from 2 to 10 [83, 84].

Hippophae rhamnoides Linn leaf extract was used to create the PdNPs, which were then investigated for heterogeneous catalytic activity in the Suzuki-Miyaura coupling process. The catalyst in the Suzuki-Miyaura coupling process is made of PdNPs (Fig. **6**). After the reaction is finished, the catalyst may be readily extracted from the reaction mixture by centrifugation, which lowers the cost of the procedure. For four new runs, the recovered catalyst was employed effectively with no apparent activity loss. By evaluating the leaching processes using inductively coupled plasma atomic emission spectroscopy, the heterogeneity of the catalyst was investigated. Only 0.2% of palladium was destroyed during the reaction.

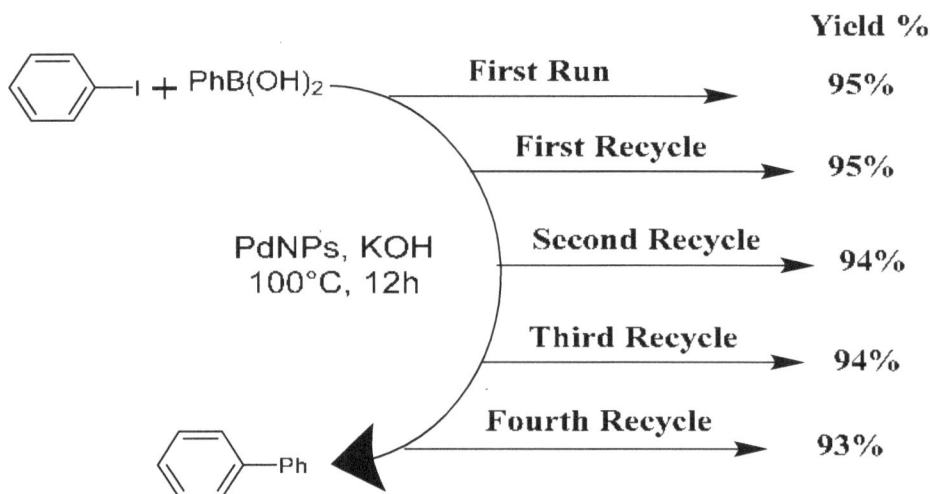

Fig. (6). Pd NP reusability in the Suzuki-Miyaura coupling process and the catalyst's yield % after each run.

In addition to acting as nanocatalysts for environmental remediation by demonstrating catalytic activity in the reduction of dyes like methylene blue, methyl orange, coomassie brilliant blue G-250, and reduction of 4-nitrophenol, some PdNPs also demonstrated exceptional antioxidant properties at a lower dose of the NPs [85, 86].

CONCLUSION AND FUTURE RECOMMENDATIONS

In recent decades, there has been a rising interest in combining green chemistry with nanotechnology, leading to a demand for environmentally friendly synthetic approaches to produce nanomaterials using plants, microorganisms, and other living organisms. Scientists have dedicated their efforts to developing eco-friendly methods for synthesizing NPs, with a particular focus on utilizing plant extracts. The use of plant extract-mediated NPs has gained significant attention in research due to their cost-effectiveness, non-toxic nature, easy availability, and environmental friendliness. These NPs hold great potential across various fields such as catalysis, biotechnology, water treatment, medicine, electronics, optics, degradation of dyes, and bioengineering. Moreover, plants possess unique compounds that aid in the synthesis process and enhance its speed. The field of green synthesis of NPs through plants is an exciting and rapidly expanding area in nanotechnology, contributing to environmental sustainability and future advancements in nanoscience. While the application possibilities of these NPs are exponentially growing, there are concerns about their long-term effects on human health, animal well-being, and the environment, which will require careful attention in the future. Biogenic NPs can be utilized in numerous ways, including water sterilization for environmental cleanup and the development of nano-weapons to combat plant pathogens. These NPs have the potential to drive innovation in the biomedical industry, particularly in the field of drug delivery systems.

REFERENCES

[1]   Allen EW. Process water treatment in Canada's oil sands industry: II. A review of emerging technologies. J Environ Eng Sci 2008; 7(5): 499-524.
[http://dx.doi.org/10.1139/S08-020]

[2]   Wang L, Hou D, Cao Y, *et al.* Remediation of mercury contaminated soil, water, and air: A review of emerging materials and innovative technologies. Environ Int 2020; 134: 105281.
[http://dx.doi.org/10.1016/j.envint.2019.105281] [PMID: 31726360]

[3]   Ajith MP, Aswathi M, Priyadarshini E, Rajamani P. Recent innovations of nanotechnology in water treatment: A comprehensive review. Bioresour Technol 2021; 342: 126000.
[http://dx.doi.org/10.1016/j.biortech.2021.126000] [PMID: 34587582]

[4]   Hu A, Apblett A, Eds. Nanotechnology for water treatment and purification. Switzerland: Springer International Publishing 2014; 22.
[http://dx.doi.org/10.1007/978-3-319-06578-6]

[5]   Russo T, Fucile P, Giacometti R, Sannino F. Sustainable removal of contaminants by biopolymers: a novel approach for wastewater treatment. Current state and future perspectives. Processes 2021; 9(4): 719.

[http://dx.doi.org/10.3390/pr9040719]

[6]   Lorenzo M, Picó Y. Wastewater-based epidemiology: current status and future prospects. Curr Opin Environ Sci Health 2019; 9: 77-84.
[http://dx.doi.org/10.1016/j.coesh.2019.05.007]

[7]   Shabir M, Yasin M, Hussain M, *et al.* A review on recent advances in the treatment of dye-polluted wastewater. J Ind Eng Chem 2022; 112: 1-19.
[http://dx.doi.org/10.1016/j.jiec.2022.05.013]

[8]   Priya AK, Pachaiappan R, Kumar PS, Jalil AA, Vo DVN, Rajendran S. The war using microbes: A sustainable approach for wastewater management. Environ Pollut 2021; 275: 116598.
[http://dx.doi.org/10.1016/j.envpol.2021.116598] [PMID: 33581625]

[9]   Albukhari SM, Ismail M, Akhtar K, Danish EY. Catalytic reduction of nitrophenols and dyes using silver nanoparticles @ cellulose polymer paper for the resolution of waste water treatment challenges. Colloids Surf A Physicochem Eng Asp 2019; 577: 548-61.
[http://dx.doi.org/10.1016/j.colsurfa.2019.05.058]

[10]   Agarwal N, Solanki VS, Pare B, Singh N, Jonnalagadda SB. Current trends in nanocatalysis for green chemistry and its applications- A mini-review. Curr Opin Green Sustain Chem 2023; 41(2): 100788.
[http://dx.doi.org/10.1016/J.CPGSC.2023.100788]

[11]   Wang L, Wang L, Meng X, Xiao FS. New strategies for the preparation of sinter-resistant metal-nanoparticle-based catalysts. Adv Mater 2019; 31(50): 1901905.
[http://dx.doi.org/10.1002/adma.201901905] [PMID: 31478282]

[12]   Sharma RK, Yadav S, Gupta R, Arora G. Synthesis of magnetic nanoparticles using potato extract for dye degradation: A green chemistry experiment. J Chem Educ 2019; 96(12): 3038-44.
[http://dx.doi.org/10.1021/acs.jchemed.9b00384]

[13]   Si A, Pal K, Kralj S, El-Sayyad GS, de Souza FG, Narayanan T. Sustainable preparation of gold nanoparticles *via* green chemistry approach for biogenic applications. Mater Today Chem 2020; 17: 100327.
[http://dx.doi.org/10.1016/j.mtchem.2020.100327]

[14]   Priya , Naveen , Kaur K, Sidhu AK. Green synthesis: An eco-friendly route for the synthesis of iron oxide nanoparticles. Front Nanotechnol 2021; 3: 655062.
[http://dx.doi.org/10.3389/fnano.2021.655062]

[15]   Saif S, Tahir A, Chen Y. Green synthesis of iron nanoparticles and their environmental applications and implications. Nanomaterials 2016; 6(11): 209.
[http://dx.doi.org/10.3390/nano6110209] [PMID: 28335338]

[16]   Welz PJ, Ramond JB, Braun L, Vikram S, Le Roes-Hill M. Bacterial nitrogen fixation in sand bioreactors treating winery wastewater with a high carbon to nitrogen ratio. J Environ Manage 2018; 207: 192-202.
[http://dx.doi.org/10.1016/j.jenvman.2017.11.015] [PMID: 29179109]

[17]   Iravani S. Green synthesis of metal nanoparticles using plants. Green Chem 2011; 13(10): 2638-50.
[http://dx.doi.org/10.1039/c1gc15386b]

[18]   Dikshit P, Kumar J, Das A, *et al.* Green synthesis of metallic nanoparticles: Applications and limitations. Catalysts 2021; 11(8): 902.
[http://dx.doi.org/10.3390/catal11080902]

[19]   Zhang D, Ma X, Gu Y, Huang H, Zhang G. Green synthesis of metallic nanoparticles and their potential applications to treat cancer. Front Chem 2020; 8: 799.
[http://dx.doi.org/10.3389/fchem.2020.00799] [PMID: 33195027]

[20]   Chandra H, Kumari P, Bontempi E, Yadav S. Medicinal plants: Treasure trove for green synthesis of metallic nanoparticles and their biomedical applications. Biocatal Agric Biotechnol 2020; 24: 101518.
[http://dx.doi.org/10.1016/j.bcab.2020.101518]

[21] Wu J, Wang S, Qi J, *et al.* Research advances in the light-driven conversion of CO_2 to valuable chemicals by two-dimensional nanomaterials. Mater Today Energy 2022; 28: 101065.
[http://dx.doi.org/10.1016/j.mtener.2022.101065]

[22] Luna AL, Matter F, Schreck M, *et al.* Monolithic metal-containing TiO_2 aerogels assembled from crystalline pre-formed nanoparticles as efficient photocatalysts for H_2 generation. Appl Catal B 2020; 267: 118660.
[http://dx.doi.org/10.1016/j.apcatb.2020.118660]

[23] Banerjee A. The design, fabrication, and photocatalytic utility of nanostructured semiconductors: focus on TiO_2-based nanostructures. Nanotechnol Sci Appl 2011; 4: 35-65.
[http://dx.doi.org/10.2147/NSA.S9040] [PMID: 24198485]

[24] Akhtar MS, Panwar J, Yun YS. Biogenic synthesis of metallic nanoparticles by plant extracts. ACS Sustain Chem Eng 2013; 1(6): 591-602.
[http://dx.doi.org/10.1021/sc300118u]

[25] Shende S, Ingle AP, Gade A, Rai M. Green synthesis of copper nanoparticles by *Citrus medica* Linn. (Idilimbu) juice and its antimicrobial activity. World J Microbiol Biotechnol 2015; 31(6): 865-73.
[http://dx.doi.org/10.1007/s11274-015-1840-3] [PMID: 25761857]

[26] Mittal AK, Chisti Y, Banerjee UC. Synthesis of metallic nanoparticles using plant extracts. Biotechnol Adv 2013; 31(2): 346-56.
[http://dx.doi.org/10.1016/j.biotechadv.2013.01.003] [PMID: 23318667]

[27] Song JY, Kwon EY, Kim BS. Biological synthesis of platinum nanoparticles using *Diopyros kaki* leaf extract. Bioprocess Biosyst Eng 2010; 33(1): 159-64.
[http://dx.doi.org/10.1007/s00449-009-0373-2] [PMID: 19701776]

[28] Zhang Y, Yang D, Kong Y, Wang X, Pandoli O, Gao G. Synergetic antibacterial effects of silver nanoparticles@ aloe vera prepared *via* a green method. Nano Biomed Eng 2010; 2(4): 252-7.
[http://dx.doi.org/10.5101/nbe.v2i4.p252-257]

[29] Krishnaraj C, Jagan EG, Rajasekar S, Selvakumar P, Kalaichelvan PT, Mohan N. Synthesis of silver nanoparticles using *Acalypha indica* leaf extracts and its antibacterial activity against water borne pathogens. Colloids Surf B Biointerfaces 2010; 76(1): 50-6.
[http://dx.doi.org/10.1016/j.colsurfb.2009.10.008] [PMID: 19896347]

[30] Kora AJ, Rastogi L. Green synthesis of palladium nanoparticles using gum ghatti (*Anogeissus latifolia*) and its application as an antioxidant and catalyst. Arab J Chem 2018; 11(7): 1097-106.
[http://dx.doi.org/10.1016/j.arabjc.2015.06.024]

[31] Kasthuri J, Veerapandian S, Rajendiran N. Biological synthesis of silver and gold nanoparticles using apiin as reducing agent. Colloids Surf B Bioint 2009; 68(1): 55-60.
[http://dx.doi.org/10.1016/j.colsurfb.2008.09.021] [PMID: 18977643]

[32] Logeswari P, Silambarasan S, Abraham J. Ecofriendly synthesis of silver nanoparticles from commercially available plant powders and their antibacterial properties. Sci Iran 2013; 20(3): 1049-54.

[33] Begum NA, Mondal S, Basu S, Laskar RA, Mandal D. Biogenic synthesis of Au and Ag nanoparticles using aqueous solutions of Black Tea leaf extracts. Colloids Surf B Bioint 2009; 71(1): 113-8.
[http://dx.doi.org/10.1016/j.colsurfb.2009.01.012] [PMID: 19250808]

[34] Pandey C, Khan E, Mishra A, Sardar M, Gupta M. Silver nanoparticles and its effect on seed germination and physiology in *Brassica juncea* L.(Indian mustard) plant. Adv Sci Lett 2014; 20(7): 1673-6.
[http://dx.doi.org/10.1166/asl.2014.5518]

[35] Vilchis-Nestor AR, Sánchez-Mendieta V, Camacho-López MA, Gómez-Espinosa RM, Camacho-López MA, Arenas-Alatorre JA. Solventless synthesis and optical properties of Au and Ag nanoparticles using *Camellia sinensis* extract. Mater Lett 2008; 62(17-18): 3103-5.
[http://dx.doi.org/10.1016/j.matlet.2008.01.138]

[36]   Mude N, Ingle A, Gade A, Rai M. Synthesis of silver nanoparticles using callus extract of *Carica papaya—a* first report. J Plant Biochem Biotechnol 2009; 18(1): 83-6.
[http://dx.doi.org/10.1007/BF03263300]

[37]   Dwivedi AD, Gopal K. Biosynthesis of silver and gold nanoparticles using *Chenopodium album* leaf extract. Colloids Surf A Physicochem Eng Asp 2010; 369(1-3): 27-33.
[http://dx.doi.org/10.1016/j.colsurfa.2010.07.020]

[38]   Aref MS, Salem SS. Bio-callus synthesis of silver nanoparticles, characterization, and antibacterial activities *viaCinnamomum camphora* callus culture. Biocatal Agric Biotechnol 2020; 27: 101689.
[http://dx.doi.org/10.1016/j.bcab.2020.101689]

[39]   Sathishkumar M, Sneha K, Kwak IS, Mao J, Tripathy SJ, Yun YS. Phyto-crystallization of palladium through reduction process using *Cinnamom zeylanicum* bark extract. J Hazard Mater 2009; 171(1-3): 400-4.
[http://dx.doi.org/10.1016/j.jhazmat.2009.06.014] [PMID: 19576689]

[40]   Prathna TC, Chandrasekaran N, Raichur AM, Mukherjee A. Biomimetic synthesis of silver nanoparticles by Citrus limon (lemon) aqueous extract and theoretical prediction of particle size. Colloids Surf B Biointerfaces 2011; 82(1): 152-9.
[http://dx.doi.org/10.1016/j.colsurfb.2010.08.036] [PMID: 20833002]

[41]   Rao SV. Picosecond nonlinear optical studies of gold nanoparticles synthesised using coriander leaves (*Coriandrum sativum*). J Mod Opt 2011; 58(12): 1024-9.
[http://dx.doi.org/10.1080/09500340.2011.590903]

[42]   Sathishkumar M, Sneha K, Yun YS. Green fabrication of zirconia nano-chains using novel *Curcuma longa* tuber extract. Mater Lett 2013; 98: 242-5.
[http://dx.doi.org/10.1016/j.matlet.2013.02.036]

[43]   Gupta AK, Ganjewala D. Synthesis of silver nanoparticles from *Cymbopogon flexuosus* leaves extract and their antibacterial properties. Int J Plant Sci Ecol 2015; 1: 225-30.

[44]   Kesharwani J, Yoon KY, Hwang J, Rai M. Phytofabrication of silver nanoparticles by leaf extract of Datura metel: hypothetical mechanism involved in synthesis. Journal of Bionanoscience 2009; 3(1): 39-44.
[http://dx.doi.org/10.1166/jbns.2009.1008]

[45]   Sardar M, Mishra A, Ahmad R. Biosynthesis of metal nanoparticles and their applications. Biosen Nanotech 2014; 239-66.

[46]   Ankamwar B, Damle C, Ahmad A, Sastry M. Biosynthesis of gold and silver nanoparticles using *Emblica Officinalis* fruit extract, their phase transfer and transmetallation in an organic solution. J Nanosci Nanotechnol 2005; 5(10): 1665-71.
[http://dx.doi.org/10.1166/jnn.2005.184] [PMID: 16245525]

[47]   Ravindra S, Murali Mohan Y, Narayana Reddy N, Mohana Raju K. Fabrication of antibacterial cotton fibres loaded with silver nanoparticles *via* "Green Approach". Colloids Surf A Physicochem Eng Asp 2010; 367(1-3): 31-40.
[http://dx.doi.org/10.1016/j.colsurfa.2010.06.013]

[48]   Dubey M, Bhadauria S, Kushwah BS. Green synthesis of nanosilver particles from extract of *Eucalyptus hybrida* (safeda) leaf. Dig J Nanomater Biostruct 2009; 4(3): 537-43.

[49]   Vishnukumar P, Vivekanandhan S, Muthuramkumar S. Plant-mediated biogenic synthesis of palladium nanoparticles: Recent trends and emerging opportunities. ChemBioEng Rev 2017; 4(1): 18-36.
[http://dx.doi.org/10.1002/cben.201600017]

[50]   Jia L, Zhang Q, Li Q, Song H. The biosynthesis of palladium nanoparticles by antioxidants in *Gardenia jasminoides Ellis*: long lifetime nanocatalysts for *p* -nitrotoluene hydrogenation. Nanotechnology 2009; 20(38): 385601.

[http://dx.doi.org/10.1088/0957-4484/20/38/385601] [PMID: 19713585]

[51] Petla R K, Vivekanandhan S, Misra M, Mohanty A K, Satyanarayana N. Soybean (Glycine max) leaf extract based green synthesis of palladium nanoparticles. J Biomat Nanobiotec 2011; 3(1).

[52] Philip D. Green synthesis of gold and silver nanoparticles using Hibiscus rosa sinensis. Physica E 2010; 42(5): 1417-24.
[http://dx.doi.org/10.1016/j.physe.2009.11.081]

[53] Sreelakshmi C, Datta KKR, Yadav JS, Reddy BVS. Honey derivatized Au and Ag nanoparticles and evaluation of its antimicrobial activity. J Nanosci Nanotechnol 2011; 11(8): 6995-7000.
[http://dx.doi.org/10.1166/jnn.2011.4240] [PMID: 22103111]

[54] Oza G, Reyes-Calderón A, Mewada A, *et al.* Plant-based metal and metal alloy nanoparticle synthesis: a comprehensive mechanistic approach. J Mater Sci 2020; 55(4): 1309-30.
[http://dx.doi.org/10.1007/s10853-019-04121-3]

[55] Rai M, Yadav A. Plants as potential synthesiser of precious metal nanoparticles: progress and prospects. IET Nanobiotechnol 2013; 7(3): 117-24.
[http://dx.doi.org/10.1049/iet-nbt.2012.0031] [PMID: 24028810]

[56] Mojally M, Sharmin E, Alhindi Y, *et al.* Hydrogel films of methanolic *Mentha piperita* extract and silver nanoparticles enhance wound healing in rats with diabetes Type I. J Taibah Univ Sci 2022; 16(1): 308-16.
[http://dx.doi.org/10.1080/16583655.2022.2054607]

[57] Prasad TNVKV, Elumalai EK. Biofabrication of Ag nanoparticles using *Moringa oleifera* leaf extract and their antimicrobial activity. Asian Pac J Trop Biomed 2011; 1(6): 439-42.
[http://dx.doi.org/10.1016/S2221-1691(11)60096-8] [PMID: 23569809]

[58] Anand K, Tiloke C, Phulukdaree A, *et al.* Biosynthesis of palladium nanoparticles by using Moringa oleifera flower extract and their catalytic and biological properties. J Photochem Photobiol B 2016; 165: 87-95.
[http://dx.doi.org/10.1016/j.jphotobiol.2016.09.039] [PMID: 27776261]

[59] Surendra TV, Roopan SM, Arasu MV, Al-Dhabi NA, Rayalu GM. RSM optimized Moringa oleifera peel extract for green synthesis of M. oleifera capped palladium nanoparticles with antibacterial and hemolytic property. J Photochem Photobiol B 2016; 162: 550-7.
[http://dx.doi.org/10.1016/j.jphotobiol.2016.07.032] [PMID: 27474786]

[60] Arias D, Rodríguez J, López B, Méndez P. Evaluation of the physicochemical properties of pectin extracted from *Musa paradisiaca* banana peels at different pH conditions in the formation of nanoparticles. Heliyon 2021; 7(1): 06059.
[http://dx.doi.org/10.1016/j.heliyon.2021.e06059] [PMID: 33537485]

[61] Santhoshkumar T, Rahuman AA, Rajakumar G, *et al.* Synthesis of silver nanoparticles using *Nelumbo nucifera* leaf extract and its larvicidal activity against malaria and filariasis vectors. Parasitol Res 2011; 108(3): 693-702.
[http://dx.doi.org/10.1007/s00436-010-2115-4] [PMID: 20978795]

[62] Singhal G, Bhavesh R, Kasariya K, Sharma AR, Singh RP. Biosynthesis of silver nanoparticles using *Ocimum sanctum* (Tulsi) leaf extract and screening its antimicrobial activity. J Nanopart Res 2011; 13(7): 2981-8.
[http://dx.doi.org/10.1007/s11051-010-0193-y]

[63] Ghodake GS, Deshpande NG, Lee YP, Jin ES. Pear fruit extract-assisted room-temperature biosynthesis of gold nanoplates. Colloids Surf B Biointerfaces 2010; 75(2): 584-9.
[http://dx.doi.org/10.1016/j.colsurfb.2009.09.040] [PMID: 19879738]

[64] Shankar SS, Ahmad A, Pasricha R, Sastry M. Bioreduction of chloroaurate ions by geranium leaves and its endophytic fungus yields gold nanoparticles of different shapes. J Mater Chem 2003; 13(7): 1822-6.

[http://dx.doi.org/10.1039/b303808b]

[65]   Qu J, Yuan X, Wang X, Shao P. Zinc accumulation and synthesis of ZnO nanoparticles using *Physalis alkekengi* L. Environ Pollut 2011; 159(7): 1783-8.
[http://dx.doi.org/10.1016/j.envpol.2011.04.016] [PMID: 21549461]

[66]   Mittal J, Batra A, Singh A, Sharma MM. Phytofabrication of nanoparticles through plant as nanofactories. Advances in Natural Sciences: Nanoscience and Nanotechnology 2014; 5(4): 043002.
[http://dx.doi.org/10.1088/2043-6262/5/4/043002]

[67]   Raghunandan D, Basavaraja S, Mahesh B, Balaji S, Manjunath SY, Venkataraman A. Biosynthesis of stable polyshaped gold nanoparticles from microwave-exposed aqueous extracellular anti-malignant guava (*Psidium guajava*) leaf extract. NanoBiotechnology 2009; 5(1-4): 34-41.
[http://dx.doi.org/10.1007/s12030-009-9030-8]

[68]   Khan M, Khan M, Kuniyil M, *et al.* Biogenic synthesis of palladium nanoparticles using *Pulicaria glutinosa* extract and their catalytic activity towards the Suzuki coupling reaction. Dalton Trans 2014; 43(24): 9026-31.
[http://dx.doi.org/10.1039/C3DT53554A] [PMID: 24619034]

[69]   Qu J, Luo C, Hou J. Synthesis of ZnO nanoparticles from Zn-hyperaccumulator (*Sedum alfredii* Hance) plants. Micro & Nano Lett 2011; 6(3): 174-6.
[http://dx.doi.org/10.1049/mnl.2011.0004]

[70]   Baran T, Yılmaz Baran N, Menteş A. Sustainable chitosan/starch composite material for stabilization of palladium nanoparticles: Synthesis, characterization and investigation of catalytic behaviour of Pd@chitosan/starch nanocomposite in Suzuki–Miyaura reaction. Appl Organomet Chem 2018; 32(2): 4075.
[http://dx.doi.org/10.1002/aoc.4075]

[71]   Venugopal K, Rather HA, Rajagopal K, *et al.* Synthesis of silver nanoparticles (Ag NPs) for anticancer activities (MCF 7 breast and A549 lung cell lines) of the crude extract of *Syzygium aromaticum*. J Photochem Photobiol B 2017; 167: 282-9.
[http://dx.doi.org/10.1016/j.jphotobiol.2016.12.013] [PMID: 28110253]

[72]   Jayaprakash N, Vijaya JJ, Kaviyarasu K, *et al.* Green synthesis of Ag nanoparticles using Tamarind fruit extract for the antibacterial studies. J Photochem Photobiol B 2017; 169: 178-85.
[http://dx.doi.org/10.1016/j.jphotobiol.2017.03.013] [PMID: 28347958]

[73]   Ankamwar B. Biosynthesis of gold nanoparticles (green-gold) using leaf extract of *Terminalia catappa*. E-J Chem 2010; 7(4): 1334-9.
[http://dx.doi.org/10.1155/2010/745120]

[74]   Silva LP. Gren synthesis of mea nanoparticles y lants: Curn trends and challenges. In: Bak V, Basiuk E, Eds. Green Proc Nanotech. Cham: Springer 2015; pp. 259-75.

[75]   Ovais M, Khalil AT, Raza A, *et al.* Green synthesis of silver nanoparticles *via* plant extracts: beginning a new era in cancer theranostics. Nanomedicine 2016; 11(23): 3157-77.
[http://dx.doi.org/10.2217/nnm-2016-0279] [PMID: 27809668]

[76]   Rout Y, Behera S, Ojha AK, Nayak PL. Green synthesis of silver nanoparticles using *Ocimum sanctum* (Tulashi) and study of their antibacterial and antifungal activities. J Microbiol Antimicrob 2012; 4(6): 103-9.
[http://dx.doi.org/10.5897/JMA11.060]

[77]   Tahir K, Nazir S, Li B, *et al. Sapium sebiferum* leaf extract mediated synthesis of palladium nanoparticles and *in vitro* investigation of their bacterial and photocatalytic activities. J Photochem Photobiol B 2016; 164: 164-73.
[http://dx.doi.org/10.1016/j.jphotobiol.2016.09.030] [PMID: 27689741]

[78]   Sharmila G, Farzana Fathima M, Haries S, Geetha S, Manoj Kumar N, Muthukumaran C. Green synthesis, characterization and antibacterial efficacy of palladium nanoparticles synthesized using

Filicium decipiens leaf extract. J Mol Struct 2017; 1138: 35-40.
[http://dx.doi.org/10.1016/j.molstruc.2017.02.097]

[79]   Kadam J, Madiwale S, Bashte B, Dindorkar S, Dhawal P, More P. Green mediated synthesis of palladium nanoparticles using aqueous leaf extract of *Gymnema sylvestre* for catalytic reduction of Cr (VI). SN Appl Sci 2020; 2(11): 1854.
[http://dx.doi.org/10.1007/s42452-020-03663-5]

[80]   Kandiah M, Chandrasekaran KN. Green synthesis of silver nanoparticles using *Catharanthus roseus* flower extracts and the determination of their antioxidant, antimicrobial, and photocatalytic activity. J Nanotechnol 2021; 2021: 1-18.
[http://dx.doi.org/10.1155/2021/5512786]

[81]   Sharmila G, Haries S, Farzana Fathima M, Geetha S, Manoj Kumar N, Muthukumaran C. Enhanced catalytic and antibacterial activities of phytosynthesized palladium nanoparticles using Santalum album leaf extract. Powder Technol 2017; 320: 22-6.
[http://dx.doi.org/10.1016/j.powtec.2017.07.026]

[82]   Majumdar R, Tantayanon S, Bag BG. Synthesis of palladium nanoparticles with leaf extract of *Chrysophyllum cainito* (Star apple) and their applications as efficient catalyst for C–C coupling and reduction reactions. Int Nano Lett 2017; 7(4): 267-74.
[http://dx.doi.org/10.1007/s40089-017-0220-4]

[83]   Han X, Xu K, Taratula O, Farsad K. Applications of nanoparticles in biomedical imaging. Nanoscale 2019; 11(3): 799-819.
[http://dx.doi.org/10.1039/C8NR07769J] [PMID: 30603750]

[84]   Siddique S, Chow JCL. Application of nanomaterials in biomedical imaging and cancer therapy. Nanomaterials 2020; 10(9): 1700.
[http://dx.doi.org/10.3390/nano10091700] [PMID: 32872399]

[85]   Nasrollahzadeh M, Sajadi SM, Maham M. Green synthesis of palladium nanoparticles using Hippophae rhamnoides Linn leaf extract and their catalytic activity for the Suzuki–Miyaura coupling in water. J Mol Catal Chem 2015; 396: 297-303.
[http://dx.doi.org/10.1016/j.molcata.2014.10.019]

[86]   Nasrollahzadeh M, Mohammad Sajadi S. Green synthesis, characterization and catalytic activity of the Pd/TiO$_2$ nanoparticles for the ligand-free Suzuki–Miyaura coupling reaction. J Colloid Interface Sci 2016; 465: 121-7.
[http://dx.doi.org/10.1016/j.jcis.2015.11.038] [PMID: 26674227]

CHAPTER 12

A Comprehensive Review on Applications of Different Domains of Nanotechnology in Wastewater Treatment

Annu Yadav[1], **Nirmala Kumari Jangid**[1,*], **Rekha Sharma**[1] and **Azhar Ullah Khan**[2]

[1] *Department of Chemistry, Banasthali Vidyapith, Banasthali, Rajasthan, India*

[2] *Department of Chemistry, School of Life and Basic Sciences, Jaipur National University, Jaipur, India*

Abstract: In the process of purification of water, nanotechnology provides the possibility of an effective removal of pollutants and germs. In recent times, nanoparticles (NPs), nanopowder and nanomembranes have been used for the detection and removal of chemical and biological substances that contain metals like cadmium, copper, lead, mercury, nickel, zinc, *etc.*, nutrients like phosphate, ammonia, nitrate and nitrite, algae, cyanobacterial toxins, viruses, bacteria, parasites, and antibiotics. Commonly, four classes of nanoscale materials that are being evaluated as functional materials for water purification are metal-containing nanoparticles, carbonaceous nanomaterials, dendrimers and zeolites. Carbon nanotubes and nanofibers are also used in the techniques of water purification. Nanomaterials (NMs) give the best results in water treatment in comparison to other techniques because NMs have a high surface area (surface/volume ratio). Silver NPs affect the activated sludge of microorganisms and play an important role in wastewater treatment since they restrain their activity and significantly reduce their number. Carbon nanostructures are widely used as nanoadsorbents for wastewater treatment owing to their abundant availability, cost-effectiveness, high chemical and thermal stabilities, high active surface areas, excellent adsorption capacities, and environmentally friendly nature. Due to the high utility of nanotechnology in the treatment of pollutants, this chapter further highlights various fundamental aspects of nanotechnology, such as types, synthesis, applications and future directions for a green and sustainable environment.

Keywords: Activated carbon, Carbon nanotubes, Dendrites, Graphene, Green synthesis, Nanoparticle, Membrane filtration, Nanofiber, Nano silver, Nanocomposites, Photocatalytic, Zeolite.

---

* **Corresponding author Nirmala K. Jangid:** Department of Chemistry, Banasthali Vidyapith, Banasthali, Rajasthan, India; E-mail: nirmalajangid.111@gmail.com

Neha Agarwal, Vijendra Singh Solanki and Sreekantha B. Jonnalagadda (Eds.)

INTRODUCTION

Water is essential for life, but it is very difficult to have clean and hygienic water every day due to continuous contamination of fresh water. Advanced technology is necessary to meet this challenge and provide people with clean water for a healthy life [1 - 4]. Nanotechnology can help by providing clean water for growing humans. The term nanoscience and the scale defined by nano is a combination of nanotechnology. The adverb "nano" is derived from nannos (Greek), which means "very short people" [5]. This technique refers to the technique that uses nanoscale particles. Depending on the size of the field, NPs have special physicochemical properties in which the structural components are sized (in at least one dimension) between 1 and 100nm [6]. Due to the nanoscale size of NMs, their electrical, optical and magnetic properties are different from those of conventional materials. NMs have a high surface area to volume ratio that allows them to effectively absorb and remove contaminants from water. The adsorption capacity of materials such as carbon nanotubes, graphene, and nanoparticles such as titanium dioxide or iron oxide has been extensively studied [7]. These NMs can selectively target and capture pollutants, including heavy metals, organic compounds and bacteria, thereby improving overall water quality. Nanotechnology has emerged as a promising field with its applications in many areas, such as the environment and water treatment. NPs can act as catalysts in various oxidation reactions, such as photocatalysis or electrocatalysis, to effectively decompose or convert pollutants into harmless products. For example, photocatalytic NMs such as titanium dioxide, when activated by light, produce reactive oxygen species that can degrade organic compounds. This approach shows promise in breaking down pollutants that are difficult to remove using conventional treatments. Nanotechnology also plays an important role in membrane separation processes, which are widely used in wastewater treatment. Reverse osmosis membranes with nanofiltration and nanoscale pores provide selective elimination of bacteria while retaining essential water. Functionalized NMs can be incorporated into tissues to improve their performance, such as improving fouling resistance, increasing permeability or facilitating selective ion removal and purifying water [8 - 10].

According to many studies, the use of NMs in water and wastewater treatment shows great promise. Currently, the most researched NMs for water and wastewater treatment mainly include zerovalent metal NPs, metal oxide NPs, carbon nanotubes (CNTs), and nanocomposites [11, 12]. Magnetic nanoparticles, metal zeolite, carbon nanotubes and other nanostructure materials can be used to remove Hg(II), Pb(II), Cr(III), Cr(VI), Ni(II), Co(II), Cu(II), Cd(II), Ag(I), As(V) and As(III) from wastewater because these metal ions can cause serious illness [13]. Nanoscale zero-valent ions are used as adsorbents and also catalyze photochemical oxidation. Due to their general adsorption properties, carbon nanotubes and dendrimers are often used to create advanced water systems [14, 15]. The introduction of nanotechnology into wastewater treatment has brought a revolution by providing oxidation, disinfection, membrane development and time management. With further research and development, nanotechnology holds great promise for the future of wastewater treatment, contributing to sustainable water management and environmental protection. Table **1** shows different types of NPs and the categories of pollutants that are removed by these NPs.

Green synthesis of NPs for wastewater treatment is focused on developing environmentally friendly methods to produce NPs and use them in water treatment processes. Conventional NP synthesis methods often involve chemical and energy-intensive processes. However, green synthesis aims to reduce or eliminate the use of chemicals and reduce energy consumption. Green synthetic methods use natural extracts, biomolecules or environmentally friendly reducing agents for production [16].

NEED FOR ADVANCED WASTEWATER TREATMENT TECHNOLOGIES

The need for advanced wastewater treatment technologies arises from many factors and problems associated with traditional methods. These include:

Emerging Contaminants

Conventional treatments are ineffective at removing emerging pollutants such as chemicals, personal care products, endocrine disruptors and microplastics that enter the environment and pose a threat to human health and hazardous ecosystems. Advanced treatment technologies are required to target and remove contaminants from wastewater.

Table 1. Different types of pollutants degraded by NPs.

Type of NPs	Type of Pollutants Degraded
Carbon nanotubes	Organic Contaminant
Nanoscale metal oxide	Heavy metals, radionuclides
Nanocatalyst PCB	Azodyes, pesticides, *etc.*
Ni/Pd nanoparticles	Dichrolophenol, trichlorobenzene
Bioactive nanoparticles	Removal of bacteria, fungi
Biomimetic membranes	Removal of salts
Photocatalysts	Heavy metal ions, Azo Dye and aromatic pollutants

(Table 1) cont.....

Type of NPs	Type of Pollutants Degraded
Nano-structured catalytic NPs	Decomposition of organic pollutants, inactivation of microorganisms
Zero-valent iron nanoparticles (Nzvi)	Inorganic ions, chlorinated organic compounds, heavy metals
TiO2 nanoparticles	Aromatic organic pollutants

Nutrient Removal

Nutrient pollution, particularly excess nitrogen and phosphorus, has become a major problem in water bodies, causing problems such as eutrophication and problematic algal blooms [17]. Advanced treatment technologies are needed to effectively remove nutrients from wastewater and prevent their release into the environment.

Water Reuse

There is an increasing interest in reusing wastewater for non-useful purposes such as irrigation, industrial production, and the use of groundwater. The purification technology enables the production of high-quality purified water that meets recycling standards and reduces pressure on fresh water.

Energy Use

Conventional wastewater treatment processes can be energy intensive, with aeration, utilization and treatment requiring large amounts of electricity. Advanced treatment technologies increase energy efficiency by using new techniques such as membrane filtration, anaerobic digestion and wastewater recovery.

Sludge Management

The disposal and management of sludge generated during wastewater treatment presents volume, disposal costs and potential environmental impact issues. Advanced technologies such as producing biogas for energy production and extracting valuable products from sludge focus on optimizing the sludge treatment and resource recovery capacity of conventional wastewater treatment systems. Advanced processing technologies are designed to improve machine durability, adaptability and robustness to better withstand and mitigate the effects of climate change.

Technological Innovation

Advances in science and technology provide opportunities to discover new and more effective treatments. Treatment technologies, including nanotechnology, advanced oxidation processes, membrane processes and biological systems. offer new solutions to improve treatment, reduce environmental impact and solve emerging problems.

Overall, the need for wastewater treatment technologies stems from the desire to improve water quality, protect public health, conserve resources, solve emerging problems, and adapt to environmental and regulatory changes. By integrating the latest technology into the wastewater treatment process, we provide safer, more efficient and effective wastewater management [18].

ROLE OF NANOTECHNOLOGY IN WASTEWATER TREATMENT

Enhanced Pollutant Removal

NMs such as NPs, nanofibers and nanocomposites have a balance between stability and certain physicochemical properties. These properties allow them to absorb, absorb or catalytically decompose more pollutants than any other material. Nanotechnology-based sorbents and catalysts can target a variety of pollutants, including heavy metals, organic compounds and natural pollutants.

Advanced Filtration and Membrane Processes

Nanotechnology has revolutionized the filtration and membrane processes used in wastewater treatment. Nanostructured membranes and filtration systems such as carbon nanotube membranes and graphene oxide membranes have higher separation, higher flux and better antifouling properties. This advanced process provides better water purification by removing small particles, bacteria, viruses and even some contaminants.

Recovery Services

Nanotechnology can recover valuable resources such as energy, nutrients and metals from wastewater. Nanomaterials can be used in processes such as microbial oil, where they can improve energy transfer and increase the strength of materials in wastewater. NPs can also be used to selectively adsorb and recover nutrients such as phosphorus and iron in wastewater, contributing to economic savings and a circular economy.

Water Disinfection

Nanotechnology offers new methods for water disinfection by providing an alternative to disinfectants such as chlorination. NMs such as silver NPs and photocatalytic NMs such as titanium dioxide have strong antibacterial properties and can effectively neutralize pathogens found in wastewater. These NMs can be used for antimicrobial applications or incorporated into wastewater treatment for microbial remediation.

Detection and Monitoring

Nanotechnology plays an important role in the development of sensors and monitoring devices for quick and easy detection of pollutants in wastewater. Nanosensors based on NPs or nanocomposites can detect and measure pollutants, pathogens or certain chemicals with high sensitivity and selectivity. These nanosensors provide real-time monitoring, early warning and effective management of wastewater processes.

Environmental Remediation

Nanotechnology provides solutions for environmental remediation of polluted water. NMs can be used to treat groundwater with pollutants such as heavy metals or organic compounds. NPs can be designed to selectively target and immobilize pollutants, improving their removal and reducing long-term environmental impacts [19].

FUNDAMENTALS OF NANOTECHNOLOGY

NMs: Types

Nanomaterials exhibit many types and properties due to their unique properties at the nanoscale. Below are some types of NMs and their basic properties.

NPs

NPs are nanoscale particles, usually 1 to 100 nanometers in size. They can be made of various materials such as metal (*e.g.*, gold, silver, iron), metal oxides (such as titanium dioxide, zinc oxide), carbon-based materials (such as carbon nanotubes, graphene) and polymers. NPs have a size-to-volume ratio that affects their reactivity, optical properties, electrical conductivity, and catalytic activity.

Nanofibers

Nanofibers are long fibers with diameters in the nanoscale range. They can be made from different materials such as polymers, metals and ceramics. Nanofibers have a high aspect ratio (length-to-length ratio) that gives them unique properties such as high surface area, excellent strength and flexibility. These properties make nanofibers suitable for use in vision, tissue engineering, and electronics.

Nanocomposites

Nanocomposites are materials composed of a matrix material (polymer, ceramic, or metal) with embedded NPs or nanofillers. The presence of NPs improves the mechanical, electrical, thermal and barrier properties of the composite. The strength, stability, electrical conductivity and other properties of nanocomposites make them useful in many applications such as aerospace, automotive and industry [20].

Carbon-based NMs

Carbon-based NMs include carbon nanotubes (CNTs), graphene, and fullerenes (such as buckyballs). CNTs are cylindrical structures formed by rolling sheets of graphene, a layer of carbon atoms. This material exhibits exceptional mechanical strength, electrical and thermal conductivity, and a high aspect ratio. They find applications in electronics, energy storage, composites, sensors, and nanoelectromechanical systems (NEMS) [21].

Quantum dots

Quantum dots are nanoscale semiconductor particles with special optical and electrical properties. They exhibit quantum confinement; that is, their electronic behavior depends on the three-dimensional confinement of electrons. Quantum dots can emit or absorb light at specific wavelengths depending on their size, allowing use in imaging, solar cells, biomedical imaging, and sensors.

Nanowires

Nanowires are one-dimensional structures, usually nanometers in diameter and up to several microns in length. They can be made from a variety of materials, including metals, semiconductors, and oxides. Nanowires have a high aspect ratio and can exhibit unique electrical, optical and thermal properties. It has applications in nanoelectronics, sensors, energy conversion and photovoltaics [22].

NMs: Properties

The important properties of NMs include-

Size-dependent Properties

Due to quantum effects and increased surface area/volume ratio, NMs exhibit size-dependent properties.

Enhanced Surface Reactivity

The high surface area of NMs provides enhanced surface reactivity, making them useful for catalysis and chemistry.

Optical Properties

NMs can exhibit unique optical properties such as enhanced absorption, scattering and fluorescence, making them suitable for use in optics and photonics.

Mechanical Strength

Some NMs, such as carbon nanotubes and nanocomposites, have good mechanical strength and flexibility.

Conductivity

Some NMs are highly electrical and thermal conductive, including carbon nanotubes, graphene, and metal nanoparticles [23].

NPs: Synthesis and Characterization

NPs can be synthesized by various methods and their characterization depends on their size, shape, composition, surface properties, *etc.* Below is a description of the NP synthesis and characterization process:

Chemical Precipitation

This method involves the precipitation of chemicals by mixing two or more chemical precursors. Chemical reactions such as temperature, pH, and concentration affect the size and composition of NPs.

Sol-gel Method

In the sol-gel method, NPs are produced by hydrolysis and condensation of precursor molecules. This method allows the synthesis of NPs of various compositions and can be used to form materials such as metal oxides.

Vapor Condensation

This process involves the evaporation of particles followed by rapid cooling to form NPs. Evaporated material condenses into NPs due to nucleation and subsequent growth.

Electrochemical Deposition

This is an electrical method that uses electric current to deposit NPs onto a substrate. This technique provides control over the size and composition of NPs.

Green Synthesis

The green synthesis method uses natural or environmentally friendly materials such as plant extracts or bacteria to reduce and stabilize metal ions to make NPs. This method provides a stable and environmentally friendly method for the synthesis of NP [24].

APPROACHES TO GREEN SYNTHESIS OF NPS USED IN WASTEWATER TREATMENT

Plant Extracts

Plant extracts containing natural compounds such as polyphenols, flavonoids and terpenoids can be used as reducing and stabilizing agents for NP synthesis. These compounds can be extracted from different parts of the plant, such as fruits, leaves and stems. The extract is mixed with a metal salt solution and reduces metal ions to form NPs. Plants need a shorter time than bacteria and viruses to reduce metal ions; therefore, they are considered the best candidates for NP synthesis [25].

Microorganisms

Microbial-based NP synthesis is biocompatible, economical, environmentally friendly and energy intensive. Metal and nonmetal, metal oxide and sulfide NPs are synthesized by bacteria, viruses, fungi and algae. Due to their physicochemical properties, nanometer sizes, controlled growth and surface modifications, these NPs are promising as adsorbents for water and wastewater treatment. Carbohydrates, proteins, and enzymes are present in these bacteria as surfactants and coating agents, reducing the use of toxic surfactants. NPs can be used to treat dyes, heavy metals, microbial pollutants and pesticides in wastewater [26]. These organisms use gold (Au), silver (Ag), iron (Fe), platinum (Pt), palladium (Pd), *etc.* and act as "biofactories" that can produce NPs.

Bio-waste and Agricultural Byproducts

Agriculture and food industry wastes such as rice husks, bagasse and fruits can be used as precursors for the synthesis of NPs. This method promotes the utilization of waste and reduces environmental pollution by converting waste into useful NPs.

Green Solvents

Conventional NP synthesis often involves the use of environmentally non-friendly organic solvents. Green solvents such as water, supercritical fluid and biocompatible solvents are used to reduce the environmental impact. These solvents are nontoxic, recyclable and recyclable and do not produce harmful by-products [27]. The NPs synthesized from this method can be used in a variety of wastewater applications. Different green methods of NP synthesis are depicted in Fig. (1).

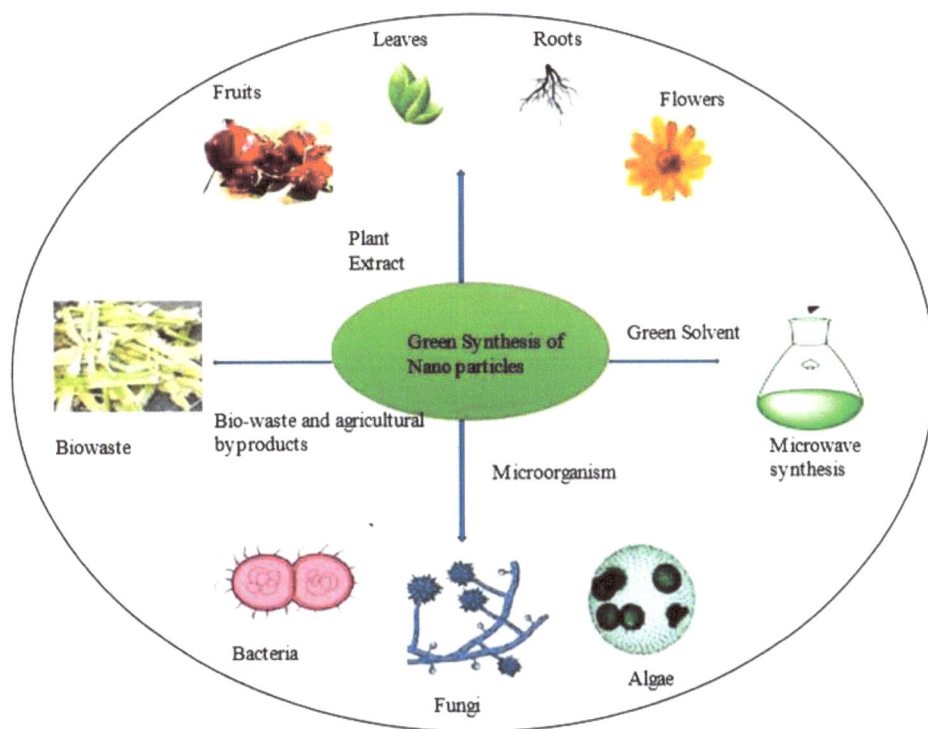

Fig. (1). Green synthesis of NPs for wastewater treatment.

Each green synthesis method has its advantages and limitations, and the choice of method depends on the requirements of NP synthesis. Biosynthesis provides reducing agents and stabilizers, but controlling particle size and shape can be difficult. Extract-mediated synthesis is simple and cost-effective, while microbial-assisted synthesis has high yields and potential functionalization options. Green reducing agents are versatile and environmentally friendly, but they may need to be optimized for certain NP materials. When choosing a green synthesis method, researchers should consider factors such as scalability, cost, type of NP required, and environmental impact. Overall, green synthesis methods provide a safe and environmentally friendly way to produce NPs for a variety of applications, including wastewater treatment. This technique has the potential to revolutionize NP synthesis by reducing reliance on traditional chemical methods, thereby promoting new and more efficient approaches [28].

CHARACTERIZATION TECHNIQUES FOR GREEN-SYNTHESIZED NPS

The characterization of NPs synthesized by the green method is important to understand their morphology, structure and composition. Many characterization techniques are widely used to identify green-synthesized NPs. Here are some common methods of characterization.

Physicochemical Characterization

Physicochemical characterization techniques provide information about the size, shape, and surface area of NPs. These techniques are discussed below.

Dynamic Light Scattering (DLS)

DLS provides information about the hydrodynamic size distribution of NPs in a solution. By this technique, hydrodynamic sizes are accurately determined for monodisperse samples in any liquid media or suitable solvents.

Zeta Potential Analysis

Zeta potential determines the surface charge and stability of NPs in solution, which offers their behavior in waste-water treatment.

Fourier Transform Infrared Spectroscopy (FTIR)

FTIR analysis identifies functional groups and conformations of biconjugate surface properties of NPs, providing information about the biomolecules responsible for their synthesis and potential interaction with bacteria.

Atomic Force Microscopy (AFM)

AFM is used to measure the 3-D sample surface mapping like Dispersion, sorption, aggregation and many other surface properties of NPs in dry, aqueous or ambient environments.

Structural Analysis

Structural analysis techniques provide information about the crystal structure and composition of NPs. This technique includes as discussed below.

X-ray Diffraction (XRD)

XRD is used to determine the crystal structure and purity of NPs. It provides information about the arrangement of atoms in NPs.

Transmission Electron Microscopy (TEM)

TEM is used for providing information about the size, size distribution, shapes and morphology of nanomaterials with larger resolution than SEM. Many analytical methods are used with TEM for the investigation of the electronic structure and chemical composition of nanomaterials.

Surface Morphology Analysis

Surface morphology analysis techniques provide information about the surface and morphology of NPs. These techniques include:

Scanning Electron Microscopy (SEM)

SEM provides detailed information about the morphology of NPs. It can give information about size, shape, and roughness.

Atomic Force Microscope (AFM)

AFM is used to analyze the surface and roughness of nanoscale NPs.

Elemental Composition Analysis

Elemental composition analysis techniques determine the elemental composition and elemental distribution of NPs. The most commonly used techniques are below.

Energy Dissipative Xray Spectroscopy (EDS)

EDS is used in conjunction with SEM or TEM to analyze the content of elements and the blueprint of NPs.

Inductively Coupled Plasma Atomic Emission Spectrometer (ICPAES)

This technique is used to determine the content of NPs and detect impurities.

It is important to use practical methods together to obtain information about NPs synthesized by green methods. The selection of the characterization method depends on the specific properties of the nanoparticles and the data required for the wastewater treatment application [29, 30].

NANOSCALE PROCESSES IN WASTEWATER TREATMENT

Nanoscale processes and NMs are increasingly being researched and used in water and wastewater treatment because of their unique properties and benefits. Some nanoscale processes that are commonly used in water and wastewater treatment are described below.

Nanofiltration and Reverse Osmosis

Nanofiltration (NF) and reverse osmosis (RO) are membrane processes that use nanoscale porous membranes to separate water pollution. These membranes have nanometer-sized pores that selectively block particles, ions, and dissolved species based on their size and charge. NF and RO remove dissolved salts, organic compounds, bacteria, viruses and other contaminants to produce purified water.

Adsorption

NMs such as activated carbon NPs, metal oxide NPs and carbon nanotubes have high adsorption capacity due to their large surface area and specific surface. They can selectively adsorb and remove various pollutants in water and wastewater, including heavy metals, organic pollutants and particulate matter. Adsorbents based on NMs have higher adsorption efficiency and can be regenerated and reused.

Photocatalysis

Photocatalysis involves the use of nanoscale photocatalysts, usually based on titanium dioxide (TiO_2) NPs that use light energy to generate reactive oxygen species that break down organic pollutants and disinfect water. When illuminated by ultraviolet or visible light, photocatalytic NPs can break down organic compounds into harmless substances. Photocatalysis can remove pollutants, dyes and some persistent pollutants [31].

Nanoscale Oxidation Process

The Advanced Oxidation Processes (AOP) use highly effective oxidizing agents to break down and mineralize pollutants in water. Nanoscale materials such as zero-valent metal NPs (nZVI), metal-doped NPs, and metal-organic frameworks can be used as catalysts in AOPs to form reactive species (such as hydroxyl radicals) that rapidly degrade bad weather. This process is especially good at removing organic compounds and bacteria from the skin.

Nanobiotechnology

Nanobiotechnology combines nanotechnology and biotechnology to develop new water treatment methods. For example, nanoscale biosorbents such as functionalized NPs or nanocomposites can be used to remove heavy metals and other pollutants by exploiting the affinity of bioactive molecules. NMs can also improve the performance of biological processes such as biofilm reactors and microbial fuel cells by improving mass transfer, substrate utilization and microbial interactions [32].

Nanosensors and Monitoring Devices

Nanotechnology-based sensors and monitoring devices provide real-time, precise detection of pollutants in water and wastewater. Nanosensors can be designed to detect specific pollutants or organisms with high sensitivity and selectivity. The sensors use NMs such as NPs, nanocomposites or nanowires as sensors that respond to target analytes and generate signals.

Nanofiltration and Enhanced Coagulation

Nanofiltration membranes and nanoscale coagulants can be used to improve coagulation and flocculation processes in water treatment. Nanoscale coagulants, such as polymer-based NPs, can improve and remove suspended solids, colloids and natural organic matter, improving the performance of the coagulation process. These nanoscale systems have the advantages of high efficiency, selectivity and versatility for water and wastewater treatment. However, in this process, it is important to consider the potential impact of NPs and NMs on the environment and health and to develop appropriate strategies for safe and stable use [32].

NMS FOR WASTEWATER TREATMENT

The elementary steps involved in the role of NMs in the process of wastewater treatment have been shown in Fig. (**2**) and explained in the following sections.

NMs in the Adsorption Process

Adsorption is a very common water treatment method that uses NMs, catalysis, adsorption, catalytic membranes, bioactive nanoparticles, biofilms, polymer and nanocomposite membranes, thin film composites, *etc*. [27]. NMs can be used as adsorbents for water treatment in many ways. The most investigated NMs for wastewater treatment are activated carbon, carbon nanotubes, graphene, Fe_3O_4,

MnO_2, Co_3O_4, TiO_2, MgO and ZnO, *etc.* Granular, tube, flake, *etc.* They can be prepared in many ways in different morphological forms, such as particles, tubes and sheets [33].

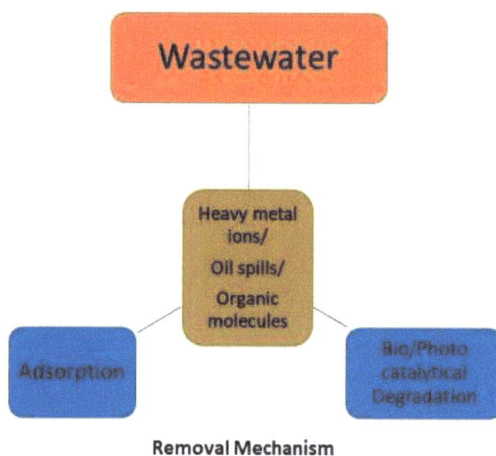

Fig. (2). Wastewater treatment process.

Activated Carbon-Based NMs

In recent years, various types of carbon-based NMs have been widely used for heavy metal and dye removal due to their non-toxicity, abundance, ease of preparation, high surface area and porosity, stable structure, and pressure. Activated charcoal was originally used as an adsorbent; however, due to the difficulty of removing heavy metals and dyes at the minute level, carbon nanotubes, fullerenes and graphene have been used as nano sorbents to overcome this difficulty. Activated carbon generally has high porosity and high surface area and can be prepared from carbonaceous precursors such as coal, wood, coconut shells and agricultural products. Activated carbon is widely used for the removal of inorganic and organic pollutants in wastewater and water treatment. For example, Activated carbon prepared from coconut shavings was used as an adsorbent for Cr (VI) removal from aqueous solutions [34].

Metal and Metal Oxide NMs

Metal and metal oxide NMs are other inorganic NMs commonly used for the removal of heavy metal ions and dyes. Nanoscale metals or metal oxides, including Fe_3O_4, MnO_2, TiO_2, MgO, CdO and ZnO, provide high surface area and high affinity. Metal oxides are less harmful to the environment, have low solubility and do not participate in the formation of secondary pollution; they are also used as adsorbents to remove heavy metals and dyes [35 - 37].

Metal NPs

Silver is the most widely used material due to its low toxicity and microbiological inactivation. Silver NPs are obtained from salts such as silver nitrate and silver chloride. Although the use of antibiotics is size dependent, smaller silver NPs (8 nm) are most effective, while smaller size (11-23 nm) results in lower bacterial activity. In addition, truncated triangular silver nanolayers exhibit better bacterial properties than spherical and rod NPs, demonstrating their shape dependence. Properties involved in the bactericidal activity of silver NPs include the generation of free radicals affecting the bacterial cell, interaction with DNA, adhesion to cell surface equipment, and enzymatic damage [38]. Silver NPs have also been incorporated into various polymers for the production of antibacterial nanofibers and nanocomposites. Various types of nanofibers containing silver NPs have been prepared for antibacterial applications and exhibit excellent antibacterial properties. Water filters made of polyurethane foam coated with silver nanofibers are effective against bacteria *Escherichia coli* (E. coli). Silver NPs are also used in water filters, such as polysulfone membranes, which are effective against different species of bacteria and viruses. Alternatively, doping silver NPs with other metal NPs or their composites can solve the problem of removing inorganic/organic compounds from water/wastewater [38, 39].

Metal Oxide NPs

Zinc Oxide NPs are excellent photocatalysts for wastewater treatment due to their special properties, such as band gap and oxidizing ability in the near ultraviolet region. The biocompatibility of these NPs makes them suitable for wastewater treatment [40]. Iron Oxide NPs, due to their simplicity and ease of synthesis, are now frequently used for heavy metal removal. This is an effective method to recover nanosorbent materials from contaminated water due to their small size, but magnetite and maghemite can be used as adsorbents due to their magnetic properties. According to the magnetic behaviour, metal oxide NPs acting as nano adsorbents can be recovered from the solution by applying an external magnet. Thus, these NPs work as nano adsorbents to remove heavy metal ions from water [41].

NMS IN THE PHOTOCATALYTIC PROCESS

The role of NMs in the photocatalytic degradation of various pollutants such as heavy metals, organic dyes, drugs, toxic organic compounds, pesticides, *etc.*, and their degradation mechanisms are shown in Fig. (**3**).

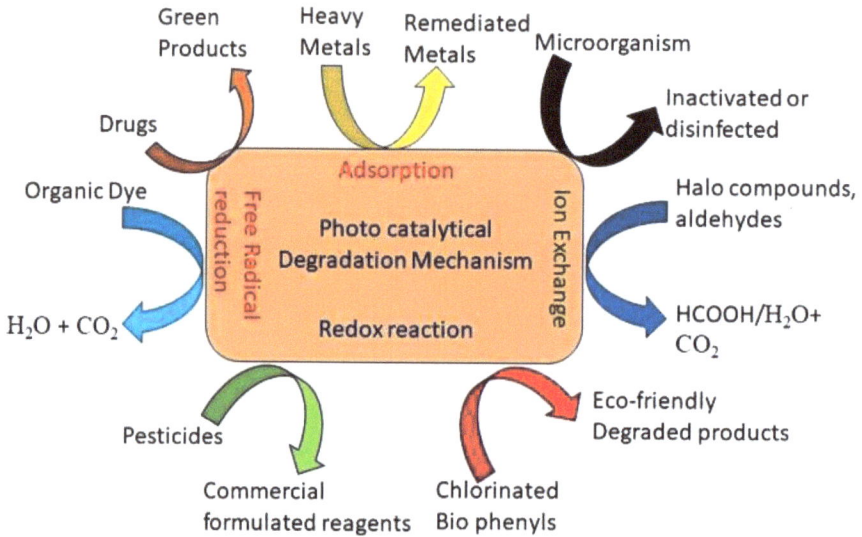

Fig. (3). Mechanism of photo catalytical degradation of heavy metals.

Titanium Oxide NPs (TiO$_2$ NPs)

TiO$_2$ NPs are one of the best photocatalysts for water purification. It can kill bacteria, insects, fungi, algae, bacteria, *etc.* The simple process of low-cost semiconductor-based photocatalysts such as TiO$_2$ is associated with the production of oxidants such as OH radicals. Ag doping of TiO$_2$ improves bacterial inactivation by eliminating or reducing the inactivation time of E. coli, thereby improving bacterial resistance to UV wavelengths and solar radiation. TiO$_2$ can be further modified and nanostructured with NPs of transition metal oxides. TiO$_2$ photocatalysts have shown great potential for water disinfection. For example, metal ion-substituted nitrogen-doped TiO$_2$ NPs photocatalysts are effective against *Escherichia coli* (Gram-negative bacteria), pseudomonas aeruginosa, insecticide disinfection efficacy, and Bacillus subtilis spores in visible light. Nitrogen-doped TiO$_2$ NPs are effective in breaking down microbial contaminants in water. Nanostructured TiO$_2$ films and membranes can kill bacteria as well as destroy pollutants under UV and visible light. Due to their stability in water, TiO$_2$ NPs can be incorporated into membranes or filters for water filtration [42].

Semiconductor NMs

Semiconductor NMs have shown great potential in wastewater treatment due to their unique properties and photocatalytic capabilities. In particular, semiconductor NMs such as TiO$_2$ and ZnO have been extensively studied for their ability to remove various pollutants from wastewater. Semiconductor NMs have been widely studied and have many applications in wastewater treatment [43].

TiO$_2$ is one of the most studied and used semiconductor NM for wastewater treatment. It has excellent photocatalytic properties and can degrade organic pollutants such as dyes, pharmaceuticals and industrial chemicals. TiO$_2$ NPs can be immobilized on the surface or used as a suspension in wastewater treatment [44]. ZnO is another photocatalytically active semiconductor NM that can effectively decompose organic pollutants and remove heavy metals. ZnO NPs can be immobilized in suspension or on supporting materials for wastewater treatment [43]. Iron oxide NPs (Fe$_2$O$_3$), especially hematite (α-Fe$_2$O$_3$) and magnetite (Fe$_3$O$_4$), show promising potential for the removal of various pollutants in wastewater. Tungsten oxide (WO$_3$) NPs showed photocatalytic activity to decompose organic pollutants and remove heavy metals from wastewater. WO$_3$ NPs can be combined in various morphologies, such as nanowires or nanolayers, to enhance their photocatalytic activity [45].

Researchers investigated the photocatalytic properties of Cadmium sulfide (CdS) NPs also in wastewater treatment. They can degrade organic pollutants such as dyes and phenols under light irradiation. However, the potential toxicity of cadmium limits the use of CdS NPs in large-scale wastewater treatment. It is important to remember that the choice of semiconductor NM depends on factors such as the nature of the disease, the treatment required, and light (UV or light). In addition, the selection of NMs should consider their stability, cost-effectiveness, and potential environmental impact. One of the main benefits of semiconductor NMs is their ability to harness solar energy and form electron-hole pairs when exposed to light. These electron-hole pairs can participate in redox reactions that lead to the degradation of organic pollutants and the removal of heavy metals from wastewater [46]. These semiconductor NMs can be used efficiently in the removal of heavy metals and the decomposition of organic pollutants from wastewater and the photocatalytic activity of these materials can lead to the degradation and extinction of bacteria and other harmful organisms [47, 48].

APPLICATIONS OF GREEN-SYNTHESIZED NPS IN WASTEWATER TREATMENT

Green synthesized NPs have many applications in wastewater treatment due to their unique properties and environmental interactions. Below are some important applications of green synthesized NPs in wastewater treatment.

Removal of Organic Pollutants

Green synthetic NPs, such as Ag NPs, are effective in removing pollutants released from wastewater. These NPs have excellent catalytic and adsorption properties, allowing them to break down and remove organic pollutants, including dyes, pesticides, chemicals and organic solvents. Various carbon-based adsorbents have been used to remove organic contaminants from raw water.

CNTs

Various types of NMs, such as nano adsorbents, CNTs, polymeric materials (such as dendrimers) and zeolites, have good adsorption properties and can be used to remove organic materials from water/wastewater. CNTs are also effective in removing polycyclic aromatic organic compounds and atrazine. Nanoporous-activated carbon fibres prepared by electrospinning carbon nanotubes showed higher organic adsorption equilibrium constants for benzene, toluene, xylene and ethylbenzene than granular activated carbons. It was found that CNTs are effective in the absorption of 1,2-dichlorobenzene. In comparison to activated carbon, both single-walled and multi-walled carbon nanotubes show higher trihalomethane adsorption capacity. Multi-walled carbon nanotubes, chlorophenols, pesticides, DDT, *etc.*, can be used as adsorbents. Finally, new polymers with carbon nanotubes are effective at removing pollutants from water.

Zero-Valent Iron

Nanocatalysts, including semiconductor materials, zero-valence metals, and bimetallic NPs, have been used for the degradation of environmental contaminants such as PCBs, pesticides, and azo dyes due to their higher surface area and shape-dependent properties. Magnetic nano sorbents have also shown their efficiency in the removal of organic contaminants. Fe_2O_3 NMs have shown better removal capabilities for organic pollutants than that has been shown by bulk materials.

Other Nanomaterials

Nanostructured ZnO semiconductor thin films are used for the degradation of organic pollutants (4-chlorocatechol). Nanocatalysts of silver and amidoxime fibers have been successfully used for the degradation of organic dyes. In a study, a bimetallic Pd-Cu/γ-alumina nanocomposite was used to reduce nitrate. The results show that halogenated organic compounds can be biodegraded using hydrogen gas and Pd-based nanoparticles. Pd NPs and bimetallic Pd/Au (gold) NPs have been successfully used for the hydrochlorination of trichloroethylene (TCE) [49]. Finally, it was found that single-enzyme NPs can be used to remove various organic pollutants [49, 50].

Removal of Heavy Metals

The green-colored synthesized NPs show high activity in the removal of heavy metal ions from wastewater. NPs such as Fe_2O_3 NPs can selectively adsorb and remove heavy metals such as lead, cadmium, mercury and arsenic. NPs can be functionalized with specific ligands or coatings to improve their affinity and selectivity for heavy metal targets. Composites of CNTs with Fe and cerium oxide (CeO_2) have also been used to remove heavy metal ions from water. Cerium oxide NPs supported on CNTs are used effectively to adsorb arsenic.

Antibacterial Domain

Some green synthesized NPs have strong antibacterial properties, making them suitable for water disinfection applications. For example, AgNPs have strong antibacterial properties against a wide variety of bacteria, including gram-positive and gram-negative types. These NPs can inhibit the growth of bacteria and prevent the spread of waterborne diseases.

Removal of Contaminants

It is difficult to remove pollutants such as personal care products and endocrine disruptors from wastewater. It has been shown that green synthesized NPs, especially those derived from plant extracts or bacteria, can break down or remove these pollutants. NPs can break down the molecular structure and turn it into non-toxic substances [51].

Enhanced Membrane Filtration for Wastewater Treatment

NPs synthesized by green methods can also be used to improve the performance of membrane filtration systems. Integrating NPs into the membrane or as a material layer can improve membrane filtration efficiency, antifouling performance and selectivity. This can improve water quality and reduce pollution during wastewater treatment.

The membrane immobilization of metal NPs has also been shown to be effective for the degradation and dichlorination of pollutants. Inorganic membranes containing nano-TiO_2 or modified nano-TiO_2 have been used effectively to reduce the degradation of pollutants, especially chlorinated chemicals. TiO_2 slurries immobilized on polyethene supports and TiO_2 slurries attached to polymer membranes proved to be effective in breaking down 1,2-dichlorobenzene and chemicals, respectively. By incorporating TiO_2 NPs into the alumina membrane, membrane fouling can be controlled by reducing the adsorption of oil droplets to the membrane during wastewater treatment. Finally, the ceramic membrane of

TiO_2 and CNTs improves the permeability and photocatalytic activity of the membrane [52, 53].

Degradation of Dyes and Elimination of Heavy Metals from Contaminated Water by Green Synthesized NPs

Water is mainly polluted by dyes such as methylene blue, methyl orange, bromophenol blue, bromocresol green, reactive dye, crystal violet, malachite green, Alizarin Yellow R (AYR), congo red, remazol yellow RR, eosin yellow, rhodamine B dye, *etc.* Nowadays, contaminated water is treated by various green synthesized NPs which is a promising and effective method to treat wastewater [54]. Table **2** shows the source of green synthesis of NPs and their application in dye degradation.

Table 2. Green synthesized NPs and their application in dye degradation.

Name of the Plant	Synthesized NPs	Degraded Dye	Refs.
Pomegranate seed (Punica granatum)	Iron oxide	UV degradation of textile dyes in wastewater	[55]
Fruit extract of *Actinidia chinensis*	Nanosized irons (FeNPs)	UV degradation of alizarin yellow R (AYR) dye	[56]
Clerodendrum infortunatum and *Clerodendrum inerme*	Cu-doped ZnO NPs	Degradation of Acid Black 234	[57]
Extract of *Hibiscus sabdariffa* flowers	Cu NPs	Removal of nitrate	[58]
P. aeruginosa JP-11	CdS	Removal of Cd	[59]
Anaerobic microbial consortium	Se	Removal of Zn(II)	[60]
Citrobacter freundii Y9	Se	Removal *of HgO*	[61]
Fruit extract of *Ficus carica*	Copper nanoparticle (CuNPs)	Photodegradation of alizarin yellow R (AYR) dye	[62]
Vitis rotundifolia (Hybrid grape pulp)	Zinc oxide (ZnO) nanoparticle	Photodegradation of malachite green dye	[63]
Leaf extract of *Moringa oleifera*	Ni/Fe_3O_4 magnetic nanoparticles	Photodegradation of malachite green dye	[64]
Leaf extract of *Parthenocissus quinquefolia* (Virginia creeper)	Fe-Cu oxides	Photodegradation of malachite green dye	[65]
Aqueous extract of *date pits*	MgO NPs	Photodegradation of methylene blue	[66]
Madhuca longifolia	Cupric oxide nanoparticles	Irradiation of methylene blue	[67]
Bagasse extract	Graphene oxide(rGO)	Photodegradation of methylene blue	[68]
Desmodesmus sp. WR1	Manganese oxides	Removal of bisphenol A	[69]

(Table 2) cont.....

Name of the Plant	Synthesized NPs	Degraded Dye	Refs.
Pseudomonas sp. G7	Nano-MnO$_x$	Oxidative degradation of 2-chlorophenol, 2,4-dichlorophenol,and 2,4,6-trichlorophenol	[70]
Vitis vinifera juice	Dy$_2$Ce$_2$O$_7$ nanostructure	Degradation of MO, RhB and 2-naphthol	[71]
Pomegranate juice	Ln$_2$Sn$_2$O$_7$ nanostructure	Degradation of EY, eriochrome black T and methyl violet	[72]
Shewanella oneidensis MR-1	Pd/Au	Dechlorination of diclofenac	[73]
Mukia maderaspatna	Silver nanocomposite	Photodegradation of methylene blue	[74]
Fruit extract of *Crataegus pentagyna*	Silver nanoparticles (AgNPs)	photodegradation of organic contaminant dyes such as rhodamine b (Rhb), eosin (EY), and methylene blue	[75]
Kernel extract of *Terminaliabellericain*	Silver nanoparticles K-AgNPs	Catalytic reduction of organic pollutants such as 4-nitrophenol, methylene blue, eosin yellow, and methyl orange.	[76]
Papaya leaf extract (*Carica papaya*)	Iron oxide nanoparticles	Photocatalytic reduction of remazol yellow RR dye	[77]
Leaf extract of *Citrofortunella microcarpa* (calamondin)	copper oxide nanoparticles	Photodegradation of rhodamine B (RhB) dye	[78]
Alpinia nigra leaves	Gold nanoparticles	Photodegradation of toxic dyes such as Rhodamine B and methyl orange	[79]
Leaf extract of *Eucalyptus spp*	Zinc oxide nanoparticles	Photodegradation of crystal violet and malachite green dye	[80]
Leaf extract of *Hibiscus Rosa-sinensis*	Zinc oxide nanoparticles	Photodegradation of congo red dye	[81]
Extract of *pomegranate* leaves	Fe$_3$O$_4$ magnetic nanoparticles	Photodegradation of congo red dye	[82]
Brassica rapa leaf extract	Cadmium tungstate (CdWO$_4$)	Photodegradation of Bismarck brown R dye	[83]
Cellulose extract of *Hibiscus sabdariffa*	Silver nanoparticles AgNPs	Catalytic reduction hazardous dyes such as methylene blue, methyl orange, bromophenol blue, eosin Y, and orange G dye	[84]
Plant extract of *Lagerstroemia speciosa*	Gold nanoparticles (AuNPs)	Photodegradation of organic pollutant dyes	[85]
Green tea extract of *Tieguanyin*	Iron nanoparticles	Photodegradation of bromophenol blue dye	[86]
Rosa damascene (RD), *Thymus vulgaris* (TV), and *Urtica dioica* (UD).	Zero-valent iron/Cu nanoparticles	Hexavalent chromium	[87]

(Table 2) cont.....

Name of the Plant	Synthesized NPs	Degraded Dye	Refs.
Coconut husk extract	Iron oxide nanoparticles	Ca (Calcium) and Cd (Cadmium	[88]
Emblica Officinalis leaf extract	Zero-valent iron nanoparticles	Removal of lead	[89]
Vaccinium floribundum	Zero-valent iron nanoparticles	Removal of petroleum oil	[90]
Leaves, namely *Azadiracta indica, Magnolia champaca, Mangifera indica, Murraya Koenigii*	Iron nanoparticle	(phosphate, chemical oxygen demands, and ammonia nitrogen	[91]
Green tea extract	Silica nanoparticles/iron oxide nanoparticles	Degradation of organochlorine pesticides	[92]

CURRENT CHALLENGES AND FUTURE PERSPECTIVES

It is important to provide good water purifying equipment to guarantee clean drinking water. Measuring the indicators against the industry level requires more effort to provide flexible water treatment technologies. NMs can offer unique advantages over other water treatment technologies, such as their ability to combine with various materials to create versatile materials such as nanocomposites and nanomembranes that can remove toxic pollutants. In addition, NMs have shown great potential due to their valuable properties, such as high surface area. However, some limitations affect the successful use of NMs [93]. These are;

1. As noted earlier, the use of products with functional NPs is potentially risky due to the possibility of NPs being released into the environment. For this reason, different laws and regulations have been created to reduce health risks.

2. Bio NPs have great potential due to the deep cultivation and sustainability of the production process and their great activity in removing environmental pollutants. However, advances in analytical and imaging techniques require different ways to evaluate and identify the grey areas in this field.

3. Surface modification and functionalization of NMs, such as polymer coating, addition of specific functional groups or adjustment of surface charge, is a challenging task, though it can improve the performance and stability of nanomaterials in wastewater treatment.

4. Nanoscale Sensors and Monitoring: the development of nanoscale sensors and monitoring systems enables timely detection and monitoring of pollutants in wastewater.

5. Nano-Enabled Advanced oxidation processes (AOP) use new semiconductor nanomaterials, nanocomposites and nanohybrids to better break down persistent organic pollutants, emerging pollutants and stubborn compounds in wastewater.

8. Environmental and safety assessment: Sustainable and safe construction, use and disposal practices to minimize risk [94].

CONCLUSION

Nanotechnology has great potential in wastewater treatment. The special properties of NMs, such as high surface area, improved reactivity and tunable surface area, allow the removal of contaminants from wastewater. In this chapter, we have explored the various applications of nanotechnology in wastewater treatment. Various NMs have shown their efficiency in the removal of contaminants, making them potentially suitable for remediation processes. However, disadvantages limit the commercialization of these devices. These disadvantages are the cost-effectiveness of the process, environmental concerns, and difficulties in scaling and installation to the commercial level. Overall, nanotechnology offers an exciting way to replace wastewater by providing better, more efficient and effective solutions to today's global water problems.

REFERENCES

[1]   Theron J, Walker JA, Cloete TE. Nanotechnology and water treatment: Applications and emerging opportunities. Crit Rev Microbiol 2008; 34(1): 43-69.
[http://dx.doi.org/10.1080/10408410701710442] [PMID: 18259980]

[2]   Riffat R. Fundamentals of Wastewater Treatment and Engineering. 1st ed. USA: CRC Press 2012; p. 359.
[http://dx.doi.org/10.1201/b12746]

[3]   Sharma R. Nanotechnology: An approach for water purification-review. IOP Conf. Ser.: Material Science and Engineering; 2020 December 18th – 19th Mathura, India. .

[4]   Sandia D. water purification roadmap-a report of the executive committee, US: Department of the interior Bureau of Reclamation and Sandia National Laboratories DWPR Progr. 2003.

[5]   Allhoff F, Lin P, Moore D. What is nanotechnology and why does it matter: From science to ethics. John Wiley & Sons 2009; pp. 1-67.

[6]   Lu H, Wang J, Stoller M, Wang T, Bao Y, Hao H. An overview of nanomaterials for water and wastewater treatment. An Overview of Nanomaterials for Water and Wastewater Treatment 2016; 1: 1-10.
[http://dx.doi.org/10.1155/2016/4964828]

[7]   Bechelany M, Eustache CG, Patrick EA, *et al.* Risks of drinking water Contamination by chemical and organic substances. Republic. Int Res J Environ Sci 2013; 2(1): 49-57.

[8]   Mussa T, Abdulla A, Alwan A, Salih D, Ali F. Removal of cadmium from wastewater using low cost natural adsorbents. Int Res J Environ Sci 2015; 4(10): 103-6.

[9]   Md H, Afroza P, Zaman A, *et al.* Qualitative and quantitative determination of the residual levels of chemical pesticides of the shrimp farm. International Science Congress Association. 11-5.

[10] Ma N, Quan X, Zhang Y, Chen S, Zhao H. Integration of separation and photocatalysis using an inorganic membrane modified with Si-doped TiO_2 for water purification. J Membr Sci 2009; 335(1-2): 58-67.
[http://dx.doi.org/10.1016/j.memsci.2009.02.040]

[11] Liu F, Yang J, Zuo J, *et al.* Graphene-supported nanoscale zero-valent iron: Removal of phosphorus from aqueous solution and mechanistic study. J Environ Sci (China) 2014; 26(8): 1751-62.
[http://dx.doi.org/10.1016/j.jes.2014.06.016] [PMID: 25108732]

[12] Kalhapure RS, Sonawane SJ, Sikwal DR, *et al.* Solid lipid nanoparticles of clotrimazole silver complex: An efficient nano antibacterial against staphylococcus aureus and MRSA. Colloids Surf B Biointerfaces 2015; 136(1): 651-8.
[http://dx.doi.org/10.1016/j.colsurfb.2015.10.003] [PMID: 26492156]

[13] Behari J, Tiwari DK, Sen P. Application of nanoparticles in waste water treatment. World Appl Sci J 2008; 3(3): 417-33.

[14] Savage N, Diallo MS. Nanomaterials and water purification: opportunities and challenges. J Nanopart Res 2005; 7(4-5): 331-42.
[http://dx.doi.org/10.1007/s11051-005-7523-5]

[15] Obare SO, Meyer GJ. Nanostructured materials for environmental remediation of organic contaminants in water. J Environ Sci Health Part A Tox Hazard Subst Environ Eng 2004; 39(10): 2549-82.
[http://dx.doi.org/10.1081/ESE-200027010] [PMID: 15509009]

[16] Yadav VK, Khan SH, *et al.* Microbial synthesis of nanoparticles and their applications for wastewater treatment First online singapore. Springer 2020; pp. 147-87.

[17] Dave S, Sharma R. Use of nanoparticles in water treatment: A review. Int Res J Environ Sci 2015; 4(10): 103-6.

[18] Gupta VK, Tyagi I, Sadegh H, Ghoshekand RS, Makhlouf ASH, Maazinejad B. Nanoparticles as adsorbent; a positive approach for removal of noxious metal ions: A review. Science, Technology and Development 2015; 34(3): 195-214.
[http://dx.doi.org/10.3923/std.2015.195.214]

[19] Madhura L, Singh S, Kanchi S, Sabela M, Bisetty K, Inamuddin . Nanotechnology-based water quality management for wastewater treatment. Environ Chem Lett 2019; 17(1): 65-121.
[http://dx.doi.org/10.1007/s10311-018-0778-8]

[20] Sadegh H, Ali GAM, Gupta VK, *et al.* The role of nanomaterials as effective adsorbents and their applications in wastewater treatment. J Nanostructure Chem 2017; 7(1): 1-14.
[http://dx.doi.org/10.1007/s40097-017-0219-4]

[21] Zare K, Gupta VK, Moradi O, *et al.* A comparative study on the basis of adsorption capacity between CNTs and activated carbon as adsorbents for removal of noxious synthetic dyes: A review. J Nanostructure Chem 2015; 5(2): 227-36.
[http://dx.doi.org/10.1007/s40097-015-0158-x]

[22] Yaqoob AA, Parveen T, Umar K, Ibrahim MNM. Role of nanomaterials in the treatment of wastewater. RE:view 2020; 12(2): 495.

[23] Jain K, Patel AS, Pardhi VP, Flora SJS. Nanotechnology in wastewater management: A new paradigm towards wastewater treatment. Molecules 2021; 26(6): 1797.
[http://dx.doi.org/10.3390/molecules26061797] [PMID: 33806788]

[24] Madhu SR, Sharma R, Bharti R. A review on the synthesis and photocatalytic application of silver nano particles. IOP Conf Ser Earth Environ Sci 2023; 1110(1)012021
[http://dx.doi.org/10.1088/1755-1315/1110/1/012021]

[25] Kalpana VN, Kataru BAS, Sravani N, Vigneshwari T, Panneerselvam A, Devi Rajeswari V.

Biosynthesis of zinc oxide nanoparticles using culture filtrates of aspergillus niger: Antimicrobial textiles and dye degradation studies. OpenNano 2018; 3: 48-55.
[http://dx.doi.org/10.1016/j.onano.2018.06.001]

[26] Kirthi AV, Rahuman AA, Rajakumar G, *et al.* Biosynthesis of titanium dioxide nanoparticles using bacterium *bacillus subtilis*. Mater Lett 2011; 65(17-18): 2745-7.
[http://dx.doi.org/10.1016/j.matlet.2011.05.077]

[27] Kuppusamy P, Yusoff MM, Maniam GP, Govindan N. Biosynthesis of metallic nanoparticles using plant derivatives and their new avenues in pharmacological applications – An updated report. Saudi Pharm J 2016; 24(4): 473-84.
[http://dx.doi.org/10.1016/j.jsps.2014.11.013] [PMID: 27330378]

[28] Malik P, Shankar R, Malik V, Sharma N, Mukherjee TK. Green chemistry based benign routes for nanoparticle synthesis. Journal of Nanoparticles 2014; 2014: 1-14.
[http://dx.doi.org/10.1155/2014/302429]

[29] Nasrollahzadeh M, Sajjadi M, Iravani S, Varma RS. Green-synthesized nanocatalysts and nanomaterials for water treatment: Current challenges and future perspectives. J Hazard Mater 2021; 401123401
[http://dx.doi.org/10.1016/j.jhazmat.2020.123401] [PMID: 32763697]

[30] Saravanan A, Kumar PS, Hemavathy RV, *et al.* A review on synthesis methods and recent applications of nanomaterial in wastewater treatment: Challenges and future perspectives 2022; 307.

[31] Solanki VS, Pare B, Agarwal N, Singh HL. Photocatalytic mineralization of pharmaceutical contaminants in water by hererogeneous photocatalysis in the presence of LED-Irradiation: A green and sustainable approach. In: Agarwal N, Ed. Pharmaceuticals:Boon or Bane. 146-75.
[http://dx.doi.org/10.52305/GPMC4427]

[32] Soni R, Pal AK, Tripathi P, Lal JA, Kesari K, Tripathi V. An overview of nanoscale materials on the removal of wastewater contaminants. Appl Water Sci 2020; 10(8): 189.
[http://dx.doi.org/10.1007/s13201-020-01275-3]

[33] Gomez-Gonzalez MA, Koronfel MA, Pullin H, *et al.* Nanoscale chemical imaging of nanoparticles under real-world wastewater treatment conditions. Adv Sustain Syst 2021; 5(7)2100023.
[http://dx.doi.org/10.1002/adsu.202100023]

[34] Patanjali P, Singh R, Kumar A, Chaudhary P. Nanotechnology forwater treatment: A green approach. Elsevier BV 2019; pp. 485-512.

[35] Khalid A, Ahmad P, Memon R., *et al.* Structural, Optical, and Renewable Energy-Assisted Photocatalytic Dye Degradation Studies of ZnO, CuZnO, and CoZnO Nanostructures for Wastewater Treatment. Separations. 2023; 10(3):184
[http://dx.doi.org/10.3390/separations10030184]

[36] Machida M, Mochimaru T, Tatsumoto H. Lead(II) adsorption onto the graphene layer of carbonaceous materials in aqueous solution. Carbon 2006; 44(13): 2681-8.
[http://dx.doi.org/10.1016/j.carbon.2006.04.003]

[37] Singh S, Barick KC, Bahadur D. Functional oxide nanomaterials and nanocomposites for the removal of heavy metals and dyes. Nanomaterials and Nanotechnology 2013; 3(1): 20.
[http://dx.doi.org/10.5772/57237]

[38] Kallman EN, Oyanedel-Craver VA, Smith JA. Ceramic filters impregnated with silver nanoparticles for point-of-use water treatment in rural guatemala. J Environ Eng 2011; 137(6): 407-15.
[http://dx.doi.org/10.1061/(ASCE)EE.1943-7870.0000330]

[39] Amin MT, Alazba AA, Manzoor U. A review of removal of pollutants from water/wastewater using different types of nanomaterials. Adv Mater Sci Eng 2014; 2014: 1-24.
[http://dx.doi.org/10.1155/2014/825910]

[40] Ghosh N, Das S, Biswas G, Haldar PK. Review on some metal oxide nanoparticles as effective

adsorbent in wastewater treatment. Water Sci Technol 2022; 85(12): 3370-95.
[http://dx.doi.org/10.2166/wst.2022.153] [PMID: 35771052]

[41] Naseem T, Durrani T. The role of some important metal oxide nanoparticles for wastewater and antibacterial applications: A review. Environmental Chemistry and Ecotoxicology 2021; 3(3): 59-75.
[http://dx.doi.org/10.1016/j.enceco.2020.12.001]

[42] Mir N, Khan A, Umar K, Muneer M. Photocatalytic study of a xanthene dye derivative, phloxine B in aqueous suspension of TiO $_2$: Adsorption isotherm and decolourization kinetics. Energy and Environment Focus 2013; 2(3): 208-16.
[http://dx.doi.org/10.1166/eef.2013.1052]

[43] Singh S, Barick K, Bahadur D. Fe$_3$O$_4$ embedded ZnO nanocomposites for the removal of toxic metal ions, organic dyes and bacterial pathogens. J Mater Chem 2013; A 1(10): 3325-33.

[44] Parva C. Using semiconductor nanomaterials as photocatalysts for water treatment. AZoNano 2023.

[45] Mbarek WB, Escoda L, Saurina J, *et al.* Nanomaterials as a sustainable choice for treating wastewater: A review. Materials (Basel) 2022; 15(23): 8576.
[http://dx.doi.org/10.3390/ma15238576] [PMID: 36500069]

[46] Yaqoob AA, Parveen T, Umar K, Mohamad Ibrahim MN. Role of nanomaterials in the treatment of wastewater: A review. Water 2020; 12(2): 495.
[http://dx.doi.org/10.3390/w12020495]

[47] Bakshi M, Abhilash P C. Nano-Materials as photocatalysts for degradation of environmental pollutants. 2020.

[48] Lee H, Kim BH, Park YK, Kim SJ, Jung SC. Application of recycled zero-Valent iron nanoparticle to the treatment of wastewater containing nitrobenzene. J Nanomater 2015; 2015: 1-8.
[http://dx.doi.org/10.1155/2015/392537]

[49] Singh NB, B H Susan MA, Guin M. Applications of green synthesized nanomaterials in water remediation. Curr Pharm Biotechnol 2021; 22(6): 733-61.
[http://dx.doi.org/10.2174/1389201021666201027160029] [PMID: 33109041]

[50] Bhatt P, Pandey SC, Joshi S, *et al.* Nanobioremediation: A sustainable approach for the removal of toxic pollutants from the environment. J Hazard Mater 2022; 427128033.
[http://dx.doi.org/10.1016/j.jhazmat.2021.128033] [PMID: 34999406]

[51] Yu S, Tang H, Zhang D, *et al.* MXenes as emerging nanomaterials in water purification and environmental remediation. Sci Total Environ 2022; 811152280.
[http://dx.doi.org/10.1016/j.scitotenv.2021.152280] [PMID: 34896484]

[52] Uthayakumar H, Panchamoorthy GK, Jayaseelan A. Chitosan based nano adsorbents and its types for heavy metal removal: A mini review. Mater Lett 2022; 312(28)131670.

[53] Saratale RG, Saratale GD, Shin HS, *et al.* New insights on the green synthesis of metallic nanoparticles using plant and waste biomaterials: current knowledge, their agricultural and environmental applications. Environ Sci Pollut Res Int 2018; 25(11): 10164-83.
[http://dx.doi.org/10.1007/s11356-017-9912-6] [PMID: 28815433]

[54] Agarwal N, Solanki VS, Pare B, Singh N, Jonnalagadda SB. Current trends in nanocatalysis for green chemistry and its applications- a mini-review. Curr Opin Green Sustain Chem 2023; 41(2)100788.
[http://dx.doi.org/10.1016/j.cogsc.2023.100788]

[55] Bibi I, Nazar N, Ata S, *et al.* Green synthesis of iron oxide nanoparticles using pomegranate seeds extract and photocatalytic activity evaluation for the degradation of textile dye. J Mater Res Technol 2019; 8(6): 6115-24.
[http://dx.doi.org/10.1016/j.jmrt.2019.10.006]

[56] Ahmed A, Usman M, Yu B, *et al.* Efficient photocatalytic degradation of toxic alizarin yellow R dye from industrial wastewater using biosynthesized Fe nanoparticle and study of factors affecting the

degradation rate. J Photochem Photobiol B 2020; 202111682.
[http://dx.doi.org/10.1016/j.jphotobiol.2019.111682] [PMID: 31731077]

[57]   Khan SA, Noreen F, Kanwal S, Iqbal A, Hussain G. Green synthesis of ZnO and Cu-doped ZnO nanoparticles from leaf extracts of Abutilon indicum, clerodendrum infortunatum, clerodendrum inerme and investigation of their biological and photocatalytic activities. Mater Sci Eng C 2018; 82: 46-59.
[http://dx.doi.org/10.1016/j.msec.2017.08.071] [PMID: 29025674]

[58]   Paixão RM, Reck IM, Bergamasco R, Vieira MF, Vieira AMS. Activated carbon of Babassu coconut impregnated with copper nanoparticles by green synthesis for the removal of nitrate in aqueous solution. Environ Technol 2018; 39(15): 1994-2003.
[http://dx.doi.org/10.1080/09593330.2017.1345990] [PMID: 28639851]

[59]   Raj R, Dalei K, Chakraborty J, Das S. Extracellular polymeric substances of a marine bacterium mediated synthesis of CdS nanoparticles for removal of cadmium from aqueous solution. J Colloid Interface Sci 2016; 462: 166-75.
[http://dx.doi.org/10.1016/j.jcis.2015.10.004] [PMID: 26454375]

[60]   Jain R, Jordan N, Schild D, *et al.* Adsorption of zinc by biogenic elemental selenium nanoparticles. Chem Eng J 2015; 260: 855-63.
[http://dx.doi.org/10.1016/j.cej.2014.09.057]

[61]   Wang X, Zhang D, Qian H, Liang Y, Pan X, Gadd GM. Interactions between biogenic selenium nanoparticles and goethite colloids and consequence for remediation of elemental mercury contaminated groundwater. Sci Total Environ 2018; 613-614: 672-8.
[http://dx.doi.org/10.1016/j.scitotenv.2017.09.113] [PMID: 28938209]

[62]   Usman M, Ahmed A, Yu B, Peng Q, Shen Y, Cong H. Photocatalytic potential of bio-engineered copper nanoparticles synthesized from ficus carica extract for the degradation of toxic organic dye from waste water: Growth mechanism and study of parameter affecting the degradation performance. Mater Res Bull 2019; 120110583.
[http://dx.doi.org/10.1016/j.materresbull.2019.110583]

[63]   Brindhadevi K, Samuel MS, Verma TN, *et al.* Zinc oxide nanoparticles (ZnONPs) -induced antioxidants and photocatalytic degradation activity from hybrid grape pulp extract (HGPE). Biocatal Agric Biotechnol 2020; 28101730.
[http://dx.doi.org/10.1016/j.bcab.2020.101730]

[64]   Prasad C, Sreenivasulu K, Gangadhara S, Venkateswarlu P. Bio inspired green synthesis of Ni/Fe$_3$O$_4$ magnetic nanoparticles using Moringa oleifera leaves extract: A magnetically recoverable catalyst for organic dye degradation in aqueous solution. J Alloys Compd 2017; 700: 252-8.
[http://dx.doi.org/10.1016/j.jallcom.2016.12.363]

[65]   Zhang P, Hou D, Li X, Pehkonen S, Varma RS, Wang X. Greener and size-specific synthesis of stable Fe-Cu oxides as earth-abundant adsorbents for malachite green. J Mater Chem A Mater Energy Sustain 2018; 6: 9229-36.
[PMID: 30147937]

[66]   Sackey J, Bashir AKH, Ameh AE, Nkosi M, Kaonga C, Maaza M. Date pits extracts assisted synthesis of magnesium oxides nanoparticles and its application towards the photocatalytic degradation of methylene blue. J King Saud Univ Sci 2020; 32(6): 2767-76.
[http://dx.doi.org/10.1016/j.jksus.2020.06.013]

[67]   Das P, Ghosh S, Ghosh R, Dam S, Baskey M. Madhuca longifolia plant mediated green synthesis of cupric oxide nanoparticles: A promising environmentally sustainable material for waste water treatment and efficient antibacterial agent. J Photochem Photobiol B 2018; 189: 66-73.
[http://dx.doi.org/10.1016/j.jphotobiol.2018.09.023] [PMID: 30312922]

[68]   Gan L, Li B, Chen Y, Yu B, Chen Z. Green synthesis of reduced graphene oxide using bagasse and its application in dye removal: A waste-to-resource supply chain. Chemosphere 2019; 219: 148-54.

[http://dx.doi.org/10.1016/j.chemosphere.2018.11.181] [PMID: 30537587]

[69]   Wang R, Wang S, Tai Y, *et al.* Biogenic manganese oxides generated by green algae Desmodesmus sp. WR1 to improve bisphenol a removal. J Hazard Mater 2017; 339: 310-9.
[http://dx.doi.org/10.1016/j.jhazmat.2017.06.026] [PMID: 28658640]

[70]   Tu J, Yang Z, Hu C. Efficient catalytic aerobic oxidation of chlorinated phenols with mixed-valent manganese oxide nanoparticles. J Chem Technol Biotechnol 2015; 90(1): 80-6.
[http://dx.doi.org/10.1002/jctb.4289]

[71]   Zinatloo-Ajabshir S, Salehi Z, Salavati-Niasari M. Green synthesis and characterization of $Dy_2Ce_2O_7$ ceramic nanostructures with good photocatalytic properties under visible light for removal of organic dyes in water. J Clean Prod 2018; 192: 678-87.
[http://dx.doi.org/10.1016/j.jclepro.2018.05.042]

[72]   Zinatloo-Ajabshir S, Morassaei MS, Salavati-Niasari M. $Nd_2Sn_2O_7$ nanostructures as highly efficient visible light photocatalyst: Green synthesis using pomegranate juice and characterization. J Clean Prod 2018; 198(4): 11-8.
[http://dx.doi.org/10.1016/j.jclepro.2018.07.031]

[73]   De Corte S, Sabbe T, Hennebel T, *et al.* Doping of biogenic Pd catalysts with Au enables dechlorination of diclofenac at environmental conditions. Water Res 2012; 46(8): 2718-26.
[http://dx.doi.org/10.1016/j.watres.2012.02.036] [PMID: 22406286]

[74]   Guruviah KD, Senthil PK, Sathish KK. Green synthesis of novel silver nanocomposite hydrogel based on sodium alginate as an efficient biosorbent for the dye wastewater treatment: Prediction of isotherm and kinetic parameters. Desalination Water Treat 2016; 57(57): 27686-99.

[75]   Ebrahimzadeh MA, Naghizadeh A, Amiri O, Shirzadi-Ahodashti M, Mortazavi-Derazkola S. Green and facile synthesis of Ag nanoparticles using crataegus pentagyna fruit extract (CP-AgNPs) for organic pollution dyes degradation and antibacterial application. Bioorg Chem 2020; 94103425.
[http://dx.doi.org/10.1016/j.bioorg.2019.103425] [PMID: 31740048]

[76]   Sherin L, Sohail A, Amjad U-S, Mustafa M, Jabeen R, Ul-Hamid A. Facile green synthesis of silver nanoparticles using Terminalia bellerica kernel extract for catalytic reduction of anthropogenic water pollutants. Colloid Interface Sci Commun 2020; 37100276.
[http://dx.doi.org/10.1016/j.colcom.2020.100276]

[77]   Bhuiyan MSH, Miah MY, Paul SC, *et al.* Green synthesis of iron oxide nanoparticle using *Carica papaya* leaf extract: application for photocatalytic degradation of remazol yellow RR dye and antibacterial activity. Heliyon 2020; 6(8)e04603.
[http://dx.doi.org/10.1016/j.heliyon.2020.e04603] [PMID: 32775754]

[78]   Rafique M, Shafiq F, Ali Gillani SS, Shakil M, Tahir MB, Sadaf I. Eco-friendly green and biosynthesis of copper oxide nanoparticles using Citrofortunella microcarpa leaves extract for efficient photocatalytic degradation of Rhodamin B dye form textile wastewater. Optik (Stuttg) 2020; 208164053.
[http://dx.doi.org/10.1016/j.ijleo.2019.164053]

[79]   Baruah D, Goswami M, Yadav RNS, Yadav A, Das AM. Biogenic synthesis of gold nanoparticles and their application in photocatalytic degradation of toxic dyes. J Photochem Photobiol B 2018; 186: 51-8.
[http://dx.doi.org/10.1016/j.jphotobiol.2018.07.002] [PMID: 30015060]

[80]   Chauhan AK, Kataria N, Garg VK. Green fabrication of ZnO nanoparticles using Eucalyptus spp. leaves extract and their application in wastewater remediation. Chemosphere 2020; 247125803.
[http://dx.doi.org/10.1016/j.chemosphere.2019.125803] [PMID: 31972482]

[81]   Debnath P, Mondal NK. Effective removal of congo red dye from aqueous solution using biosynthesized zinc oxide nanoparticles. Environ Nanotechnol Monit Manag 2020; 14100320
[http://dx.doi.org/10.1016/j.enmm.2020.100320]

[82]   Prasad C, Karlapudi S, Venkateswarlu P, Bahadur I, Kumar S. Green arbitrated synthesis of Fe 3 O 4 magnetic nanoparticles with nanorod structure from pomegranate leaves and Congo red dye degradation studies for water treatment. J Mol Liq 2017; 240: 322-8.
[http://dx.doi.org/10.1016/j.molliq.2017.05.100]

[83]   Fatima B, Siddiqui SI, Ahmed R, Chaudhry SA. Green synthesis of f-CdWO$_4$ for photocatalytic degradation and adsorptive removal of Bismarck Brown R dye from water. Water Resour Ind 2019; 22100119.
[http://dx.doi.org/10.1016/j.wri.2019.100119]

[84]   Goswami M, Baruah D, Das AM. Green synthesis of silver nanoparticles supported on cellulose and their catalytic application in the scavenging of organic dyes. New J Chem 2018; 42(13): 10868-78.
[http://dx.doi.org/10.1039/C8NJ00526E]

[85]   Choudhary BC, Paul D, Gupta T, *et al.* Photocatalytic reduction of organic pollutant under visible light by green route synthesized gold nanoparticles. J Environ Sci (China) 2017; 55: 236-46.
[http://dx.doi.org/10.1016/j.jes.2016.05.044] [PMID: 28477818]

[86]   Xin H, Yang X, Liu X, Tang X, Weng L, Han Y. Biosynthesis of iron nanoparticles using tie guanyin tea extract for degradation of bromothymol blue. J Nanotechnol 2016; 5: 1-8.
[http://dx.doi.org/10.1155/2016/4059591]

[87]   Zhu F, Ma S, Liu T, Deng X. Green synthesis of nano zero-valent iron/Cu by green tea to remove hexavalent chromium from groundwater. J Clean Prod 2018; 174: 184-90.
[http://dx.doi.org/10.1016/j.jclepro.2017.10.302]

[88]   Sebastian A, Nangia A, Prasad MNV. A green synthetic route to phenolics fabricated magnetite nanoparticles from coconut husk extract: Implications to treat metal contaminated water and heavy metal stress in Oryza sativa L. J Clean Prod 2018; 174: 355-66.
[http://dx.doi.org/10.1016/j.jclepro.2017.10.343]

[89]   Kumar R, Singh N, Pandey SN. Potential of green synthesized zero-valent iron nanoparticles for remediation of lead-contaminated water. Int J Environ Sci Technol 2015; 12(12): 3943-50.
[http://dx.doi.org/10.1007/s13762-015-0751-z]

[90]   Murgueitio E, Cumbal L, Abril M, Izquierdo A, Debut A, Tinoco O. Green synthesis of iron nanoparticles: Application on the removal of petroleum oil from contaminated water and soils. J Nanotechnol 2018; 2018: 1-8.
[http://dx.doi.org/10.1155/2018/4184769]

[91]   Devatha CP, Thalla AK, Katte SY. Green synthesis of iron nanoparticles using different leaf extracts for treatment of domestic waste water. J Clean Prod 2016; 139: 1425-35.
[http://dx.doi.org/10.1016/j.jclepro.2016.09.019]

[92]   El-Said WA, Fouad DM, Ali MH, El-Gahami MA. Green synthesis of magnetic mesoporous silica nanocomposite and its adsorptive performance against organochlorine pesticides. International Journal of Environmental Science ans Technology 2018; 15: 1731-44.

[93]   Kishore S, Malik S, Shah MP, *et al.* A comprehensive review on removal of pollutants from wastewater through microbial nanobiotechnology -based solutions. Biotechnol Genet Eng Rev 2022; 1-26.
[http://dx.doi.org/10.1080/02648725.2022.2106014] [PMID: 35923085]

[94]   Mojiri A, Bashir MJK. Wastewater treatment: Current and future techniques. Water 2022; 14(3): 448.
[http://dx.doi.org/10.3390/w14030448]

CHAPTER 13

Application of Nanomaterials in the Degradation of Micro and Nano Plastics

V. J. Maodiswari[1], E. Rajalakshmi[2], S. Ambika[2], J. Princymerlin[2] and Y. Manojkumar[2,*]

[1] *Department of Botany, Bishop Heber College, Tiruchirappalli, Tamil Nadu, India*

[2] *Department of Chemistry, Bishop Heber College, Tiruchirappalli, Tamil Nadu, India*

Abstract: In recent years, microplastics (MPs) and nanoplastics (NPs) have become significant environmental concerns due to their persistent nature and potentially harmful effects on ecosystems and human health. Most of the reported materials and methods for the degradation of such toxic pollutants show limitations such as low recovery, high energy consumption and environmental impacts. As a result, more efficient green materials and methods are the need of the hour. Recently, researchers have reported efficient materials for the degradation of MPs and NPs. Hence, in this chapter, a comprehensive overview of eco-friendly initiatives and preventive measures is highlighted. It covers detailed information about the sources of MPs and NPs and their toxic impact on the environment and human health. It also highlights the existing techniques for processing and degradation of MPs and NPs and the potential of green nanomaterials in the degradation of plastics. The authors believe that this information will pave the way for the design and development of new alternate methods for further implementation.

Keywords: Eco-friendly, Environmental impacts, Degradation, Green nanomaterials, Microplastics, Nanoplastics, Preventive measures.

INTRODUCTION

Plastic consumption is constantly increasing worldwide, and last year, 390.7 million tons were produced [1]. Plastics have many useful properties, such as lightness, affordability, adaptability, durability and resistance to corrosion and flame. These materials significantly improve the quality of life of millions of people, making them safe to use. However, the disposal and mishandling of plastics have caused serious environmental and health issues that require urgent global attention [2]. Plastics have a remarkable resistance to degradation, which

---

* **Corresponding author Y. Manojkumar:** PG and Research Department of Chemistry, Bishop Heber College, Tiruchirappalli, Tamil Nadu, India; E-mail: yrmanojkumar@gmail.com

Neha Agarwal, Vijendra Singh Solanki and Sreekantha B. Jonnalagadda (Eds.)

allows them to remain in the environment for centuries. Plastics are classified by size into macroplastics and microplastics. Macro plastics are larger and have been the main focus of environmentalists for many years.

However, currently, more attention has been paid to MPs in the scientific community. Thompson coined the term "microplastic" in 2004 to describe the tiny fragments of plastic found in the marine environment [3]. The pieces of MPs, known as NPs, are the biggest threat to the environment and human health as they are even smaller in size than MPs. Polymers are classified commonly as macroplastics (>25 mm), mesoplastics (5-25 mm), MPs (<5 mm) and NPs (>100 nm) (Fig. **1**) [4].

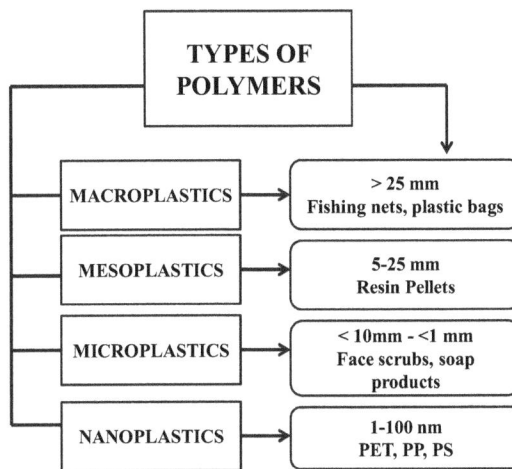

```
                    ┌─────────────────┐
                    │    TYPES OF     │
                    │    POLYMERS     │
                    └─────────────────┘

  ┌──────────────────┐     ┌──────────────────────────┐
  │  MACROPLASTICS   │ ──▶ │        > 25 mm           │
  └──────────────────┘     │  Fishing nets, plastic bags │
                           └──────────────────────────┘

  ┌──────────────────┐     ┌──────────────────────────┐
  │  MESOPLASTICS    │ ──▶ │        5-25 mm           │
  └──────────────────┘     │       Resin Pellets      │
                           └──────────────────────────┘

  ┌──────────────────┐     ┌──────────────────────────┐
  │  MICROPLASTICS   │ ──▶ │      < 10mm - <1 mm      │
  └──────────────────┘     │     Face scrubs, soap    │
                           │         products         │
                           └──────────────────────────┘

  ┌──────────────────┐     ┌──────────────────────────┐
  │  NANOPLASTICS    │ ──▶ │        1-100 nm          │
  └──────────────────┘     │       PET, PP, PS        │
                           └──────────────────────────┘
```

Fig. (1). Classification of plastics with their dimension.

MPs and NPs have a large surface area that facilitates the absorption and release of pollutants and chemicals and their possible entry into the food chain and subsequent migration to trophic levels [5]. Compared to MPs, NPs have been relatively less studied in the field of marine litter, which continues to pose serious threats to both the environment and human health. The widespread use of plastics in consumer goods and industrial applications, along with their slow rate of degradation, has led to the accumulation of MPs and NPs in various ecosystems, including oceans, rivers, soil and even the atmosphere. These small particles can persist in the environment for hundreds of years and pose a long-term threat to biodiversity and ecosystem functioning. These particles can end up in water bodies, where they pose a threat to aquatic organisms when they are ingested, causing physical damage, reducing feeding efficiency and disrupting reproductive processes. In addition, the accumulation of MPs and NPs in sediments affects benthic organisms and bottom-dwelling species, while in terrestrial ecosystems, they pollute the soil, affecting soil health, nutrient cycling and plant growth. Further, MPs and NPs can become airborne, polluting the air, which can have an impact on air quality and, subsequently, on human health. From a human health perspective, there is increasing evidence that MPs can enter the human body through the consumption of contaminated food and water, as well as inhalation [6]. The combined environmental and health concerns associated with MPs and NPs emphasize the need to intervene and mitigate their spread in our ecosystems. Due to the global and constant presence of MPs and NP pollution, there is an urgent need to develop effective strategies for their mitigation and elimination [7].

Traditional plastic degradation methods are not effective in processing MPs and NPs due to their small size and fragmentation. Green NMs offer a promising solution in this regard, as they are derived from renewable natural resources and have unique properties suitable for the degradation of MPs and NPs. These materials can enhance degradation processes through surface modification, enzymatic reactions, photocatalysis, chemical reactions and biodegradation [8]. By harnessing the power of green NMs, we can effectively mitigate the harmful effects of MPs and NPs and at the same time, avoid the additional environmental burden. The introduction of sustainable degradation methods is essential to prevent the accumulation of these particles in the ecosystems.

SOURCES OF MICROPLASTICS AND NANOPLASTICS

Fragmentation of Larger Plastics

Large plastic items such as bottles, bags and packaging materials can break down over time due to weathering, UV radiation and mechanical stress. This fragmentation process leads to smaller plastic particles such as MPs and NPs [9].

Synthetic Textiles

Synthetic fibers such as polyester, nylon and acrylic, commonly used in clothing, carpets and other textile products, can release microfibers during washing, use as well as disposal. Reports indicate that washing processes for synthetic textiles are the major source of MPs in the oceans. These microfibers can enter water bodies through sewage systems and contribute to MP pollution [10, 11].

Industrial Processes

Various industrial operations can produce MPs and NPs. For example, the production and processing of plastics, including the manufacturing, molding and extrusion of plastics, can lead to the release of plastic particles into the environment [12].

Personal Care Products

Some personal care products and cosmetics contain microbeads, which are tiny plastic particles added for exfoliation. When these products are used and washed, the microbeads can end up in drains and eventually in the aquatic environment. Reports indicate that facial cleansers contain NPs (polyethylene microbeads with a diameter of 0.2 mm) [13].

Packaging Materials

Plastic packaging materials, including films, wraps and foams, can degrade and form MPs. Improper disposal and littering contribute to the release of these particles into the environment. Over time, they degrade into MPs and NPs [14].

Paints and Coatings

Some paints, coatings, and varnishes contain MP particles for structural or other functional purposes. As these coatings wear down over a period, they can release MPs into the environment [15].

Agricultural Activity

Plastic mulch films used in agriculture for weed control and moisture retention can break down over a period and release MPs into the soil. In addition, the use of plastic-based fertilizers, such as polymer-coated fertilizers, can contribute to MP pollution [16, 17].

Aquaculture and Fishing

Aquaculture activities, including fish farming and shellfish farming, can release plastic particles into the aquatic environment. Plastic fishing gear, such as nets and lines, can also contribute to MP pollution through wear and tear [18].

Urban Drainage

Rainwater in urban areas can carry MPs and NPs from streets, sidewalks and other surfaces into the water bodies. This runoff can carry plastic debris, including trash and plastic waste that has been improperly disposed of. Sources and pathways of MPs and NPs can vary depending on geographic location, human activities and environmental conditions. Efforts to reduce MPs and NP pollution require a holistic approach aimed at reducing the generation of plastic waste, proper waste management practices and the development of sustainable alternatives to plastic [19].

IMPACT OF MPS AND NPS ON THE ENVIRONMENT

MPs and NPs can accumulate in marine and freshwater environments, where they can be consumed by many organisms. These particles can cause physical damage, clog the digestive tract and disrupt the feeding and reproductive processes of marine organisms such as fish, seabirds, turtles and marine mammals [20]. They can enter the food chain when lower organisms consume them and then larger animals consume those contaminated organisms. This process, called biomagnification, can lead to higher levels of MPs, including the kind consumed by humans [21]. This raises concerns about potential health effects. The presence of MPs in ecosystems can disrupt ecological processes. For example, the accumulation of MPs on the surface of water bodies can reduce light penetration, which affects photosynthesis and the productivity of aquatic plants. It can also alter microbial communities and nutrient cycling, affecting the overall ecosystem and its functions. It can release harmful chemicals and impurities into the environment. Plastic particles can adsorb and concentrate pollutants such as polychlorinated biphenyls (PCBs), polycyclic aromatic hydrocarbons (PAHs) and heavy metals [22]. If ingested, these chemicals can cause toxic effects on organisms and enter the food chain. It can change habitats, especially in marine and coastal areas. They can accumulate in sediments, affecting sediment-dwelling organisms and changing sediment properties. This can disrupt the structure and function of benthic ecosystems and affect species that depend on these habitats for shelter and food. It can pollute water bodies and soil. In freshwater systems, they can degrade water quality and affect the health of aquatic organisms. MPs in the soil can affect nutrient cycling, soil structure and microbial activity, which can affect plant growth and agricultural productivity. It has been found in remote places such as the Arctic and Antarctic, indicating their ability to travel long distances by air and water currents [23]. Thus, the widespread distribution shows the global nature of the problem of MP pollution.

IMPACT OF MPS AND NPS ON HUMAN HEALTH

The impact of MPs and NPs on human health is an area of ongoing research, and our understanding of the potential risks is still evolving (Fig. **2**). Although the direct impact of MPs on human health is not yet fully understood, there are several problems and possible ways through which these particles can pose a threat; they can enter the human body by consumption of contaminated food and water. Studies have shown that MPs have been found in various foods, such as seafood, salt and drinking water [24].

The potential health risks of ingesting MPs, including physical tissue damage, inflammatory reactions and transfer of harmful chemicals found in MPs, continue to be explored. It is also possible to inhale MP, especially in environments with high levels of MP pollution [25]. These particles can be in the air we breathe, including indoor air, and can cause respiratory health risks [26]. However, the extent to which inhalation of MPs affects human exposure and potential health effects is not yet fully understood. It can adsorb and condense chemicals from the surrounding environment, including persistent organic pollutants (POPs) and other toxic substances. When MPs are ingested, these chemicals can be released and absorbed into the body [27].

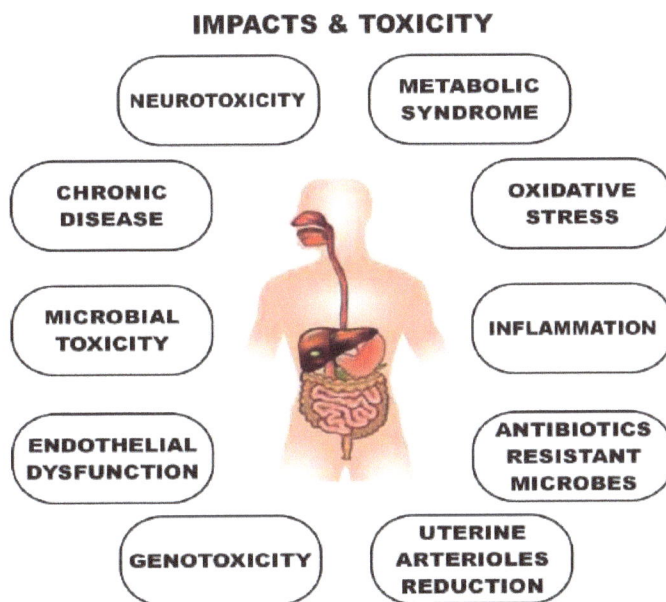

Fig. (2). Pictorial representation of the impact of MPs and NPs on human health.

The potential health effects of these chemicals, especially in long-term exposure scenarios, are the subject of ongoing research. There is growing evidence that MPs can interact with the gut microbiome, the collection of microorganisms that live in our digestive system [28]. Research has shown that MPs can alter the composition and diversity of animal gut microbiota, with implications for human health, as the gut microbiome plays an important role in many aspects of our well-being, including digestion, immune function and mental health. Some studies have shown that MPs can cross biological barriers, such as intestinal walls and enter tissues and organs [29]. While the extent of this translocation and the health implications are still being studied, there is great concern that the accumulation of MPs in tissues may cause local inflammation, cellular damage and potential long-term health effects. Table **1** summarizes a few important studies on the degradation of various types of plastics.

Table 1. Previous studies on the degradation of various types of plastics.

Polymer	Micro-Organisms / Nano Materials	Source / Degradation Time / Efficiency (weight loss)	Refs.
Low-density polyethylene (LDPE)	Microbulbifer	Marine pulp mill waste / 30 days / 79.18%	[30]
Polyethylene terephthalate (PET)	Streptomyces	Drinking Water Bottles / 18 days / 68.8%	[31]
Polyethylene (PE)	Bacillus	Municipal Landfill / 60 days / 14.7%	[32]
Polystyrene (PS)	Pseudomonas aeruginosa	The gut of the super worms Zophobasaratus / 15 days / 2.6%	[33]
Polyvinyl Chloride (PVC)	*Tenebrio molitor*	PVC Microplastic / 16 days / 2.9%	[34]
Polyhydroxybutyrate (PHB) and polycyclic aromatic hydrocarbons (PAHs)	Micrococcus luteus,	Marine water and oil spilt area of the northeast to down south seashore regions / 30 / 11.4%	[35]
Low-density polyethylene (LDPE)	Palladium	Purchased / 30 days/49%	[36]
Polyethylene	Silver	Purchased/5 weeks 64.5% (LDPE) and 44.4% of HDPE	[37]
Low-density polyethylene (LDPE)	Iron-Zinc Oxide	Purchased/120 hours / 41.3%	[38]
Low-density polyethylene (LDPE)	Nickel oxide	Purchased/240 hours / 33%	[39]

IDENTIFICATION OF MPS AND NPS

MPs and NPs can be difficult to detect due to their small size and various shapes and compositions. However, several techniques have been developed to detect and analyze these particles. Here are some commonly used MPs and NP detection techniques.

Optical Microscopy

Optical microscopy has proven to be a versatile tool in the detection and characterization of MPs and NPs. This method uses visible light to visualize and analyze the plastic particles found in various samples [40]. Using optical microscopy, scientists can observe the size, shape and color of plastic particles, which allows them to be identified and distinguished from other materials (Fig. **3**). One common technique used in optical microscopy is bright-field microscopy, which produces high-resolution images of MPs and NPs. By adjusting the microscope settings, researchers can improve the contrast and clarity of the plastic particles, making them easier to see [41]. In addition, techniques such as polarized light microscopy and differential interference contrast microscopy can provide additional information about the physical properties and texture of plastic particles [42]. To facilitate the detection process, researchers often use staining techniques or fluorescent dyes that selectively bind to plastics and improve their visibility under the microscope. These methods can improve the detection and quantification of MPs and NPs in complex samples. However, it is important to note that optical microscopy has limitations in detecting very small NPs as their size approaches or falls below the wavelength of visible light [43]. In such cases, additional methods such as electron microscopy, spectroscopy, or chemical analysis may be required to enable accurate identification and characterization.

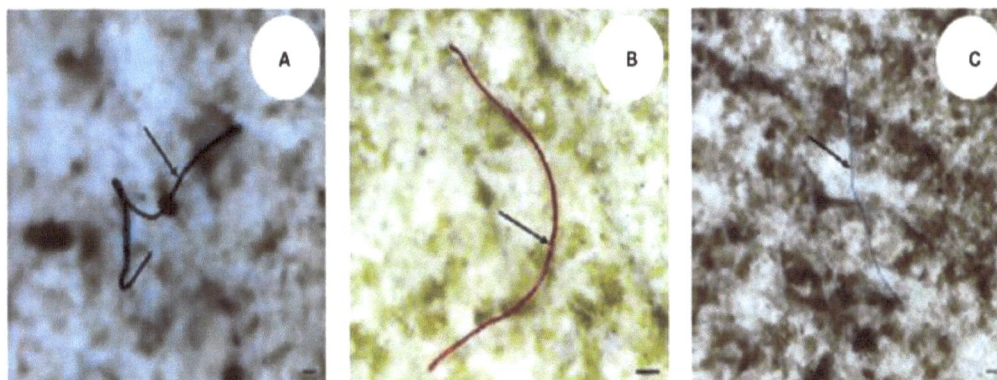

Fig. (3). Images of different shapes of microplastics in biological samples. The arrows indicate fibers (A-C) [43].

Fourier Transform Infrared Spectroscopy (FT-IR)

It is a powerful technique for the detection and analysis of MPs and NPs. It works by measuring the absorption of infrared light due to the effect of chemical bonds in the sample and provides information about the functional groups and molecular composition of the material. For MPs and NPs, it can help to distinguish between different polymers commonly found in plastic materials. Each polymer has a unique infrared absorption spectrum that allows them to be identified and distinguished. Among these three IR spectral regions, the higher energy near-infrared region (14,000– 4000 cm^{-1}), the mid-infrared region (4000 – 400 cm^{-1}), the far infrared region (400 – 10 cm^{-1}), the mid-infrared region are very useful to characterize the MPs and NPs. By comparing the spectrum obtained from the plastic sample to a database of known spectra, researchers can determine the type of polymer present (Fig. **4**).

It can be performed on whole samples as well as on individual particles [45]. For bulk samples, a small amount of plastic material is analyzed to obtain the average spectrum of the mixture. Single particle analysis involves microsampling techniques such as attenuated total reflectance (ATR) or micro-FTIR, which allow the study of individual MPs and NPs [46]. One of the advantages of FTIR spectroscopy is its ability to analyze MPs and NPs in complex matrices, such as environmental samples or biological tissues [47]. This technique can be used to detect and quantify MPs and NPs in water, sediments, soil and even biological samples. However, it alone cannot provide information about the size or shape of MPs and NPs [48]. Other techniques, such as microscopy or particle size classification methods, are often used in conjunction with FTIR to obtain a more comprehensive characterization of plastic particles.

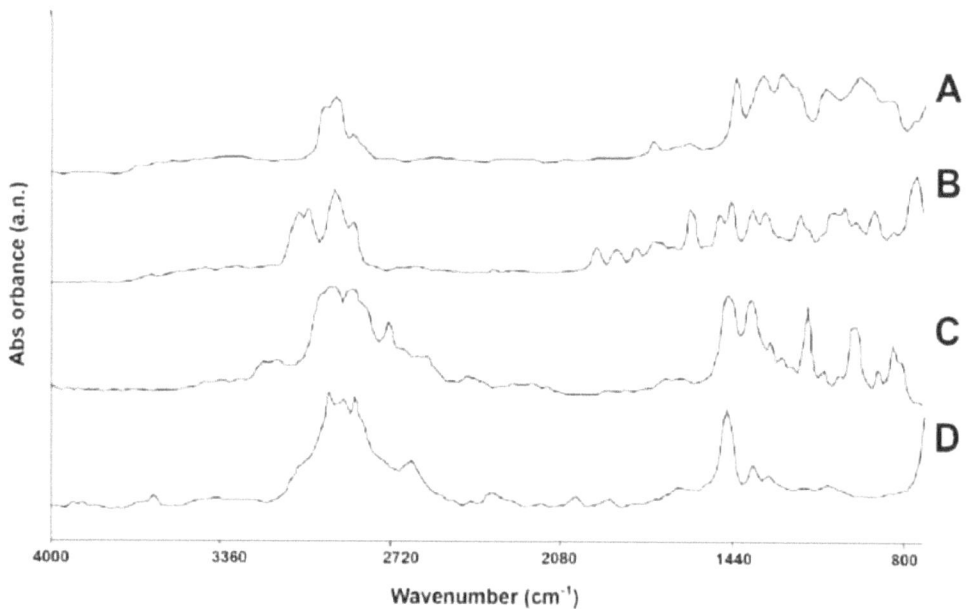

Fig. (4). FT-IR spectra of selected and magnified microplastic fragments. A: Polyvinylchloride, B: Polystyrene; C: Polypropylene; D: Polyethylene [44].

Scanning Electron Microscopy (SEM)

It is a powerful technique for the detection and characterization of MPs and NPs. It provides high-resolution visualization of the surface morphology and structure of particles and further provides valuable information about their size, shape and surface properties [49]. When MPs and NPs can reveal the physical properties of the particles, such as their surface structure, irregularities and structural properties. This information helps to distinguish plastic particles from other materials and helps identify them. In SEM analysis, the sample is placed in a vacuum chamber and a focused electron beam is scanned over the surface of the particles. When electrons interact with the sample, various signals are emitted, including secondary electrons and backscattered electrons. These signals are then detected and used to create an image of the sample. It can provide detailed images of individual MPs and NP particles, allowing their size and shape to be observed (Fig. **5**). By measuring the dimensions of the particles, it is possible to classify them according to size classes, for example, MPs (>5 mm) or NPs (>1 µm). In addition to morphology, it can be coupled with energy-dispersive X-ray spectroscopy (EDS) to obtain information on elemental composition [50]. EDS analysis allows the detection of specific elements in plastic particles, which can further help in their characterization and separation. It is particularly useful for studying the physical properties of MPs and NPs and their interactions with other materials or organisms [43].

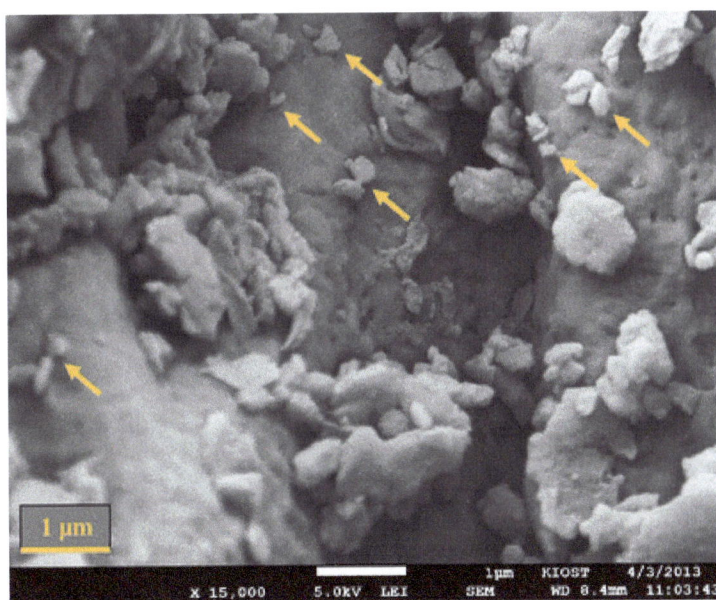

Fig. (5). Scanning electron microscopy image of micro- and nano-sized polystyrene particles attached on the surface of polystyrene spherule. Nanometre-sized particles are indicated by yellow arrows [51].

It can be applied to various sample types, including environmental samples, biological tissues and laboratory preparations. However, the analysis requires sample preparation, which may include methods such as filtering, drying and coating with conductive materials.

Micro-Fourier Transform Infrared Spectroscopy (Micro-FTIR)

It is a powerful technique for the detection of MPs and NPs. It uses infrared radiation to analyze the chemical composition of particles and provides valuable information about the types of plastics present [52]. In micro-FTIR analysis, a small sample containing MPs and NP particles is placed in a sample holder.

Infrared light is then passed through the sample and the resulting absorption spectrum is measured. Plastics are characterized by an absorption band in the infrared region, which allows them to be identified by their unique molecular vibration. By comparing the resulting spectrum with reference spectra of known plastic polymers, the type of plastic in the sample can be determined. This technique can detect a variety of plastic materials, including commonly used polymers such as polyethylene, polypropylene, polystyrene, and polyethylene terephthalate. It can also provide additional information about the chemical structure and functional groups of the plastic [53]. This can be particularly useful for differentiating different types of plastic, or for identifying impurities or coatings that may be present. One of the advantages of micro-FTIR is its ability to analyze particles directly without extensive sample processing (Fig. **6**). This allows rapid analysis and minimizes the risk of contamination or particle alteration during sample handling. However, it is important to note that it has limitations in terms of spatial resolution. Although it can analyze individual particles, the particle size must be in the micrometer range for accurate detection. In addition, micro-FTIR can have difficulty detecting transparent or low infrared-absorbing plastics.

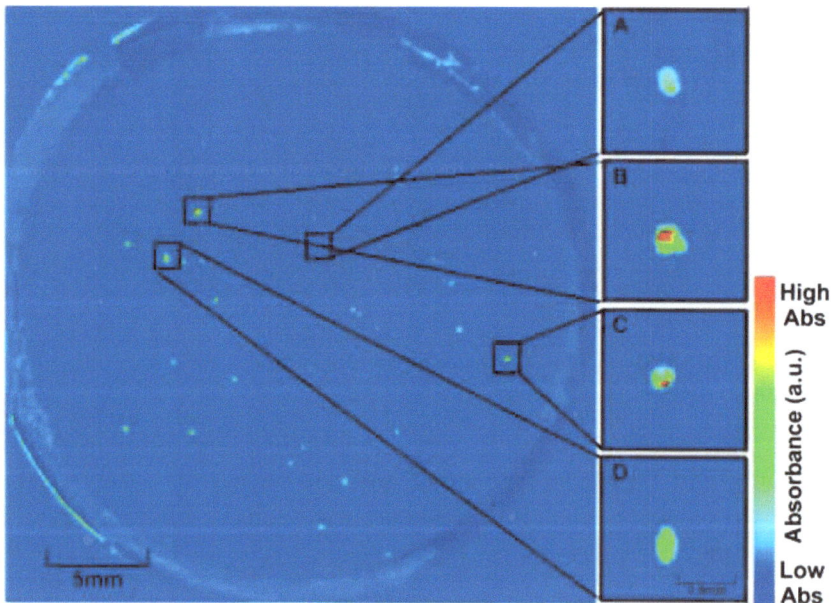

Fig. (6). Combined and overlapped false-color images showing 4 different microplastic types: A: Polyvinylchloride; B: Polystyrene; C: polypropylene; D: Polyethylene. Four fragments of each polymer type have been selected and magnified [54].

Raman Spectroscopy

It is an effective technique for the detection of micro- and nanoplastics. It uses Raman scattering to analyze the molecular structure and composition of the particles and provides valuable information about the types of plastics available. In Raman analysis, a laser beam is directed at the sample containing MPa and NP particles [54]. The laser light interacts with the particles, causing the light to disperse at different wavelengths. The scattered light, known as Raman scattering, is collected and analyzed to determine the vibrational modes and molecular bonds of the particles. Different plastics have unique Raman spectra due to differences in their molecular structure and chemical composition. By comparing the obtained Raman spectrum with reference spectra of known plastic polymers, the specific type of plastic in the sample can be identified. This technique can reliably identify common plastic materials such as polyethylene, polypropylene, polystyrene and polyethylene terephthalate [45]. It also provides additional information about the crystallinity, orientation and presence of additives or fillers in plastics. This makes it possible to distinguish between different types of plastics and to characterize certain material properties. One of the advantages of micro-Raman spectroscopy is its high spatial resolution, which allows the analysis of individual particles ranging in size from micrometers to nanometers [56]. This enables the identification and characterization of MPs and NPs in complex samples. However, it is important to note that it has limitations in terms of its sensitivity to fluorescent impurities and possible heating of the sample under the laser beam. In addition, transparent or low-emissivity plastics can cause problems with accurate detection.

Flow Cytometry

It is an effective method for the detection and characterization Of MPs and NPs. Although traditionally used in the field of cell analysis, flow cytometry has been adapted to analyze and quantify particles, including MPs and NPs, based on their physical and optical properties [57]. In flow cytometry, a sample containing MPs and NP particles is suspended in a liquid and passed through a flow cell. As the particles flow in a single file through the flow cell, they pass through the laser beam. The laser light interacts with the particles, causing them to break apart and fluoresce. It can provide information about particle size, shape and optical properties by detecting and measuring scattered light and fluorescence emitted by particles [58]. Parameters such as forward scatter (FSC) and side scatter (SSC) can be used to determine particle size and granularity [59].

In addition, fluorescent dyes or labels can be used to distinguish different types of plastics based on their fluorescence signals. To facilitate the detection of MPs and

NPs, special gating strategies and fluorescence channels can be used to distinguish plastic particles from other types of particles or background noise. By comparing the data obtained with reference standards or calibration beads, a certain type of plastic can be determined in the sample. Flow cytometry offers several advantages for the detection of micro- and NPs [60]. It offers a high-throughput analysis that enables rapid screening and quantification of particles in large sample volumes. It also allows different types of plastics to be separated based on their optical properties, which facilitates their identification and classification. However, it is important to consider the limitations of flow cytometry in MPs and NP analysis. The technique is based on the optical properties of the particles, and therefore, transparent or low-emission plastics cannot be easily detected or distinguished. In addition, the accuracy of the analysis can be affected by the presence of autofluorescent substances or background noise.

NEED FOR EFFECTIVE DECOMPOSITION METHODS

The presence of MPs and NPs in various parts of the environment, such as oceans, soil and freshwater, has raised concerns about their ecological and health effects. These tiny plastic particles, often invisible to the naked eye, can accumulate in organisms, disrupt ecosystems and possibly enter the food chain [61]. Therefore, the development of efficient degradation methods is crucial to reduce the environmental burden caused by MPs and NPs. Several important reasons emphasize the importance of effective degradation methods. Firstly, MPs and NPs remain in the environment for a long time, causing them to accumulate in ecosystems and endanger marine, terrestrial and human health [62]. By reducing their abundance and the associated ecological risks, decomposition methods can mitigate these harmful effects. Secondly, MPs and NPs can enter the food chain, causing physical damage, reproductive disruption and physiological disruption at various trophic levels [63].

Developing decomposition methods can help mitigate these ecological consequences and protect biodiversity. In addition, evidence has revealed the presence of MPs and NPs in food, drinking water and even air, raising concerns about potential health risks. Effective disposal methods can help reduce human exposure and protect public health. Conventional waste management methods, such as landfills and incineration, are not effective in eliminating MPs and NPs due to their small size and the limitations of existing filtration systems [64]. Therefore, it is necessary to find innovative and sustainable degradation methods to meet the challenges of plastic waste management. The development of such methods complies with the principles of sustainable development, as they use renewable natural resources, minimize energy consumption and reduce harmful by-products, thus promoting a more sustainable and circular economy. In addition, recognition of the environmental and health risks associated with MPs and NPs has led to the creation of regulatory frameworks and policies aimed at reducing plastic pollution [65]. Effective degradation methods can help meet the requirements and goals set by these regulations, supporting the transition to a cleaner and more sustainable future.

METHODS OF MPS AND NPS DISINTEGRATION

Mechanical Disintegration

The process of mechanical disintegration is used to break MPs and NPs into smaller particles using physical forces and mechanical procedures [56]. This method involves applying mechanical energy to plastic materials, causing them to break, splinter or break into smaller pieces. A variety of methods are used for mechanical disintegration, including milling, grinding, grinding and crushing. During the milling and grinding processes, the plastic particles are subjected to high shear force and impact, which reduces their size. These techniques often use specialized equipment, such as ball mills or flour mills, which rotate or vibrate to generate strong mechanical forces [66]. The plastic particles are introduced into the grinding chamber, where they collide with each other or with the grinding materials, resulting in their fragmentation. Milling is an alternative mechanical degradation method where mechanical forces are used to break and grind MPs and NPs into smaller particles [67]. This process is often used in industries such as recycling and waste management, where plastic materials need to be reduced in size for further processing or disposal.

Shredding is a mechanical breaking technique where plastic materials are cut or torn into smaller pieces using sharp blades or other cutting mechanisms. This method is often used to reduce larger plastic items to their original size before further processing. The mechanical degradation of MPs and NPs offers several advantages [68]. It is a relatively simple and cost-effective method that does not require the use of additional chemicals or advanced equipment. However, this approach also has some limitations. The efficiency of mechanical degradation depends on factors such as plastic-type, morphology and size distribution. Certain plastics, such as those with high molecular weight or cross-linked structures, may be more resistant to mechanical forces, requiring additional pretreatment or alternative degradation methods.

Chemical Degradation

The chemical degradation method offers the possibility of controlled and efficient degradation of plastic materials into smaller, less harmful components. It can be used for different types and sizes of plastics. However, the selection of appropriate chemicals, reaction conditions and degradation pathways is critical to ensure the desired degradation results while minimizing the generation of toxic byproducts. In addition, chemical degradation methods may require additional steps to separate and purify the degraded products from the reaction mixture [69]. The selection of suitable solvents, catalysts and reaction conditions is important to ensure environmental compatibility and minimize the generation of hazardous waste. Different chemical degradation techniques are used for MPs and NPs. Some of the more common methods include hydrolysis, oxidation and catalytic degradation.

Hydrolysis

Hydrolysis is a chemical degradation method that breaks down MPs and NPs using water or aqueous solutions. It involves breaking the chemical bonds of plastic polymer chains by reacting with water molecules. The process can be accelerated by adjusting pH, temperature and reaction time [70]. Water molecules interact with the polymer chains, breaking the bonds between the monomer units and resulting in smaller plastic particles. The efficiency of hydrolysis depends on the type of polymer, chemical structure and reaction conditions. It is particularly effective for polyesters and polyamides with easily hydrolyzable ester and amide bonds [71]. When MPs and NPs made from these polymers come into contact with water or aqueous solutions, the polymer chains can be split, which breaks down the plastic particles. The pH of the solution can be adjusted to improve the hydrolysis reaction of certain polymers and the efficiency is also affected by temperature and reaction time [72].

Hydrolysis has advantages such as mild reaction conditions, making it an environmentally friendly way to degrade MPs and NPs. However, the possible release of harmful by-products or degradation intermediates must be considered. Ongoing research is focused on optimizing the hydrolysis conditions and exploring the use of catalysts or additives to develop more efficient and sustainable methods to degrade MPs and NPs, thus reducing plastic pollution.

Oxidation

Oxidation is a chemical degradation process that can effectively degrade MPs and NPs. It involves a reaction between plastic particles and oxidizing agents such as oxygen, ozone or hydrogen peroxide, which leads to the breaking of the polymer chains and the degradation of the plastic material. This reaction causes the chemical bonds to break and smaller fragments to form, reducing the size of the plastic particles and promoting their degradation. The extent of oxidation depends on factors such as polymer type, concentration, type of oxidant and reaction conditions such as temperature and reaction time. There are various methods and techniques for carrying out oxidation reactions [73]. Exposure to ozone or hydrogen peroxide can initiate the process or oxidizing agents can be incorporated into the polymer matrix or combined with catalysts to improve reaction rates. One advantage of oxidation is its ability to degrade many different plastic polymers, such as polyethylene, polypropylene, polystyrene and polyvinyl chloride. However, the oxidation efficiency can vary depending on the composition and structure of the particular polymer [74]. There are important environmental considerations for oxidation processes. Reactive intermediates and byproducts formed during oxidation reactions can have toxicological effects or contribute to secondary pollution. Proper management and handling of these by-products are critical to the overall environmental sustainability of the oxidation process. Oxidation of MPs and NPs is an active area of research, where researchers are exploring different approaches to optimize efficiency and selectivity. By studying different oxidizers and reaction conditions catalysts, the researchers aim to develop efficient and environmentally friendly oxidation methods to break down MPs and NPs, which help reduce plastic pollution in the environment.

Catalytic Degradation

Metal catalysts such as iron, copper, and manganese facilitate oxidation reactions, while nanocatalysts offer high surface area and reactivity [75]. Photocatalysis, which uses semiconductor catalysts such as titanium dioxide and zinc oxide, uses light to break down plastic to form reactive substances [76]. Enzyme catalysts such as lipases and esterases selectively hydrolyze plastics [77]. Further optimization and evaluation of these catalyst systems are necessary for large-scale implementation, considering their potential environmental impact and cost-effectiveness.

Photodegradation

Photodegradation is the process by which MPs and NPs are broken down by exposure to ultraviolet (UV) radiation from sunlight [78]. UV radiation produces energy that breaks the polymer chains, leading to the fragmentation of the plastic material. UVA and UVB wavelengths of sunlight trigger photochemical reactions in plastic polymers [79]. High-energy photons interact with the polymer chains, resulting in the formation of reactive species such as free radicals [80]. These reactive parts initiate chain-end reactions by breaking polymer bonds and forming smaller plastic parts. The efficiency of photodegradation depends on factors such as the type of polymer, the presence of additives or pigments, and the intensity and duration of UV radiation [81].

Different polymers have different sensitivities to photodegradation, and certain plastics are more prone to degradation. For example, polyethylene and polypropylene undergo significant photodegradation due to their molecular structure and chemical composition [82]. UV lamps can also be used as artificial UV sources to accelerate photodegradation under controlled laboratory conditions by emitting UV radiation at specific wavelengths to induce degradation reactions in plastic samples [83]. Photodegradation of MPs and NPs has several advantages. It is a natural process that occurs in the environment and does not require the addition of external chemicals. In addition, it can degrade many types of plastic polymers. However, process efficiency can be affected by factors such as particle thickness, surface area, and other environmental conditions (*e.g.*, temperature and humidity) that affect the rate of degradation [84].

DEGRADATION OF MPS AND NPS USING GREEN METHODS

Biodegradation

Enzymes produced by certain microorganisms can break down plastics. Scientists have identified microorganisms in various environments, such as soil and sewage, that can degrade plastics, such as polyethylene and polypropylene [84]. These microorganisms produce enzymes such as lipases and esterases that can break down plastic polymers into smaller fragments [85]. Developing and promoting the use of biodegradable polymers can help reduce the accumulation of microplastics in the environment. Biodegradable plastics are designed to break down more easily through natural processes such as microbial action or enzymatic degradation.

Natural Degradation

Exposure to sunlight and weather can gradually degrade the plastic over time, but with microplastics, this process is often slow [86]. However, increasing natural degradation by using additives or modifying the plastic surface can accelerate degradation [87].

Environmentally Friendly Materials and Recycling

Promoting the use of environmentally friendly materials and improving recycling systems can help prevent MPs from entering the environment. By increasing the degree of reduction and recycling of plastic waste, the total burden of MPs can be minimized [88].

Biochar-assisted Degradation

Biochar, a carbon-rich material made from biomass, has shown the potential to promote MP degradation [89]. Biochar can adsorb plastic particles, which increases their surface area and facilitates microbial degradation [90]. It can also change environmental conditions such as pH and nutrient availability to support the growth of plastic-degrading microorganisms [91].

Degradation Based on Agricultural Waste

Using agricultural waste, such as agricultural waste or food waste, to degrade micro- and nanoplastics is a sustainable approach [92]. Agricultural waste can be processed into bioactive compounds or used as a carbon source for microorganisms that can degrade plastics. This approach promotes circular economy principles and reduces waste production.

Green Solvents

Traditional solvents used to break down plastics can have an environmental impact. As an alternative, green solvents such as supercritical carbon dioxide (scCO2) or ionic liquids are being investigated [93]. These solvents are non-toxic, have a low environmental impact and can be used in environmentally friendly degradation processes. It is important to note that environmentally sustainable methods to degrade MPs and NPs are still being developed, and further research is needed to optimize their efficiency and scalability [94]. Therefore, widespread adoption of these methods will help reduce plastic pollution and promote sustainable management practices.

ADVANTAGES OF GREEN METHODS

One of the benefits of using green methods to degrade MPs and NPs is their potential to improve environmental sustainability. This could visually summarize the advantages of using bacteria for microplastic removal, including selectivity, efficiency, and environmental friendliness. Green methods often use ecological processes and materials, which minimizes the negative impact on ecosystems and human health. Here are some specific benefits of using green methods to break down MPs and NPs.

Reduce Environmental Damage

Green methods favor the use of non-toxic and biodegradable substances, minimizing the generation of harmful by-products and pollution risks. Their goal is to prevent or minimize the release of dangerous chemicals and substances into the environment, which contributes to the preservation of the ecosystem [95]. It is important to note that some biodegradation processes produce methane, a greenhouse gas; they typically occur at slower rates and with less overall impact compared to burning fossil fuels. Biodegradation generally has a lower environmental impact than combustion.

Lower Energy Requirements

Green methods often emphasize energy-efficient processes that reduce total energy consumption during the decomposition process [96]. Several technologies offer promising solutions for microplastic (MP) and nanoparticle (NP) pollution, but biological degradation stands out for its environmental benefits. This natural process utilizes microorganisms to break down MPs and NPs, requiring minimal energy input, releasing virtually no harmful emissions, and contributing to carbon sequestration. This can lead to lower operating costs and a smaller carbon footprint [97]. This eco-friendly approach not only tackles plastic pollution but also contributes to solving the energy problem by generating renewable energy sources like biogas. Compared to other techniques like coagulation/flocculation and photodegradation, biological degradation offers a sustainable and potentially cost-effective solution for long-term environmental remediation. The use of magnetic composite catalysts as micromotors/microrobots for microplastic (MP) and nanoparticle (NP) degradation holds significant potential for environmental remediation [98].

Compatibility with Principles of Circular Economy

Green methods are compatible with the principles of circular economy, which aim to reduce waste, promote recycling and reuse, and minimize the consumption of resources. Using green methods, the degradation of MPs and NPs can help close the plastic life cycle and promote a more sustainable approach to plastic waste management [99]. Biodegradable plastics can indeed contribute to this concept by returning some materials to the environment, though their complete breakdown and reintegration into the cycle can vary depending on specific types and environmental conditions. Biological recycling through naturally occurring microbes offers an environmentally friendly option, but its effectiveness for different materials and contexts requires further exploration. To support these advancements, encouraging research and development of efficient biodegradation processes alongside tailored sustainability policies that promote responsible plastic use and address potential downsides of biodegradability can pave the way for a more sustainable plastic waste management system [100].

Health and Safety Benefits

Generally, the degradation of MPs/NPs is not complete, which raises the concern of generating hazardous/toxic by-products. Such by-products may lead to new pollutants and pose potential risks to the aquatic environment and public health. So, it is necessary to figure out the degradation pathway, monitor potential products, and assess the toxicity of the degradation intermediates [100]. Green practices prioritize the health and safety of both workers and the public by minimizing exposure to hazardous chemicals and substances [101]. This reduces the risks associated with the use and disposal of toxic substances and improves occupational health and general environmental safety.

DISADVANTAGES OF GREEN METHODS

Although green methods to degrade MPs and NPs offer promising opportunities, they also have certain disadvantages and limitations. Here are some of the disadvantages of decomposing MPs and NPs using green methods.

Efficiency and Effectiveness

Green degradation methods may not be as efficient or effective as traditional MPs and NP degradation methods [102]. Some green methods, such as enzymatic degradation or microbial degradation, may be slower or less efficient than chemical or physical methods [103]. Achieving complete degradation of all types of MPs and NPs using green methods can be difficult. The biodegradation products of MPs have not been collected effectively and detected.

Specificity and Selectivity

Enterobacter and Pseudomonas bacteria were isolated from cow dung to outperform individual strains in degrading PE and PP microplastics. However, the mechanism behind this enhanced efficiency remains unclear. Therefore, carefully selecting suitable microbial consortia for biodegrading MPs and NPs is crucial, as only specific combinations possess the desired degrading activity [104]. Green degradation methods may have limitations in specificity and selectivity, meaning that they may be effective at degrading certain types of plastics but less effective at others. For example, enzymes or microorganisms may have substrate preferences that limit their ability to degrade all plastic polymers [105]. The development of green methods that can effectively target a wide range of MPs and NPs is still an ongoing area of research.

Optimization and Scalability

Green decomposition methods often require further optimization to improve efficiency and scalability [106]. Factors such as temperature, pH, enzyme concentrations, or microbial conditions must be carefully controlled to achieve optimal degradation. Scaling these methods for large applications can present challenges in terms of cost, infrastructure, and process optimization.

Environmental Impacts of Enzyme Production

Enzymes used in green degradation methods are often produced by biotechnological processes [107]. Enzyme production can have environmental impacts, including energy and resource consumption, use of fermentation substrates, and potential waste generation. Assessing the life cycle impacts of enzyme production is crucial in assessing the overall environmental sustainability of green degradation methods. However, using enzymes to degrade MP/NPs is safe; the procedure is only suitable for small amounts of samples due to its high cost. This procedure is inapplicable to large-scale treatment since each enzyme needs an optimal pH to function and digest materials [108]. Polymers affected by enzymes did not exhibit deterioration when compared to other materials through chemical degradation. Enzymes are also not harmful. The key drawback is that processing samples requires an extensive period. There may be variations in the appropriate pH level and optimum temperature range for each enzyme [109].

Limited Applicability to Complex Matrices

Green degradation methods may have limitations when applied to complex matrices or mixed plastic wastes. Environmental samples or waste streams often contain various impurities, additives, or other materials that can interfere with the degradation process. Ensuring the effectiveness of green methods in real scenarios with complex plastic mixtures is a challenge that requires further research and optimization [110].

Costs and Economic Feasibility

Large-scale implementation of green decomposition methods can have cost implications. Some methods, such as enzymatic digestion, are expensive due to the production and purification cost of the enzymes [111]. According to Cole's enzymatic digestion investigation, proteinase-K broke down 0.2 g of the material with a degradation rate of 97%. However, the implementation of the enzyme for bigger samples is rather expensive due to its high cost [112]. The economic feasibility of green methods must be carefully evaluated to ensure their practical implementation and widespread use.

GREEN NANOMATERIALS FOR DEGRADATION

Clay Nanoparticles

Clay nanoparticles, such as montmorillonite or kaolin, are naturally occurring minerals that can be used as nanofillers in biodegradable polymer composites [113]. These clay nanoparticles can improve the mechanical properties of the composite and increase the degradation of the polymer matrix. They can act as nucleation sites for microbial colonization and enzymatic degradation of MPs and NPs [114].

Carbon-Based Nanomaterials

Carbon-based nanomaterials such as carbon nanotubes (CNTs) or graphene oxide (GO) have shown potential in the degradation of MPs and NPs [115]. CNTs can mechanically degrade plastic particles, facilitating their degradation, while GO can act as a carrier or support for plastic-degrading enzymes or microorganisms [116]. These carbon-based nanomaterials offer durability and low environmental impact.

Bio-Inspired Nanomaterials

Bio-inspired nanomaterials are inspired by natural systems and use sustainable materials to break down plastics. For example, NMs can be developed that mimic enzymes or their active sites [117 - 119]. These are nanozymes (NMs mimics of an enzyme) used in biomedical fields such as tumor therapy and antibacterial, antioxidant and bioorthogonal reactions. These NMs can have the catalytic activity of natural enzymes that promote the degradation of MPs and NPs in a green and sustainable way.

Nanoenzymes

Nanoenzymes are nanomaterials designed to mimic the activity of natural enzymes [120]. These NMs can have catalytic properties similar to enzymes and can be designed to target specific plastic polymers for degradation [121]. Nanoenzymes offer advantages such as stability, reusability, and versatility in substrate specificity, making them promising for the degradation of MPs and NPs [122].

Natural NMs

Natural NMs, such as cellulose, chitosan, lignin, starch, *etc.*, can be used to break down MPs and NPs [123]. These materials come from renewable sources and have properties suitable for the degradation of plastics [124]. They can act as adsorbents or carriers that enhance the biological degradation process or facilitate the attachment of plastic-degrading microorganisms. Some important natural nanomaterials are described below.

Cellulose-Based Nanomaterials

Cellulose nanocrystals (CNCs) have emerged as promising green nanomaterials for the degradation of micro- and nanoplastics [125]. CNCs are derived from cellulose, the most abundant biopolymer on earth found in plants and trees. CNCs are highly biodegradable because they are naturally derived from cellulose. Microorganisms, enzymes, and other natural degradation processes can easily destroy them by reducing their environmental impact. They have a large aspect ratio, which means that their length is much greater than their width. This unique shape allows a better interaction with plastic surfaces, which increases the efficiency of the degradation processes. The surface of CNC machines can be chemically modified to improve their degradability. Functional groups can be added to the surface of CNCs to improve their compatibility and reactivity with micro- and nanoplastics. CNCs can exhibit catalytic activity by promoting the degradation of plastics through various chemical reactions. For example, they can act as catalysts in hydrolysis or oxidation reactions, which facilitates the breakdown of plastic polymers into smaller fragments. It also has excellent mechanical properties, such as high stiffness and strength. These properties make them useful for reinforcing plastic composites, reducing total plastic content and improving the recyclability of plastic materials. CNCs can be activated with special enzymes or catalysts to improve their adhesion to plastic surfaces [126]. This allows enzymes or catalysts to come into direct contact with the plastics, promoting their degradation. CNCs can act as carriers for enzymes involved in the degradation of plastics [127].

The presence of CNCs improves the degradation properties of the compound, making it more effective in degrading plastic materials. Cellulose nanocrystals can be modified for photocatalytic properties by adding photocatalytic materials such as TiO_2 or ZnO [127]. These photocatalytic cellulose nanocrystals can facilitate the degradation of micro- and nanoplastics under the influence of UV light, which increases the degradation process and reduces the persistence of plastic particles.

Chitosan-Based Nanomaterials

It has also shown great potential in the degradation of micro- and nanoplastics [128]. Their unique properties and interaction with plastic particles enable efficient and environmentally friendly degradation processes [129, 130]. Chitosan nanoparticles have mucoadhesive properties that allow them to adhere to micro and nanoplastic surfaces. This stickiness facilitates the encapsulation of plastic particles, prevents their aggregation and facilitates their degradation. Chitosan nanoparticles can act as carriers for enzymes involved in the degradation of plastics [130]. Enzymes such as lipases, esterases, and proteases can be immobilized on the surface of chitosan nanoparticles or encapsulated in them [131]. When these enzymes come into contact with plastic particles, they can catalyze the breaking of the polymer chains, leading to their degradation. Chitosan nanoparticles can be modified and functionalized to enhance their interactions with micro- and nanoplastics. Surface modifications can include the attachment of specific target ligands or antibodies to plastic particles, which improves their selectivity and efficiency in degrading plastic [132]. Chitosan nanoparticles can be designed to encapsulate degrading agents such as enzymes, reactive oxygen species or other chemical compounds involved in plastic degradation.

These nanoparticles can release degradable substances in a controlled manner, which prolongs their activity and increases the degradation of the plastic. Chitosan nanoparticles can be combined with other green nanomaterials or degradation strategies to improve their performance in degradable micro- and nanoplastics. For example, combining chitosan nanoparticles with metal or metal oxide nanoparticles can produce synergistic effects where the nanoparticles work together to facilitate the degradation process. Chitosan nanoparticles can be modified to contain photocatalytic material, TiO_2 [132]. These photocatalytic chitosan nanoparticles can degrade micro- and nanoplastics under the influence of UV light. The photocatalytic properties of TiO_2 improve the degradation process, while chitosan provides encapsulation and adhesion properties that facilitate interaction with plastic particles.

Lignin-Based Nanomaterials

Lignin nanoparticles have recently attracted attention as a potential material for the degradation of micro- and nanoplastics. Lignin, a natural polymer found in plant cell walls, is an abundant and renewable natural resource [133]. Its unique properties, such as high reactivity and surface area, make it a promising candidate for various applications, including the degradation of plastic materials. Lignin nanoparticles can be used to coat the surfaces of micro- and nanoplastics, promoting their degradation [134]. The high reactivity and adhesion properties of lignin facilitate encapsulation and interaction with plastic particles, leading to improved degradation processes. Lignin nanoparticles can act as carriers for enzymes involved in the degradation of plastics [135]. Enzymes such as esterases or lipases can be immobilized on the surface of lignin nanoparticles. Enzyme-rich lignin nanoparticles, when they come in contact with micro- and nano-plastics, release enzymes that catalyze the degradation of plastic polymers and allow their degradation [136]. Lignin nanoparticles can be incorporated into composite materials to degrade micro- and nanoplastics. For example, lignin nanoparticles can be combined with biodegradable polymers such as polyhydroxyalkanoates (PHA) to form nanocomposites [137]. The presence of lignin nanoparticles improves the degradation properties of the compound, making it more effective in degrading plastic materials. Lignin nanoparticles can be combined with other green nanomaterials or degradation strategies to improve their efficiency in degrading MPs and NPs [138, 139]. Lignin nanoparticles can be modified to obtain photocatalytic properties by adding photocatalytic materials such as TiO_2 or ZnO [140].

Starch-Based Nanomaterials

Starch, a biodegradable and renewable polymer derived from plant sources, offers several advantages in breaking down plastics [141]. Starch nanoparticles can be used to create coatings on the surfaces of micro and nanoplastics. The large surface area and film-forming properties of starch nanoparticles allow them to encapsulate and interact with plastic particles, facilitating their degradation over time. These coatings can increase the exposure of plastic surfaces to environmental factors such as moisture or enzymes that promote their degradation [142]. Enzymes such as lipases or esterases can be immobilized on the surface of starch nanoparticles. Upon contact with micro- and nanoplastics, enzyme-rich starch nanoparticles release enzymes that catalyze the breakdown of plastic polymers, leading to their degradation. Starch nanocomposites: Starch nanoparticles can be incorporated into nanocomposite structures to degrade micro- and nanoplastics. For example, starch nanoparticles can be combined with other biodegradable polymers, such as polylactic acid (PLA) or polyhydroxyalkanoate (PHA), to form nanocomposites. The presence of starch nanoparticles improves the degradation properties of the compound, making it more effective in degrading plastic materials. Starch nanoparticles can be modified for photocatalytic properties by adding photocatalytic materials such as titanium dioxide (TiO_2) or zinc oxide (ZnO) [143]. These photocatalytic starch nanoparticles can facilitate the degradation of micro- and nanoplastics by UV light. Starch nanoparticles have been studied as micro- and nanoplastic adsorbents [144].

INORGANIC GREENS OF NANOMATERIALS FOR DEGRADATION

Iron nanoparticles, especially zero-valent iron nanoparticles, have shown promise in the degradation of micro- and nanoplastics [145]. Iron nanoparticles Titanium dioxide and Zinc oxide nanoparticles have been extensively studied for their photocatalytic properties. When exposed to UV light, these nanoparticles can form electron-hole pairs that drive photocatalytic reactions [146, 147]. These reactions cause the production of reactive oxygen species that can degrade plastic polymers, promoting their degradation [148].

Silver nanoparticles have antimicrobial properties and have been studied for micro- and nanoplastic degradation along with other degradation strategies [149]. The presence of silver nanoparticles can prevent the growth of microorganisms on plastic surfaces, preventing the formation of biofilm and promoting the subsequent degradation of plastic materials. Copper nanoparticles have shown potential in the degradation of micro- and nanoplastics [150]. Their large surface area and reactivity allow them to interact with plastic particles and initiate degradation processes [151]. Copper nanoparticles can catalyze oxidation reactions, causing and promoting the degradation of plastic polymers. It is important to note that the use of inorganic green metal nanoparticles to degrade micro- and nanoplastics is still an active area of research. Further research is needed to optimize their performance, understand degradation mechanisms and assess potential environmental impacts [152]. However, these examples highlight the potential of inorganic green metal nanoparticles as effective micro- and nanoplastic-degrading agents, offering a promising way to combat plastic pollution.

OUTLOOK AND FUTURE DIRECTIONS

The use of nanomaterials to break down micro- and nanoplastics is promising in the fight against plastic pollution. Future research will focus on developing nanomaterials with better degradation efficiency. This includes research on new types of nanocatalysts, nano enzymes and nanocomposites that can effectively degrade various micro- and nanoplastics. Optimizing the properties of nanomaterials, such as surface chemistry, size and morphology, is crucial to improve their catalytic or enzymatic activity. Selectivity and specificity: One of the challenges of plastic degradation is selectivity and specificity for different types of plastics. Researchers are investigating the synergistic effects of combining different nanomaterials or degradation strategies to improve overall degradation efficiency. For example, combining nanocatalysts with microbial degradation or using nanomaterials as carriers for enzymes or microorganisms can lead to better degradation results. Such combinatorial approaches can offer advantages in terms of degradability, selectivity and degradation of complex plastic mixtures. The development of green and sustainable nanomaterials will be central in the future. The goal of the researchers is to use nanomaterials obtained from renewable natural resources or waste materials and reduce the environmental impacts associated with their production and use. This includes research on bioactive nanomaterials or environmentally friendly nanomaterial synthesis methods. As nanomaterials are used to degrade plastics, it is important to assess their safety and potential environmental impacts. Future research will focus on understanding the fate and behavior of nanomaterials in the environment, including their potential toxicity and long-term effects. This ensures that the use of nanomaterials to break down plastics remains sustainable and does not create new environmental problems. The scalability and practical application of nanomaterial degradation methods are another focus. The development of cost-effective and scalable processes for the production of nanomaterials and their integration into plastic waste management systems is crucial. The research aims to overcome the challenges associated with large-scale applications, including cost, infrastructure requirements and process optimization.

CONCLUSION

Degradation of micro- and nanoplastics with green nanomaterials holds promise for mitigating the plastic pollution crisis. Green metal nanoparticles have been found to have the potential for efficient plastic degradation. However, further research is needed to optimize the degradation processes, assess the environmental effects of the degradation products, and develop practical solutions for the actual use of these materials. By combining the power of nanotechnology with sustainable development, we can work towards a cleaner and healthier future free of micro and nanoplastic pollution. The chapter provides a comprehensive overview of the use of green nanomaterials in the degradation of micro- and nanoplastics. It underlines the potential of these materials to solve the growing problem of plastic pollution and emphasizes the importance of sustainable and environmentally friendly solutions.

ACKNOWLEDGEMENT

The authors, S.A. and Y.M., thank the management of Bishop Heber College (Autonomous), Tiruchirappalli, Tamil Nadu, India, for the support. (Ref. No. 02.05.2022 MRP/1009/2022 (BHC)) and (Ref. No. 18.04.2023 MRP/1005/2023 (BHC)).

REFERENCES

[1]   Montanari W, Antonini D, Avella R, Frioni V, Giffoni M, Masi M, *et al.* Position Paper: The sustainability of plastics. Chem Eng Trans 2023; 98: 1-8.

[2]   Kundu A, Shetti NP, Basu S, Raghava Reddy K, Nadagouda MN, Aminabhavi TM. Identification and removal of micro- and nano-plastics: Efficient and cost-effective methods. Chem Eng J 2021; 421(1): 129816.
[http://dx.doi.org/10.1016/j.cej.2021.129816] [PMID: 34504393]

[3]   Thompson RC, Olsen Y, Mitchell RP, *et al.* Lost at sea: where is all the plastic? Science 2004; 304(5672): 838.
[http://dx.doi.org/10.1126/science.1094559] [PMID: 15131299]

[4]   Allen S, Allen D, Karbalaei S, Maselli V, Walker TR. Micro(nano)plastics sources, fate, and effects: What we know after ten years of research. J Hazard Mater Adv 2022; 6: 100057.
[http://dx.doi.org/10.1016/j.hazadv.2022.100057]

[5]   Peng L, Fu D, Qi H, Lan CQ, Yu H, Ge C. Micro- and nano-plastics in marine environment: Source, distribution and threats — A review. Sci Total Environ 2020; 698: 134254.
[http://dx.doi.org/10.1016/j.scitotenv.2019.134254] [PMID: 31514025]

[6]   Yee MSL, Hii LW, Looi CK, *et al.* Impact of microplastics and nanoplastics on human health. Nanomaterials (Basel) 2021; 11(2): 496.
[http://dx.doi.org/10.3390/nano11020496] [PMID: 33669327]

[7]   Silva P, AL Patrício Silva, JC Prata, *et al.* Rocha-Santos Increased 2020.

[8]   Liu L, Xu M, Ye Y, Zhang B. On the degradation of (micro)plastics: Degradation methods, influencing factors, environmental impacts. Sci Total Environ 2022; 806(Pt 3): 151312.
[http://dx.doi.org/10.1016/j.scitotenv.2021.151312] [PMID: 34743885]

[9]   Koelmans AA, Besseling E, Shim WJ. Nanoplastics in the aquatic environment. Critical review Marine anthropogenic litter 2015; 325-40.

[10]   De Falco F, Di Pace E, Cocca M, Avella M. The contribution of washing processes of synthetic clothes to microplastic pollution. Sci Rep 2019; 9(1): 6633.
[http://dx.doi.org/10.1038/s41598-019-43023-x] [PMID: 31036862]

[11]   Acharya S, Rumi SS, Hu Y, Abidi N. Microfibers from synthetic textiles as a major source of microplastics in the environment: A review. Text Res J 2021; 91(17-18): 2136-56.
[http://dx.doi.org/10.1177/0040517521991244]

[12]   Wang F, Wang B, Duan L, *et al.* Occurrence and distribution of microplastics in domestic, industrial, agricultural and aquacultural wastewater sources: A case study in Changzhou, China. Water Res 2020; 182: 115956.
[http://dx.doi.org/10.1016/j.watres.2020.115956] [PMID: 32622124]

[13]   Hernandez LM, Yousefi N, Tufenkji N. Are there nanoplastics in your personal care products? Environ Sci Technol Lett 2017; 4(7): 280-5.
[http://dx.doi.org/10.1021/acs.estlett.7b00187]

[14]   Jadhav EB, Sankhla MS, Bhat RA, Bhagat DS. Microplastics from food packaging: An overview of human consumption, health threats, and alternative solutions. Environ Nanotechnol Monit Manag 2021; 16: 100608.
[http://dx.doi.org/10.1016/j.enmm.2021.100608]

[15]   Turner A. Paint particles in the marine environment: An overlooked component of microplastics. Water Res X 2021; 12: 100110.
[http://dx.doi.org/10.1016/j.wroa.2021.100110] [PMID: 34401707]

[16]   Kim SK, Kim JS, Lee H, Lee HJ. Abundance and characteristics of microplastics in soils with different agricultural practices: Importance of sources with internal origin and environmental fate. J Hazard Mater 2021; 403: 123997.
[http://dx.doi.org/10.1016/j.jhazmat.2020.123997] [PMID: 33265033]

[17]   van Schothorst B, Beriot N, Huerta Lwanga E, Geissen V. Sources of light density microplastic related to two agricultural practices: the use of compost and plastic mulch. Environments (Basel) 2021; 8(4): 36.
[http://dx.doi.org/10.3390/environments8040036]

[18]   Iheanacho S, Ogbu M, Bhuyan MS, Ogunji J. Microplastic pollution: An emerging contaminant in aquaculture. Aquac Fish 2023; 8(6): 603-16.
[http://dx.doi.org/10.1016/j.aaf.2023.01.007]

[19]   Müller A, Österlund H, Marsalek J, Viklander M. The pollution conveyed by urban runoff: A review

of sources. Sci Total Environ 2020; 709: 136125.
[http://dx.doi.org/10.1016/j.scitotenv.2019.136125] [PMID: 31905584]

[20]   Urli S, Corte Pause F, Crociati M, Baufeld A, Monaci M, Stradaioli G. Impact of microplastics and nanoplastics on livestock health: An emerging risk for reproductive efficiency. Animals (Basel) 2023; 13(7): 1132.
[http://dx.doi.org/10.3390/ani13071132] [PMID: 37048387]

[21]   Benson NU, Agboola OD, Fred-Ahmadu OH, De-la-Torre GE, Oluwalana A, Williams AB. Micro (nano) plastics prevalence, food web interactions, and toxicity assessment in aquatic organisms: A review. Front Mar Sci 2022; 9: 851281.
[http://dx.doi.org/10.3389/fmars.2022.851281]

[22]   Sprovieri M, Feo ML, Prevedello L, *et al.* Heavy metals, polycyclic aromatic hydrocarbons and polychlorinated biphenyls in surface sediments of the naples harbour (southern Italy). Chemosphere 2007; 67(5): 998-1009.
[http://dx.doi.org/10.1016/j.chemosphere.2006.10.055] [PMID: 17157354]

[23]   Pakhomova S, Berezina A, Lusher AL, *et al.* Microplastic variability in subsurface water from the arctic to antarctica. Environ Pollut 2022; 298: 118808.
[http://dx.doi.org/10.1016/j.envpol.2022.118808] [PMID: 35007674]

[24]   Smith M, Love DC, Rochman CM, Neff RA. Microplastics in seafood and the implications for human health. Curr Environ Health Rep 2018; 5(3): 375-86.
[http://dx.doi.org/10.1007/s40572-018-0206-z] [PMID: 30116998]

[25]   Chen G, Feng Q, Wang J. Mini-review of microplastics in the atmosphere and their risks to humans. Sci Total Environ 2020; 703: 135504.
[http://dx.doi.org/10.1016/j.scitotenv.2019.135504] [PMID: 31753503]

[26]   Ageel HK, Harrad S, Abdallah MAE. Occurrence, human exposure, and risk of microplastics in the indoor environment. Environ Sci Process Impacts 2022; 24(1): 17-31.
[http://dx.doi.org/10.1039/D1EM00301A] [PMID: 34842877]

[27]   Prata JC, João P. Environmental exposure to microplastics: an overview on possible human health. Sci Total Environ 2019; 44: 1-32.
[PMID: 31733547]

[28]   Sun M, Chao H, Zheng X, Deng S, Ye M, Hu F. Ecological role of earthworm intestinal bacteria in terrestrial environments: A review. Sci Total Environ 2020; 740: 140008.
[http://dx.doi.org/10.1016/j.scitotenv.2020.140008] [PMID: 32562986]

[29]   Ribeiro F, O'Brien JW, Galloway T, Thomas KV. Accumulation and fate of nano- and micro-plastics and associated contaminants in organisms. Trends Analyt Chem 2019; 111: 139-47.
[http://dx.doi.org/10.1016/j.trac.2018.12.010]

[30]   Li Z, Wei R, Gao M, *et al.* Biodegradation of low-density polyethylene by microbulbifer hydrolyticus IRE-31. J Environ Manage 2020; 263: 110402.
[http://dx.doi.org/10.1016/j.jenvman.2020.110402] [PMID: 32174537]

[31]   Farzi A, Dehnad A, Fotouhi AF. Biodegradation of polyethylene terephthalate waste using Streptomyces species and kinetic modeling of the process. Biocatal Agric Biotechnol 2019; 17: 25-31.
[http://dx.doi.org/10.1016/j.bcab.2018.11.002]

[32]   Park SY, Kim CG. Biodegradation of micro-polyethylene particles by bacterial colonization of a mixed microbial consortium isolated from a landfill site. Chemosphere 2019; 222: 527-33.
[http://dx.doi.org/10.1016/j.chemosphere.2019.01.159] [PMID: 30721811]

[33]   Kim HR, Lee HM, Yu HC, *et al.* Biodegradation of polystyrene by Pseudomonas sp. isolated from the gut of superworms (larvae of *Zophobas atratus*). Environ Sci Technol 2020; 54(11): 6987-96.
[http://dx.doi.org/10.1021/acs.est.0c01495] [PMID: 32374590]

[34]   Peng BY, Chen Z, Chen J, *et al.* Biodegradation of polyvinyl chloride (PVC) in *Tenebrio molitor*

(*Coleoptera: Tenebrionidae*) larvae. Environ Int 2020; 145: 106106.
[http://dx.doi.org/10.1016/j.envint.2020.106106] [PMID: 32947161]

[35]   Mohanrasu K, Premnath N, Siva Prakash G, Sudhakar M, Boobalan T, Arun A. Exploring multi
       potential uses of marine bacteria; an integrated approach for PHB production, PAHs and polyethylene
       biodegradation. J Photochem Photobiol B 2018; 185: 55-65.
       [http://dx.doi.org/10.1016/j.jphotobiol.2018.05.014] [PMID: 29864727]

[36]   Olajire AA, Mohammed AA. Green synthesis of palladium nanoparticles using *Ananas comosus* leaf
       extract for solid-phase photocatalytic degradation of low density polyethylene film. J Environ Chem
       Eng 2019; 7(4): 103270.
       [http://dx.doi.org/10.1016/j.jece.2019.103270]

[37]   Jayaprakash V, Palempalli UMD. Studying the effect of biosilver nanoparticles on polyethylene
       degradation. Appl Nanosci 2019; 9(4): 491-504.
       [http://dx.doi.org/10.1007/s13204-018-0922-6]

[38]   Lam SM, Sin JC, Zeng H, *et al.* Green synthesis of Fe-ZnO nanoparticles with improved sunlight
       photocatalytic performance for polyethylene film deterioration and bacterial inactivation. Mater Sci
       Semicond Process 2021; 123: 105574.
       [http://dx.doi.org/10.1016/j.mssp.2020.105574]

[39]   Olajire AA, Mohammed AA. Green synthesis of nickel oxide nanoparticles and studies of their
       photocatalytic activity in degradation of polyethylene films. Adv Powder Technol 2020; 31(1): 211-8.
       [http://dx.doi.org/10.1016/j.apt.2019.10.012]

[40]   Wang ZM, Wagner J, Ghosal S, Bedi G, Wall S. SEM/EDS and optical microscopy analyses of
       microplastics in ocean trawl and fish guts. Sci Total Environ 2017; 603-604: 616-26.
       [http://dx.doi.org/10.1016/j.scitotenv.2017.06.047] [PMID: 28646780]

[41]   Jonkman J, Brown CM, Wright GD, Anderson KI, North AJ. Tutorial: guidance for quantitative
       confocal microscopy. Nat Protoc 2020; 15(5): 1585-611.
       [http://dx.doi.org/10.1038/s41596-020-0313-9] [PMID: 32235926]

[42]   Samanta P, Dey S, Kundu D, Dutta D, Jambulkar R, Mishra R, *et al.* An insight on sampling,
       identification, quantification and characteristics of microplastics in solid wastes. Trends in
       Environmental Analytical Chemistry 2022.
       [http://dx.doi.org/10.1016/j.teac.2022.e00181]

[43]   Mariano S, Tacconi S, Fidaleo M, Rossi M, Dini L. Micro and nanoplastics identification: classic
       methods and innovative detection techniques. Front Toxicol 2021; 3: 636640.
       [http://dx.doi.org/10.3389/ftox.2021.636640] [PMID: 35295124]

[44]   Ojeda J, Tagg A, Sapp M, Harrison J. Identification and quantification of microplastics in wastewater
       using focal plane array-based reflectance micro-FT-IR imaging. 2015.

[45]   Popovicheva O, Kireeva E, Persiantseva N, *et al.* Microscopic characterization of individual particles
       from multicomponent ship exhaust. J Environ Monit 2012; 14(12): 3101-10.
       [http://dx.doi.org/10.1039/c2em30338h] [PMID: 23090431]

[46]   Shim WJ, Hong SH, Eo SE. Identification methods in microplastic analysis: a review. Anal Methods
       2017; 9(9): 1384-91.
       [http://dx.doi.org/10.1039/C6AY02558G]

[47]   Adhikari S, Kelkar V, Kumar R, Halden RU. Methods and challenges in the detection of microplastics
       and nanoplastics: a mini-review. Polym Int 2022; 71(5): 543-51.
       [http://dx.doi.org/10.1002/pi.6348]

[48]   Strungaru SA, Jijie R, Nicoara M, Plavan G, Faggio C. Micro- (nano) plastics in freshwater
       ecosystems: Abundance, toxicological impact and quantification methodology. Trends Analyt Chem
       2019; 110: 116-28.
       [http://dx.doi.org/10.1016/j.trac.2018.10.025]

[49]   Gniadek M, Dąbrowska A. The marine nano- and microplastics characterisation by SEM-EDX: The potential of the method in comparison with various physical and chemical approaches. Mar Pollut Bull 2019; 148: 210-6.
[http://dx.doi.org/10.1016/j.marpolbul.2019.07.067] [PMID: 31437623]

[50]   Scimeca M, Bischetti S, Lamsira HK, Bonfiglio R, Bonanno E. Energy Dispersive X-ray (EDX) microanalysis: A powerful tool in biomedical research and diagnosis. Euro J Histochem 2018; 62: (1).

[51]   Bergmann M, Gutow L, Klages M. Nanoplastics in the Aquatic Environment.Critical Review. Nanoplastics in the Aquatic Environment. Springer Berlin/Heidelberg, Germany 2015; pp. 325-40.

[52]   Chen Y, Wen D, Pei J, *et al.* Identification and quantification of microplastics using Fourier-transform infrared spectroscopy: Current status and future prospects. Curr Opin Environ Sci Health 2020; 18: 14-9.
[http://dx.doi.org/10.1016/j.coesh.2020.05.004]

[53]   Jiang C, Chen Z, Lu B, *et al.* Hydrothermal pretreatment reduced microplastics in sewage sludge as revealed by the combined micro-Fourier transform infrared (FTIR) and Raman imaging analysis. Chem Eng J 2022; 450: 138163.
[http://dx.doi.org/10.1016/j.cej.2022.138163]

[54]   Vélez-Escamilla LY, Contreras-Torres FF. Latest advances and developments to detection of micro- and nanoplastics using surface-enhanced Raman spectroscopy. Part Part Syst Charact 2022; 39(3): 2100217.
[http://dx.doi.org/10.1002/ppsc.202100217]

[55]   Mecozzi M, Pietroletti M, Monakhova YB. FTIR spectroscopy supported by statistical techniques for the structural characterization of plastic debris in the marine environment: Application to monitoring studies. Mar Pollut Bull 2016; 106(1-2): 155-61.
[http://dx.doi.org/10.1016/j.marpolbul.2016.03.012] [PMID: 26997255]

[56]   Unnimaya S, Mithun N, Lukose J, Nair MP, Gopinath A, Chidangil S, Eds. Identification of microplastics using a custom built micro-raman spectrometer Journal of physics: Conference series. IOP Publishing 2023.

[57]   Caputo F, Vogel R, Savage J, *et al.* Measuring particle size distribution and mass concentration of nanoplastics and microplastics: addressing some analytical challenges in the sub-micron size range. J Colloid Interface Sci 2021; 588: 401-17.
[http://dx.doi.org/10.1016/j.jcis.2020.12.039] [PMID: 33422789]

[58]   Zhu C, Zhang X, Xu R, *et al.* Starch granular size and multi-scale structure determine population patterns in bivariate flow cytometry sorting. Int J Biol Macromol 2023; 231: 123306.
[http://dx.doi.org/10.1016/j.ijbiomac.2023.123306] [PMID: 36669629]

[59]   Bianco A, Carena L, Peitsaro N, Sordello F, Vione D, Passananti M. Rapid detection of nanoplastics and small microplastics by nile-red staining and flow cytometry. Environ Chem Lett 2023; 21(2): 647-53.
[http://dx.doi.org/10.1007/s10311-022-01545-3]

[60]   Baumgarth N, Roederer M. A practical approach to multicolor flow cytometry for immunophenotyping. J Immunol Methods 2000; 243(1-2): 77-97.
[http://dx.doi.org/10.1016/S0022-1759(00)00229-5] [PMID: 10986408]

[61]   Horton AA, Walton A, Spurgeon DJ, Lahive E, Svendsen C. Microplastics in freshwater and terrestrial environments: Evaluating the current understanding to identify the knowledge gaps and future research priorities. Sci Total Environ 2017; 586: 127-41.
[http://dx.doi.org/10.1016/j.scitotenv.2017.01.190] [PMID: 28169032]

[62]   Bradney L, Wijesekara H, Palansooriya KN, *et al.* Particulate plastics as a vector for toxic trace-element uptake by aquatic and terrestrial organisms and human health risk. Environ Int 2019; 131: 104937.

[http://dx.doi.org/10.1016/j.envint.2019.104937] [PMID: 31284110]

[63] Wang W, Do ATN, Kwon JH. Ecotoxicological effects of micro- and nanoplastics on terrestrial food web from plants to human beings. Sci Total Environ 2022; 834: 155333.
[http://dx.doi.org/10.1016/j.scitotenv.2022.155333] [PMID: 35452728]

[64] Chang X, Xue Y, Li J, Zou L, Tang M. Potential health impact of environmental micro- and nanoplastics pollution. J Appl Toxicol 2020; 40(1): 4-15.
[http://dx.doi.org/10.1002/jat.3915] [PMID: 31828819]

[65] Oliveira M, Almeida M, Miguel I. A micro(nano)plastic boomerang tale: A never ending story? Trends Analyt Chem 2019; 112: 196-200.
[http://dx.doi.org/10.1016/j.trac.2019.01.005]

[66] Corcoran PL. Degradation of microplastics in the environment Handbook of microplastics in the environment. Springer 2022; pp. 531-42.
[http://dx.doi.org/10.1007/978-3-030-39041-9_10]

[67] Johnson RW, Scopelliti HR, Herrold NT, Wakabayashi K. Solid-state shear pulverization of post-industrial ultra-high molecular weight polyethylene: Particle morphology and molecular structure modifications toward conventional mechanical recycling. Polym Eng Sci 2023; 63(2): 319-30.
[http://dx.doi.org/10.1002/pen.26207]

[68] El Hadri H, Gigault J, Maxit B, Grassl B, Reynaud S. Nanoplastic from mechanically degraded primary and secondary microplastics for environmental assessments. NanoImpact 2020; 17: 100206.
[http://dx.doi.org/10.1016/j.impact.2019.100206]

[69] Zeenat , Elahi A, Bukhari DA, Shamim S, Rehman A. Plastics degradation by microbes: A sustainable approach. J King Saud Univ Sci 2021; 33(6): 101538.
[http://dx.doi.org/10.1016/j.jksus.2021.101538]

[70] Yang Q, Luo K, Li X, *et al.* Enhanced efficiency of biological excess sludge hydrolysis under anaerobic digestion by additional enzymes. Bioresour Technol 2010; 101(9): 2924-30.
[http://dx.doi.org/10.1016/j.biortech.2009.11.012] [PMID: 20045636]

[71] Heumann S, Eberl A, Pobeheim H, *et al.* New model substrates for enzymes hydrolysing polyethyleneterephthalate and polyamide fibres. J Biochem Biophys Methods 2006; 69(1-2): 89-99.
[http://dx.doi.org/10.1016/j.jbbm.2006.02.005] [PMID: 16624419]

[72] Tollini F, Occhetta A, Broglia F, *et al.* Influence of pH on the kinetics of hydrolysis reactions: the case of epichlorohydrin and glycidol. React Chem Eng 2022; 7(10): 2211-23.
[http://dx.doi.org/10.1039/D2RE00191H]

[73] Lambert S, Wagner M. Characterisation of nanoplastics during the degradation of polystyrene. Chemosphere 2016; 145: 265-8.
[http://dx.doi.org/10.1016/j.chemosphere.2015.11.078] [PMID: 26688263]

[74] Andrady AL, Barnes PW, Bornman JF, *et al.* Oxidation and fragmentation of plastics in a changing environment; from UV-radiation to biological degradation. Sci Total Environ 2022; 851(Pt 2): 158022.
[http://dx.doi.org/10.1016/j.scitotenv.2022.158022] [PMID: 35970458]

[75] Goswami C, Hazarika KK, Bharali P. Transition metal oxide nanocatalysts for oxygen reduction reaction. Mater Sci Energy Technol 2018; 1(2): 117-28.
[http://dx.doi.org/10.1016/j.mset.2018.06.005]

[76] Ibhadon A, Fitzpatrick P. Heterogeneous photocatalysis: recent advances and applications. Catalysts 2013; 3(1): 189-218.
[http://dx.doi.org/10.3390/catal3010189]

[77] Hou Q, Zhen M, Qian H, *et al.* Upcycling and catalytic degradation of plastic wastes. Cell Rep Phys Sci 2021; 2(8): 100514.
[http://dx.doi.org/10.1016/j.xcrp.2021.100514]

[78] Sun J, Zheng H, Xiang H, Fan J, Jiang H. The surface degradation and release of microplastics from plastic films studied by UV radiation and mechanical abrasion. Sci Total Environ 2022; 838(Pt 3): 156369.
[http://dx.doi.org/10.1016/j.scitotenv.2022.156369] [PMID: 35654205]

[79] Yousif E, Haddad R. Photodegradation and photostabilization of polymers, especially polystyrene: Review. Springerplus 2013; 2(1): 398.
[http://dx.doi.org/10.1186/2193-1801-2-398] [PMID: 25674392]

[80] Muzzarelli R, Biagini G, Pugnaloni A, *et al.* Reconstruction of parodontal tissue with chitosan. Biomaterials 1989; 10(9): 598-603.
[http://dx.doi.org/10.1016/0142-9612(89)90113-0] [PMID: 2611308]

[81] Andrady AL, Hamid H, Torikai A. Effects of solar UV and climate change on materials. Photochem Photobiol Sci 2011; 10(2): 292-300.
[http://dx.doi.org/10.1039/c0pp90038a] [PMID: 21253664]

[82] Carlsson DJ, Wiles DM. The photodegradation of polypropylene films. III. Photolysis of polypropylene hydroperoxides. Macromolecules 1969; 2(6): 597-606.
[http://dx.doi.org/10.1021/ma60012a007]

[83] Andrady AL, Heikkilä AM, Pandey KK, *et al.* Effects of UV radiation on natural and synthetic materials. Photochem Photobiol Sci 2023; 22(5): 1177-202.
[http://dx.doi.org/10.1007/s43630-023-00377-6] [PMID: 37039962]

[84] Mekhilef S, Saidur R, Kamalisarvestani M. Effect of dust, humidity and air velocity on efficiency of photovoltaic cells. Renew Sustain Energy Rev 2012; 16(5): 2920-5.
[http://dx.doi.org/10.1016/j.rser.2012.02.012]

[85] Sharma I. An advance tool in biological research. Open J Environ Biol 2020.

[86] Zhu K, Jia H, Sun Y, *et al.* Long-term phototransformation of microplastics under simulated sunlight irradiation in aquatic environments: Roles of reactive oxygen species. Water Res 2020; 173: 115564.
[http://dx.doi.org/10.1016/j.watres.2020.115564] [PMID: 32028245]

[87] Kyrikou I, Briassoulis D. Biodegradation of agricultural plastic films: A critical review. J Polym Environ 2007; 15(2): 125-50.
[http://dx.doi.org/10.1007/s10924-007-0053-8]

[88] Enfrin M, Myszka R, Giustozzi F. Paving roads with recycled plastics: Microplastic pollution or eco-friendly solution? J Hazard Mater 2022; 437: 129334.
[http://dx.doi.org/10.1016/j.jhazmat.2022.129334] [PMID: 35716564]

[89] Wang S, Zhao Y, Wang J, *et al.* The efficiency of long-term straw return to sequester organic carbon in Northeast China's cropland. J Integr Agric 2018; 17(2): 436-48.
[http://dx.doi.org/10.1016/S2095-3119(17)61739-8]

[90] Kumar R, Verma A, Rakib MRJ, *et al.* Adsorptive behavior of micro(nano)plastics through biochar: Co-existence, consequences, and challenges in contaminated ecosystems. Sci Total Environ 2023; 856(Pt 1): 159097.
[http://dx.doi.org/10.1016/j.scitotenv.2022.159097] [PMID: 36179840]

[91] Qi C, Wang R, Jia S, *et al.* Biochar amendment to advance contaminant removal in anaerobic digestion of organic solid wastes: A review. Bioresour Technol 2021; 341: 125827.
[http://dx.doi.org/10.1016/j.biortech.2021.125827] [PMID: 34455247]

[92] Maraveas C. Production of sustainable and biodegradable polymers from agricultural waste. Polymers 2020; 12(5): 1127.
[http://dx.doi.org/10.3390/polym12051127] [PMID: 32423073]

[93] Skouta R. Selective chemical reactions in supercritical carbon dioxide, water, and ionic liquids. Green Chem Lett Rev 2009; 2(3): 121-56.

[http://dx.doi.org/10.1080/17518250903230001]

[94]   Picó Y, Barceló D. Micro(Nano)plastic analysis: A green and sustainable perspective. J Hazard Mater Adv 2022; 6: 100058.
[http://dx.doi.org/10.1016/j.hazadv.2022.100058]

[95]   Ncube A, Mtetwa S, Bukhari M, Fiorentino G, Passaro R. Circular economy and green chemistry: the need for radical innovative approaches in the design for new products. Energies 2023; 16(4): 1752.
[http://dx.doi.org/10.3390/en16041752]

[96]   Monteiro SS, Pinto da Costa J. Methods for the extraction of microplastics in complex solid, water and biota samples. Trends in Environmental Analytical Chemistry 2022; 33: e00151.
[http://dx.doi.org/10.1016/j.teac.2021.e00151]

[97]   Alvarez S, Rubio A. Carbon footprint in Green Public Procurement: A case study in the services sector. J Clean Prod 2015; 93: 159-66.
[http://dx.doi.org/10.1016/j.jclepro.2015.01.048]

[98]   Chen Z, Wei W, Ni BJ, Chen H. Plastic wastes derived carbon materials for green energy and sustainable environmental applications. Environmental Functional Materials 2022; 1(1): 34-48.
[http://dx.doi.org/10.1016/j.efmat.2022.05.005]

[99]   Sheldon RA, Norton M. Green chemistry and the plastic pollution challenge: Towards a circular economy. Green Chem 2020; 22(19): 6310-22.
[http://dx.doi.org/10.1039/D0GC02630A]

[100]   Song JH, Murphy RJ, Narayan R, Davies GBH. Biodegradable and compostable alternatives to conventional plastics. Philos Trans R Soc Lond B Biol Sci 2009; 364(1526): 2127-39.
[http://dx.doi.org/10.1098/rstb.2008.0289] [PMID: 19528060]

[101]   Zimmerman JB, Anastas PT, Erythropel HC, Leitner W. Designing for a green chemistry future. Science 2020; 367(6476): 397-400.
[http://dx.doi.org/10.1126/science.aay3060] [PMID: 31974246]

[102]   Korde P, Ghotekar S, Pagar T, Pansambal S, Oza R, Mane D. Plant extract assisted eco-benevolent synthesis of selenium nanoparticles-a review on plant parts involved, characterization and their recent applications. J Chem Rev 2020; 2(3): 157-68.

[103]   Gu JD. Microbiological deterioration and degradation of synthetic polymeric materials: Recent research advances. Int Biodeterior Biodegradation 2003; 52(2): 69-91.
[http://dx.doi.org/10.1016/S0964-8305(02)00177-4]

[104]   Skariyachan S, Taskeen N, Kishore AP, Krishna BV, Naidu G. Novel consortia of Enterobacter and Pseudomonas formulated from cow dung exhibited enhanced biodegradation of polyethylene and polypropylene. J Environ Manage 2021; 284: 112030.
[http://dx.doi.org/10.1016/j.jenvman.2021.112030] [PMID: 33529882]

[105]   Roohi , Bano K, Kuddus M, *et al.* Microbial enzymatic degradation of biodegradable plastics. Curr Pharm Biotechnol 2017; 18(5): 429-40.
[PMID: 28545359]

[106]   Newman SG, Jensen KF. The role of flow in green chemistry and engineering. Green Chem 2013; 15(6): 1456-72.
[http://dx.doi.org/10.1039/c3gc40374b]

[107]   Adrio JL, Demain AL. Recombinant organisms for production of industrial products. Bioeng Bugs 2010; 1(2): 116-31.
[http://dx.doi.org/10.4161/bbug.1.2.10484] [PMID: 21326937]

[108]   Stock F, Kochleus C, Bänsch-Baltruschat B, Brennholt N, Reifferscheid G. Sampling techniques and preparation methods for microplastic analyses in the aquatic environment – A review. Trends Analyt Chem 2019; 113: 84-92.
[http://dx.doi.org/10.1016/j.trac.2019.01.014]

[109]   Lusher AL, Welden NA, Sobral P, Cole M. Sampling, isolating and identifying microplastics ingested by fish and invertebrates. Anal Methods 2017; 9(9): 1346-60.
[http://dx.doi.org/10.1039/C6AY02415G]

[110]   Vilaplana F, Karlsson S. Quality concepts for the improved use of recycled polymeric materials: A review. Macromol Mater Eng 2008; 293(4): 274-97.
[http://dx.doi.org/10.1002/mame.200700393]

[111]   Sinha A, Mishra S, Sharif A, Yarovaya L. Does green financing help to improve environmental & social responsibility? Designing SDG framework through advanced quantile modelling. J Environ Manage 2021; 292: 112751.
[http://dx.doi.org/10.1016/j.jenvman.2021.112751] [PMID: 33991831]

[112]   Cole M, Lindeque P, Halsband C, Galloway TS. Microplastics as contaminants in the marine environment: A review. Mar Pollut Bull 2011; 62(12): 2588-97.
[http://dx.doi.org/10.1016/j.marpolbul.2011.09.025] [PMID: 22001295]

[113]   Dharini V, Periyar Selvam S, Jayaramudu J, Sadiku Emmanuel R. Functional properties of clay nanofillers used in the biopolymer-based composite films for active food packaging applications - Review. Appl Clay Sci 2022; 226: 106555.
[http://dx.doi.org/10.1016/j.clay.2022.106555]

[114]   Huang H, Yang Y. Preparation of silver nanoparticles in inorganic clay suspensions. Compos Sci Technol 2008; 68(14): 2948-53.
[http://dx.doi.org/10.1016/j.compscitech.2007.10.003]

[115]   Deng J, You Y, Sahajwalla V, Joshi RK. Transforming waste into carbon-based nanomaterials. Carbon 2016; 96: 105-15.
[http://dx.doi.org/10.1016/j.carbon.2015.09.033]

[116]   Khan ZU, Kausar A, Ullah H. A review on composite papers of graphene oxide, carbon nanotube, polymer/GO, and polymer/CNT: Processing strategies, properties, and relevance. Polym Plast Technol Eng 2016; 55(6): 559-81.
[http://dx.doi.org/10.1080/03602559.2015.1098693]

[117]   Song M, Ju J, Luo S, *et al.* Controlling liquid splash on superhydrophobic surfaces by a vesicle surfactant. Sci Adv 2017; 3(3): e1602188.
[http://dx.doi.org/10.1126/sciadv.1602188] [PMID: 28275735]

[118]   Zhang X, Xu X, Yao S, *et al.* Boosting Electrocatalytic Activity of Single Atom Catalysts Supported on Nitrogen-Doped Carbon through N Coordination Environment Engineering. Small 2022; 18(10): 2105329.
[http://dx.doi.org/10.1002/smll.202105329] [PMID: 35023622]

[119]   Shang Y, Liu F, Wang Y, Li N, Ding B. Enzyme mimic nanomaterials and their biomedical applications. ChemBioChem 2020; 21(17): 2408-18.
[http://dx.doi.org/10.1002/cbic.202000123] [PMID: 32227615]

[120]   Zhang R, Yan X, Fan K. Nanozymes inspired by natural enzymes. Acc Mater Res 2021; 2(7): 534-47.
[http://dx.doi.org/10.1021/accountsmr.1c00074]

[121]   Bilal M, Khaliq N, Ashraf M, *et al.* Enzyme mimic nanomaterials as nanozymes with catalytic attributes. Colloids Surf B Biointerfaces 2023; 221: 112950.
[http://dx.doi.org/10.1016/j.colsurfb.2022.112950] [PMID: 36327773]

[122]   Samal S, Dey P, Baral SS, Rangarajan V. Plastic degradation—contemporary enzymes versus nanozymes-based technologies Advances in nano and biochemistry. Elsevier 2023; pp. 127-49.

[123]   Mishra RK, Ha SK, Verma K, Tiwari SK. Recent progress in selected bio-nanomaterials and their engineering applications: An overview. J Sci Adv Mater Devices 2018; 3(3): 263-88.
[http://dx.doi.org/10.1016/j.jsamd.2018.05.003]

[124] Ates B, Koytepe S, Ulu A, Gurses C, Thakur VK. Chemistry, structures, and advanced applications of nanocomposites from biorenewable resources. Chem Rev 2020; 120(17): 9304-62.
[http://dx.doi.org/10.1021/acs.chemrev.9b00553] [PMID: 32786427]

[125] Ferreira FV, Dufresne A, Pinheiro IF, *et al.* How do cellulose nanocrystals affect the overall properties of biodegradable polymer nanocomposites: A comprehensive review. Eur Polym J 2018; 108: 274-85.
[http://dx.doi.org/10.1016/j.eurpolymj.2018.08.045]

[126] Dufresne A. Nanocellulose processing properties and potential applications. Curr For Rep 2019; 5(2): 76-89.
[http://dx.doi.org/10.1007/s40725-019-00088-1]

[127] Erlantz L, Goikuria U, Vilas JL, Cristofaro F, Bruni G, Fortunati E, *et al.* Metal Nanoparticles Embedded in Cellulose Nanocrystal Based Films: Material Properties and Postuse Analysis. 2018.

[128] Pandey VK, Upadhyay SN, Niranjan K, Mishra PK. Antimicrobial biodegradable chitosan-based composite Nano-layers for food packaging. Int J Biol Macromol 2020; 157: 212-9.
[http://dx.doi.org/10.1016/j.ijbiomac.2020.04.149] [PMID: 32339572]

[129] Cheaburu-Yilmaz CN, Yilmaz O, Vasile C. Eco-friendly Polymer Nanocomposites: Chemistry and Applications 2015; 341-86.
[http://dx.doi.org/10.1007/978-81-322-2473-0_11]

[130] Xiong Chang X, Mujawar Mubarak N, Ali Mazari S, *et al.* A review on the properties and applications of chitosan, cellulose and deep eutectic solvent in green chemistry. J Ind Eng Chem 2021; 104: 362-80.
[http://dx.doi.org/10.1016/j.jiec.2021.08.033]

[131] Tang ZX, Qian JQ, Shi LE. Preparation of chitosan nanoparticles as carrier for immobilized enzyme. Appl Biochem Biotechnol 2007; 136(1): 77-96.
[http://dx.doi.org/10.1007/BF02685940] [PMID: 17416979]

[132] Nadesh R, Narayanan D, P r S, *et al.* Hematotoxicological analysis of surface-modified and -unmodified chitosan nanoparticles. J Biomed Mater Res A 2013; 101(10): 2957-66.
[http://dx.doi.org/10.1002/jbm.a.34591] [PMID: 23613460]

[133] Hatakeyama H, Hatakeyama T. Lignin structure, properties, and applications. Biopolymers: lignin, proteins, bioactive nanocomposites 2010; 1-63.

[134] Rout PR, Mohanty A, Aastha , *et al.* Micro- and nanoplastics removal mechanisms in wastewater treatment plants: A review. J Hazard Mater Adv 2022; 6: 100070.
[http://dx.doi.org/10.1016/j.hazadv.2022.100070]

[135] Sipponen MH, Lange H, Crestini C, Henn A, Österberg M. Lignin for Nano-and Microscaled Carrier Systems 2019.

[136] Alnoch RC, Alves dos Santos L, Marques de Almeida J, Krieger N, Mateo C. Recent trends in biomaterials for immobilization of lipases for application in non-conventional media. Catalysts 2020; 10(6): 697.
[http://dx.doi.org/10.3390/catal10060697]

[137] Keskin G, Kızıl G, Bechelany M, Pochat-Bohatier C, Öner M. Potential of polyhydroxyalkanoate (PHA) polymers family as substitutes of petroleum based polymers for packaging applications and solutions brought by their composites to form barrier materials. Pure Appl Chem 2017; 89(12): 1841-8.
[http://dx.doi.org/10.1515/pac-2017-0401]

[138] Li W, Sun H, Wang G, Sui W, Dai L, Si C. Lignin as a green and multifunctional alternative to phenol for resin synthesis. Green Chem 2023; 25(6): 2241-61.
[http://dx.doi.org/10.1039/D2GC04319J]

[139] Lizundia E, Armentano I, Luzi F, *et al.* Synergic effect of nanolignin and metal oxide nanoparticles

into Poly (l-lactide) bionanocomposites: Material properties, antioxidant activity, and antibacterial performance. ACS Appl Bio Mater 2020; 3(8): 5263-74.
[http://dx.doi.org/10.1021/acsabm.0c00637] [PMID: 35021701]

[140] Mohamad Idris NH, Cheong KY, Kennedy BJ, Ohno T, Lee HL. Buoyant titanium dioxide (TiO_2) as high performance photocatalyst and peroxide activator: A critical review on fabrication, mechanism and application. J Environ Chem Eng 2022; 10(3): 107549.
[http://dx.doi.org/10.1016/j.jece.2022.107549]

[141] Mohanty AK, Misra M, Hinrichsen G. Biofibres, biodegradable polymers and biocomposites: An overview. Macromol Mater Eng 2000; 276-277(1): 1-24.
[http://dx.doi.org/10.1002/(SICI)1439-2054(20000301)276:1<1::AID-MAME1>3.0.CO;2-W]

[142] Shah U, Gani A, Ashwar BA, *et al.* A review of the recent advances in starch as active and nanocomposite packaging films. Cogent Food Agric 2015; 1(1): 1115640.
[http://dx.doi.org/10.1080/23311932.2015.1115640]

[143] Mochane MJ, Motloung MT, Mokhena TC, Mofokeng TG. Morphology and photocatalytic activity of zinc oxide reinforced polymer composites: A mini review. Catalysts 2022; 12(11): 1439.
[http://dx.doi.org/10.3390/catal12111439]

[144] Kim HY, Park SS, Lim ST. Preparation, characterization and utilization of starch nanoparticles. Colloids Surf B Biointerfaces 2015; 126: 607-20.
[http://dx.doi.org/10.1016/j.colsurfb.2014.11.011] [PMID: 25435170]

[145] Reddy AVB, Yusop Z, Jaafar J, *et al.* Recent progress on Fe-based nanoparticles: Synthesis, properties, characterization and environmental applications. J Environ Chem Eng 2016; 4(3): 3537-53.
[http://dx.doi.org/10.1016/j.jece.2016.07.035]

[146] LL HRLJC, Liu DQ, Li HZ, Zheng Y. Ding J Powder Technol 2009; 189: 426-32.

[147] Nabi I, Bacha A-U-R, Ahmad F, Zhang L. Application of titanium dioxide for the photocatalytic degradation of macro- and micro-plastics: A review. J Environ Chem Eng 2021; 9(5): 105964.
[http://dx.doi.org/10.1016/j.jece.2021.105964]

[148] Wang Z, Saadé NK, Ariya PA. Advances in ultra-trace analytical capability for micro/nanoplastics and water-soluble polymers in the environment: fresh falling urban snow. Environ Pollut 2021; 276: 116698.
[http://dx.doi.org/10.1016/j.envpol.2021.116698] [PMID: 33611197]

[149] Domenech J, Cortés C, Vela L, Marcos R, Hernández A. Polystyrene nanoplastics as carriers of metals. Interactions of polystyrene nanoparticles with silver nanoparticles and silver nitrate, and their effects on human intestinal caco-2 cells. Biomolecules 2021; 11(6): 859.
[http://dx.doi.org/10.3390/biom11060859] [PMID: 34207836]

[150] Din MI, Rehan R. Synthesis, characterization, and applications of copper nanoparticles. Anal Lett 2017; 50(1): 50-62.
[http://dx.doi.org/10.1080/00032719.2016.1172081]

[151] Muthulakshmi L, Varada Rajalu A, Kaliaraj GS, Siengchin S, Parameswaranpillai J, Saraswathi R. Preparation of cellulose/copper nanoparticles bionanocomposite films using a bioflocculant polymer as reducing agent for antibacterial and anticorrosion applications. Compos, Part B Eng 2019; 175: 107177.
[http://dx.doi.org/10.1016/j.compositesb.2019.107177]

[152] Ricardo IA, Alberto EA, Silva Júnior AH, *et al.* A critical review on microplastics, interaction with organic and inorganic pollutants, impacts and effectiveness of advanced oxidation processes applied for their removal from aqueous matrices. Chem Eng J 2021; 424: 130282.
[http://dx.doi.org/10.1016/j.cej.2021.130282]

SUBJECT INDEX

A

Acid(s) 34, 79, 99, 103, 106, 120, 154, 189,
    193, 209, 241, 264
  carboxylic 79, 120
  chlorobenzene 34
  deoxyribose nucleic (DNA) 34, 103, 106,
    154, 189, 241, 264
  glutamic 103
  organic 99, 193, 209
  phenolic 154
  thioctic 103
Adsorbents 267, 303
  carbon-based 267
  nanoplastic 303
Adsorption 104, 107, 186, 190, 211, 214, 217,
    218, 219, 222, 230, 261, 262, 268
  activated carbon-based 230
   -based techniques 104
  heavy metal ion 104
Advanced 6, 7, 8, 251, 252, 253, 261, 272
  oxidation processes (AOPs) 6, 7, 8, 253,
    261, 272
  processing technologies 252
  treatment technologies 251, 252
Aldehyde dehydrogenase 26, 218
Anaerobic degradation 27
Anti-human lung carcinoma 53
Antibacterial 53, 54, 119, 133, 141, 161, 237,
    238, 240
  activity 53, 54, 141, 237
  agents 119, 133, 238
  effects 238, 240
  properties 161
Anticancer activity 53, 54
Antimicrobial 53, 106, 107, 123
  activity 53, 123
  agents 106, 107
Antioxidant 44, 52, 53, 54, 121, 140, 209,
    234, 300
  antimicrobial 54
Arsenate, toxic 214

Artificial electroplating wastewater 195
Atomic force microscopy (AFM) 154, 259,
    260
Attenuated total reflectance (ATR) 286

B

Bacteria 75, 76, 161, 163, 214
  iron-reducing 161
  marine 75, 76
  reducing 163, 214
Bacterial metabolism 79
Bacteriophages 33, 68
Bio-based technique 152
Bioaugmentation treatments 33
Biodegradable 11, 295, 297
  plastics 295, 297
  waste 11
Biodegradation 7, 9, 19, 20, 23, 24, 26, 30, 31,
    32, 34, 36, 65, 66, 68, 72, 78, 81, 85,
    221
  anaerobic 26
  mechanism of cycloalkane 78
  microbial 24
  nanotechnology-based enzyme 221
  of dyes 81
  of petroleum-based hydrocarbon 24
  reactions 72
Bioremediate 84
  uranium 84
Bioremediation 3, 9, 24, 33, 34, 66, 67, 68, 70,
    73, 119, 153, 209, 210, 211, 212, 214,
    216, 219, 221, 222
  activity 212
  agents 66, 209
  of wastewater 3, 119, 214, 219, 221, 222
  process 9, 24, 33, 34, 66, 67, 68, 70, 73,
    153, 216
  techniques 9, 67, 70, 210, 211
By-products 4, 5, 19, 152, 293, 296
  disinfection 4
  harmful 152, 293, 296

www.ingramcontent.com/pod-product-compliance
Lightning Source LLC
Chambersburg PA
CBHW050807220326
41598CB00006B/146